U.S. ARMY
INTELLIGENCE AND
INTERROGATION
HANDBOOK

T0058146

U.S. ARMY
INTELLIGENCE AND
INTERROGATION
HANDBOOK

DEPARTMENT OF THE ARMY

SKYHORSE PUBLISHING

All inquiries should be addressed to Skyhorse Publishing, 307 West 36th Street, 11th Floor, New York, NY 10018.

Skyhorse Publishing books may be purchased in bulk at special discounts for sales promotion, corporate gifts, fund-raising, or educational purposes. Special editions can also be created to specifications. For details, contact the Special Sales Department, Skyhorse Publishing, 307 West 36th Street, 11th Floor, New York, NY 10018 or info@skyhorsepublishing.com.

Skyhorse® and Skyhorse Publishing® are registered trademarks of Skyhorse Publishing, Inc.®, a Delaware corporation.

Visit our website at www.skyhorsepublishing.com.

10 9 8 7 6 5 4 3 2

Library of Congress Cataloging-in-Publication Data is available on file.

ISBN: 978-1-62636-098-3

Printed in the United States of America

Field Manual
No. 2-22.3

*FM 2-22.3 (FM 34-52)
Headquarters
Department of the Army

Human Intelligence Collector Operations

Contents

*This publication supersedes FM 34-52, 28 September 1992, and ST 2-22.7, Tactical Human Intelligence and Counterintelligence Operations, April 2002.

Preface

This manual provides doctrinal guidance, techniques, and procedures governing the employment of human intelligence (HUMINT) collection and analytical assets in support of the commander's intelligence needs. It outlines—

- HUMINT operations.
- The HUMINT collector's role within the intelligence operating system.
- The roles and responsibilities of the HUMINT collectors and the roles of those providing the command, control, and technical support of HUMINT collection operations.

This manual expands upon the information contained in FM 2-0. It supersedes FM 34-52 and rescinds ST 2-22.7. It is consistent with doctrine in FM 3-0, FM 5-0, FM 6-0, and JP 2-0. In accordance with the Detainee Treatment Act of 2005, the only interrogation approaches and techniques that are authorized for use against any detainee, regardless of status or characterization, are those authorized and listed in this Field Manual. Some of the approaches and techniques authorized and listed in this Field Manual also require additional specified approval before implementation.

This manual will be reviewed annually and may be amended or updated from time to time to account for changes in doctrine, policy, or law, and to address lessons learned.

This manual provides the doctrinal guidance for HUMINT collectors and commanders and staffs of the MI organizations responsible for planning and executing HUMINT operations. This manual also serves as a reference for personnel developing doctrine, tactics, techniques, and procedures (TTP); materiel and force structure; institutional and unit training; and standing operating procedures (SOPs), for HUMINT operations at all army echelons. In accordance with TRADOC Regulation 25-36, the doctrine in this field manual is not policy (in and of itself), but is "...a body of thought on how Army forces operate....[It] provides an authoritative guide for leaders and soldiers, while allowing freedom to adapt to circumstances."

This manual applies to the Active Army, the Army National Guard/Army National Guard of the United States, and the United States Army Reserve unless otherwise stated. This manual also applies to DOD civilian employees and contractors with responsibility to engage in HUMINT collection activities. It is also intended for commanders and staffs of joint and combined commands, and Service Component Commands (SCC). Although this is Army doctrine, adaptations will have to be made by other Military Departments, based on each of their organizations and specific doctrine.

Material in this manual applies to the full range of military operations. Principles outlined also are valid under conditions involving use of electronic warfare (EW) or nuclear, biological, or chemical (NBC) weapons.

This manual is intended for use by military, civilian, and civilian contractor HUMINT collectors, as well as commanders, staff officers, and military intelligence (MI) personnel charged with the responsibility of the HUMINT collection effort.

HUMINT operations vary depending on the source of the information. It is essential that all HUMINT collectors understand that, whereas operations and sources may differ, the handling and treatment of sources must be accomplished in accordance with applicable law and policy. Applicable law and policy include US law; the law of war; relevant international law; relevant directives including DOD Directive 3115.09, "DOD Intelligence Interrogations, Detainee Debriefings, and Tactical Questioning"; DOD Directive 2310.1E, "The Department of Defense Detainee Program"; DOD instructions; and military execute orders including fragmentary orders (FRAGOs).

Interrogation, the HUMINT subdiscipline responsible for MI exploitation of enemy personnel and their documents to answer the supported specific information requirements (SIRs), requires the HUMINT collector to be fully familiar with both the classification of the source and applicable law. The principles and techniques of HUMINT collection are to be used within the constraints established by US law including the following:

- The Uniform Code of Military Justice (UCMJ).
- Geneva Convention for the Amelioration of the Condition of the Wounded and Sick in Armed Forces in the Field (including Common Article III), August 12, 1949; hereinafter referred to as GWS.
- Geneva Convention Relative to the Treatment of Prisoners of War (including Common Article III), August 12, 1949; hereinafter referred to as GPW.
- Geneva Convention Relative to the Protection of Civilian Persons in Time of War (including Common Article III), August 12, 1949; hereinafter referred to as GC.
- Detainee Treatment Act of 2005, Public Law No. 109-163, Title XIV.

HUMINT collectors must understand specific terms used to identify categories of personnel when referring to the principles and techniques of interrogation. Determination of a detainee's status may take a significant time and may not be completed until well after the time of capture. Therefore, there will be no difference in the treatment of a detainee of any status from the moment of capture until such a determination is made. The following terms are presented here and in the glossary.

- Civilian Internee: A person detained or interned in the United States or in occupied territory for security reasons, or for protection, or because he or she has committed an offense against the detaining power, and who is entitled to "protected person" status under the GC.
- Enemy Prisoner of War (EPW): A detained person, as defined in Articles 4 and 5 of the GPW. In particular, one who, while engaged in combat under orders of his or her government, is captured by the armed forces of the enemy. As such, he or she is entitled to the combatant's privilege of immunity from the municipal law of the capturing state for warlike acts that do not amount to breaches of the law of armed conflict. For example, an EPW may be, but is not limited to, any person belonging to one of the following categories of personnel who have fallen into the power of the enemy; a member of the armed forces, organized militia or volunteer corps; a person who accompanies the armed forces, without actually being a member thereof; a member of a merchant marine or civilian aircraft crew not qualifying for more favorable treatment; or individuals who, on the approach of the enemy, spontaneously take up arms to resist invading forces.
- Other Detainees: Persons in the custody of the US Armed Forces who have not been classified as an EPW (Article 4, GPW), retained personnel (Article 33, GPW), and Civilian Internee (Articles 27, 41, 48, and 78, GC) shall be treated as EPWs until a legal status is ascertained by competent authority; for example, by Article 5 Tribunal.
- Retained Personnel: (See Articles 24 and 26, GWS.)

- Official medical personnel of the armed forces exclusively engaged in the search for, or the collection, transport or treatment of wounded or sick, or in the prevention of disease, and staff exclusively engaged in the administration of medical units and facilities.
- Chaplains attached to the armed forces.

- Staff of National Red Cross Societies and that of other Volunteer Aid Societies, duly recognized and authorized by their governments to assist Medical Service personnel of their own armed forces, provided they are exclusively engaged in the search for, or the collection, transport or treatment of wounded or sick, or in the prevention of disease, and provided that the staff of such societies are subject to military laws and regulations.

• Protected Persons: Include civilians entitled to protection under the GC, including those we retain in the course of a conflict, no matter what the reason.

• Enemy Combatant: In general, a person engaged in hostilities against the United States or its coalition partners during an armed conflict. The term "enemy combatant" includes both "lawful enemy combatants" and "unlawful enemy combatants." All captured or detained personnel, regardless of status, shall be treated humanely, and in accordance with the Detainee Treatment Act of 2005 and DOD Directive 2310.1E, "Department of Defense Detainee Program", and no person in the custody or under the control of DOD, regardless of nationality or physical location, shall be subject to torture or cruel, inhuman, or degrading treatment or punishment, in accordance with and as defined in US law.

- **Lawful Enemy Combatant:** Lawful enemy combatants, who are entitled to protections under the Geneva Conventions, include members of the regular armed forces of a State Party to the conflict; militia, volunteer corps, and organized resistance movements belonging to a State Party to the conflict, which are under responsible command, wear a fixed distinctive sign recognizable at a distance, carry their arms openly, and abide by the laws of war; and members of regular armed forces who profess allegiance to a government or an authority not recognized by the detaining power.

- **Unlawful Enemy Combatant:** Unlawful enemy combatants are persons not entitled to combatant immunity, who engage in acts against the United States or its coalition partners in violation of the laws and customs of war during an armed conflict. For the purposes of the war on terrorism, the term "unlawful enemy combatant" is defined to include, but is not limited to, an individual who is or was part of or supporting Taliban or al Qaeda forces, or associated forces that are engaged in hostilities against the United States or its coalition partners.

Headquarters, U.S. Army Training and Doctrine Command (TRADOC) is the proponent for this publication. The preparing agency is the US Army Intelligence Center and Fort Huachuca, Fort Huachuca, AZ. Send written comments and recommendations on DA Form 2028 (Recommended Changes to Publications and Blank Forms) directly to Commander, ATZS-CDI-D (FM 2-22.3), U.S. Army Intelligence Center and Fort Huachuca, 550 Cibeque Street, Fort Huachuca, AZ 85613-7017. Send comments and recommendations by e-mail to ATZS-FDT-D@hua.army.mil. Follow the DA Form 2028 format or submit an electronic DA Form 2028.

Unless otherwise stated, masculine nouns and pronouns do not refer exclusively to men. Use of the terms "he" and "him" in this manual should be read as referring to both males and females unless otherwise expressly noted.

PART ONE

HUMINT Support, Planning, and Management

HUMINT collection activities include three general categories: screening, interrogation, and debriefing. In some cases these may be distinguished by legal distinctions between source categories such as between interrogation and debriefing. In others, the distinction is in the purpose of the questioning. Regardless of the type of activity, or goal of the collection effort, HUMINT collection operations must be characterized by effective support, planning, and management.

Chapter 1

Introduction

INTELLIGENCE BATTLEFIELD OPERATING SYSTEM

1-1. The Intelligence battlefield operating system (BOS) is one of seven operating systems—Intelligence, maneuver, fire support, air defense, mobility/countermobility/survivability, combat service support (CSS), and command and control—that enable commanders to build, employ, direct, and sustain combat power. The Intelligence BOS is a flexible force of Intelligence personnel, organizations, and equipment. Individually and collectively, these assets generate knowledge of and products portraying the enemy and the environmental features required by a command planning, preparing, executing, and assessing operations. Inherent within the Intelligence BOS is the capability to plan, direct, and synchronize intelligence, surveillance, and reconnaissance (ISR) operations; collect and process information; produce relevant intelligence; and disseminate intelligence and critical information in an understandable and presentable form to those who need it, when they need it. As one of the seven disciplines of the Intelligence BOS, HUMINT provides a capability to the supported commander in achieving information superiority on the battlefield.

INTELLIGENCE PROCESS

1-2. Intelligence operations consist of the functions that constitute the intelligence process: **plan, prepare, collect, process, produce**, and the common tasks of **analyze, disseminate**, and **assess** that occur throughout the intelligence process. Just as the activities of the operations process overlap and recur as circumstances demand, so do the functions of the intelligence process. Additionally, the analyze, disseminate, and assess tasks of the intelligence process occur continuously throughout the intelligence process. (See Figure 1-1.)

- **Plan.** This step of the intelligence process consists of activities that include assessing the situation, envisioning a desired outcome (also known as setting the vision), identifying pertinent information and intelligence requirements, developing a strategy for ISR operations to satisfy those requirements, directing intelligence operations, and synchronizing the ISR effort. The commander's intent, planning guidance, and commander's critical information requirements (CCIRs) (priority information requirements [PIRs] and friendly force information requirements [FFIRs]) drive the planning of intelligence operations. Commanders must involve their supporting staff judge advocate (SJA) when planning intelligence operations (especially HUMINT operations). Planning, managing, and coordinating these operations are continuous activities necessary to obtain information and produce intelligence essential to decisionmaking.

- **Prepare.** This step includes those staff and leader activities that take place upon receiving the operations plan (OPLAN), operations order (OPORD), warning order (WARNO), or commander's intent to improve the unit's ability to execute tasks or missions and survive on the battlefield.

- **Collect.** Recent ISR doctrine necessitates that the entire staff, especially the G3/S3 and G2/S2, must change their reconnaissance and surveillance (R&S) mindset to conducting ISR. The staff must carefully focus ISR on the CCIR but also enable the quick re-tasking of units and assets as the situation changes. This doctrinal requirement ensures that the enemy situation, not just our OPLAN, "drives" ISR operations. Well-developed procedures and carefully planned flexibility to support emerging targets, changing requirements, and the need to support combat assessment are critical. The G3/S3 and G2/S2 play a critical role in this challenging task that is sometimes referred to as "fighting ISR" because it is so staff intensive during planning and execution (it is an operation within the operation). Elements of all units on the battlefield obtain information and data about enemy forces, activities, facilities, and resources as well as information concerning the environmental and geographical characteristics of a particular area.

- **Process.** This step converts relevant information into a form suitable for analysis, production, or immediate use by the commander. Processing also includes sorting through large amounts of collected information and intelligence (multidiscipline reports from the unit's ISR assets, lateral and higher echelon units and organizations, and non-MI elements in the battlespace). Processing identifies and exploits that information which is pertinent to the commander's intelligence requirements and facilitates situational understanding. Examples of processing include developing film, enhancing imagery, translating a document from a foreign language, converting electronic data into a standardized report that can be analyzed by a system operator, and correlating dissimilar or jumbled information by assembling like elements before the information is forwarded for analysis.

- **Produce.** In this step, the G2/S2 integrates evaluated, analyzed, and interpreted information from single or multiple sources and disciplines into finished intelligence products. Like collection operations, the G2/S2 must ensure the unit's information processing and intelligence production are prioritized and synchronized to support answering the collection requirements.

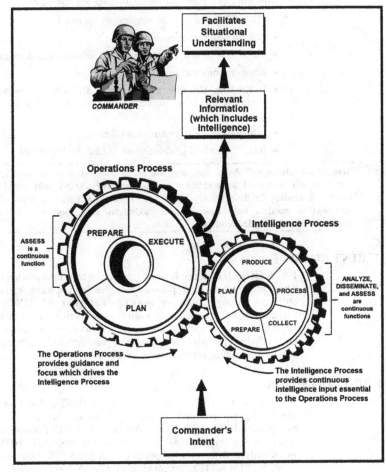

Figure 1-1. Intelligence Process.

1-3. For more information on the Intelligence process, see FM 2-0.

_____

HUMAN INTELLIGENCE

1-4. HUMINT is the collection of information by a trained HUMINT collector (military occupational specialties [MOSs] 97E, 351Y [formerly 351C], 351M [formerly 351E], 35E, and 35F), from people and their associated documents and media sources to identify elements, intentions, composition, strength, dispositions, tactics, equipment, personnel, and capabilities. It uses human sources as a tool and a variety of collection methods, both passively and actively, to gather information to satisfy the commander's intelligence requirements and cross-cue other intelligence disciplines.

1-5. HUMINT tasks include but are not limited to—

- Conducting source operations.
- Liaising with host nation (HN) officials and allied counterparts.
- Eliciting information from select sources.
- Debriefing US and allied forces and civilian personnel including refugees, displaced persons (DPs), third-country nationals, and local inhabitants.
- Interrogating EPWs and other detainees.
- Initially exploiting documents, media, and materiel.

Note. In accordance with Army regulatory and policy guidance, a select set of intelligence personnel may be trained and certified to conduct certain HUMINT tasks outside of those which are standard for their primary MOS. Such selection and training will qualify these personnel to conduct only those specific additional tasks, and will not constitute qualifications as a HUMINT collector.

HUMINT SOURCE

1-6. A HUMINT source is a person from whom information can be obtained. The source may either possess first- or second-hand knowledge normally obtained through sight or hearing. Potential HUMINT sources include threat, neutral, and friendly military and civilian personnel. Categories of HUMINT sources include but are not limited to detainees, refugees, DPs, local inhabitants, friendly forces, and members of foreign governmental and non-governmental organizations (NGOs).

HUMINT COLLECTOR

1-7. For the purpose of this manual, a HUMINT collector is a person who is specifically trained and certified for, tasked with, and engages in the collection of information from individuals (HUMINT sources) for the purpose of answering intelligence information requirements. HUMINT collectors specifically include enlisted personnel in MOS 97E, Warrant Officers (WOs) in MOS 351M (351E) and MOS 351Y (351C), commissioned officers in MOS 35E and MOS 35F, select other specially trained MOSs, and their Federal civilian employee and civilian contractor counterparts. These specially trained and certified individuals are the **only** personnel authorized to conduct HUMINT collection operations, although CI agents also use HUMINT collection techniques in the conduct of CI operations. HUMINT collection operations must be conducted in accordance with applicable law and policy. Applicable law and policy include US law; the law of war; relevant international law; relevant directives including DOD Directive 3115.09, "DOD Intelligence Interrogations, Detainee Debriefings, and Tactical Questioning"; DOD Directive 2310.1E, "The Department of Defense Detainee Program"; DOD instructions; and military execute orders including FRAGOs. Additional policies and regulations apply to management of contractors engaged in HUMINT collection. (See Bibliography for additional references on contractor management.) HUMINT collectors are not to be confused with CI agents, MOS 97B and WO MOS 351L (351B). CI agents are trained and certified for, tasked with, and carry out the mission of denying the enemy the ability to collect information on the activities and intentions of friendly forces. Although personnel in 97E and 97B MOSs may use similar methods to carry out their missions, commanders should not use them interchangeably. See Figure 1-2 for HUMINT and CI functions.

PHASES OF HUMINT COLLECTION

1-8. Every HUMINT questioning session, regardless of the methodology used or the type of operation, consists of five phases. The five phases of HUMINT collection are planning and preparation, approach, questioning, termination, and reporting. They are generally sequential; however, reporting may occur at any point within the process when critical information is obtained and the approach techniques used will be reinforced as required through the questioning and termination phases.

Planning and Preparation

1-9. During this phase, the HUMINT collector conducts the necessary research and operational planning in preparation for a specific collection effort with a specific source. Chapter 7 discusses this phase in detail.

Approach

1-10. During the approach phase, the HUMINT collector establishes the conditions of control and rapport to gain the cooperation of the source and to facilitate information collection. Chapter 8 discusses approach and termination strategies in detail.

Questioning

1-11. During the questioning phase, the HUMINT collector uses an interrogation, debriefing, or elicitation methodology to ask a source questions systematically on relevant topics, collect information in response to the intelligence tasking, and ascertain source veracity. Chapter 9 discusses questioning techniques in detail. (See Appendix B for a source and reliability matrix.)

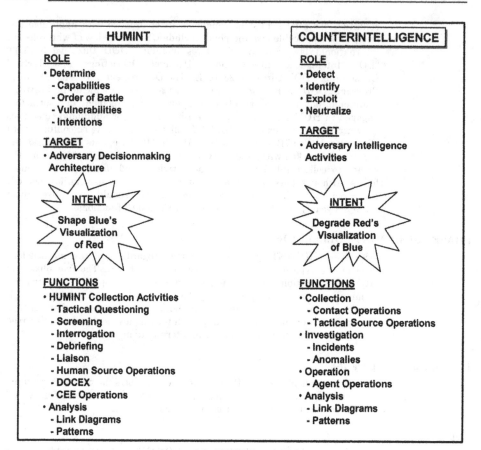

Figure 1-2. HUMINT and CI Functions.

Termination

1-12. During the termination phase, the HUMINT collector completes a questioning session and establishes the necessary conditions for future collection from the same source by himself or another HUMINT collector. (See Chapter 8.)

Reporting

1-13. During the reporting phase, the HUMINT collector writes, edits, and submits written, and possibly oral, reports on information collected in the course of a HUMINT collection effort. These reports will be reviewed, edited, and analyzed as they are forwarded through the appropriate channels. Chapter 10 discusses reporting in detail.

HUMINT COLLECTION AND RELATED ACTIVITIES

1-14. HUMINT collection activities include these categories: tactical questioning, screening, interrogation, debriefing, liaison, human source contact operations (SCOs), document exploitation (DOCEX), and captured enemy equipment (CEE) operations. DOCEX and CEE operations are activities supported by HUMINT collection but usually are only conducted by HUMINT collectors when the CEE or captured enemy document (CED) is associated with a source being questioned. In some cases, these determinations may depend on legal distinctions between collection methods such as interrogation and debriefing. In others, the distinction is in the purpose of the questioning. For example, screening is used to identify the knowledgeability and cooperation of a source, as opposed to the other activities that are used to collect information for intelligence purposes.

1-15. The activities may be conducted interactively. For example, a HUMINT collector may be screening a potential source. During the course of the screening, the HUMINT collector identifies that the individual has information that can answer requirements. He might at that point debrief or interrogate the source on that specific area. He will then return to screening the source to identify other potential areas of interest.

1-16. HUMINT collection activities vary depending on the source of the information. Once the type of activity has been determined, leaders use the process of plan, prepare, execute, and assess to conduct the activity. The following are the different types of HUMINT collection activities.

TACTICAL QUESTIONING

1-17. Tactical questioning is expedient initial questioning for information of immediate tactical value. Tactical questioning is generally performed by members of patrols, but can be done by any DOD personnel. (See ST 2-91.6.)

SCREENING

1-18. Screening is the process of identifying and assessing the areas of knowledge, cooperation, and possible approach techniques for an individual who has information of intelligence value. Indicators and discriminators used in screening can range from general appearance, possessions, and attitude to specific questions to assess areas of knowledge and degree of cooperation to establish if an individual matches a predetermined source profile. Screening is not in itself an intelligence collection technique but a timesaving measure that identifies those individuals most likely to have information of value.

1-19. Screening operations are conducted to identify the level of knowledge, level of cooperation, and the placement and access of a given source. Screening operations can also assist in the determination of which discipline or agency can best conduct the exploitation. Chapter 6 discusses screening in detail. Screening operations include but are not limited to—

- Mobile and static checkpoint screening, including screening of refugees and DPs.
- Locally employed personnel screening.

- Screening as part of a cordon and search operation.
- EPW and detainee screening.

INTERROGATION

1-20. Interrogation is the systematic effort to procure information to answer specific collection requirements by direct and indirect questioning techniques of a person who is in the custody of the forces conducting the questioning. Some examples of interrogation sources include EPWs and other detainees. Interrogation sources range from totally cooperative to highly antagonistic. Interrogations may be conducted at all echelons in all operational environments. Detainee interrogation operations conducted at a Military Police (MP) facility, coalition-operated facility, or other agency-operated collection facility are more robust and require greater planning, but have greater logistical support. Interrogations may only be conducted by personnel trained and certified in the interrogation methodology, including personnel in MOSs 97E, 351M (351E), or select others as may be approved by DOD policy. Interrogations are always to be conducted in accordance with the Law of War, regardless of the echelon or operational environment in which the HUMINT collector is operating.

DEBRIEFING

1-21. Debriefing is the process of questioning cooperating human sources to satisfy intelligence requirements, consistent with applicable law. The source usually is not in custody and usually is willing to cooperate. Debriefing may be conducted at all echelons and in all operational environments. The primary categories of sources for debriefing are refugees, émigrés, DPs, and local civilians; and friendly forces.

- **Refugees, Émigrés, DPs, and Local Civilians Debriefing Operations.** Refugee, émigré, and DP debriefing operations are the process of questioning cooperating refugees and émigrés to satisfy intelligence requirements. The refugee may or may not be in custody, and a refugee or émigré's willingness to cooperate need not be immediate or constant. Refugee debriefings are usually conducted at refugee collection points or checkpoints and may be conducted in coordination with civil affairs (CA) or MP operations. Local civilian debriefing operations are the process of questioning cooperating local civilians to satisfy intelligence requirements. As with refugees and émigrés, the local civilians being debriefed may or may not be in custody and the civilian's willingness to cooperate may not be immediate or constant. Debriefing operations must be conducted consistent with applicable law and policy. Applicable law and policy include US law; the law of war; relevant international law; relevant directives including DOD Directive 3115.09, "DOD Intelligence Interrogations, Detainee Debriefings, and Tactical Questioning"; DOD Directive 2310.1E, "The Department of Defense Detainee Program"; DOD instructions; and military execute orders including FRAGOs.
- **Friendly Force Debriefing Operations.** Friendly force debriefing operations are the systematic debriefing of US forces to answer

collection requirements. These operations must be coordinated with US units. (See Chapter 6.)

LIAISON OPERATIONS

1-22. Liaison operations are programs to coordinate activities and exchange information with host country and allied military and civilian agencies and NGOs.

HUMAN SOURCE CONTACT OPERATIONS

1-23. Human SCO are operations directed toward the establishment of human sources who have agreed to meet and cooperate with HUMINT collectors for the purpose of providing information. Within the Army, SCO are conducted by trained personnel under the direction of military commanders. The entire range of HUMINT collection operations can be employed. SCO sources include one-time contacts, continuous contacts, and formal contacts from debriefings, liaison, and contact operations. SCO consist of collection activities that utilize human sources to identify attitude, intentions, composition, strength, dispositions, tactics, equipment, target development, personnel, and capabilities of those elements that pose a potential or actual threat to US and coalition forces. SCO are also employed to develop local source or informant networks that provide early warning of imminent danger to US and coalition forces and contribute to the Military Decision-Making Process (MDMP). See Chapter 5 for discussion of approval, coordination, and review for each type of activity.

DOCEX OPERATIONS

1-24. DOCEX operations are the systematic extraction of information from open, closed, published, and electronic source documents. These documents may include documents or data inside electronic communications equipment, including computers, telephones, Personal Digital Assistants (PDAs), and Global Positioning System (GPS) terminals. This operation is not solely a HUMINT function, but may be conducted by any intelligence personnel with appropriate language support.

1-25. Many CEDs are associated with EPWs and other human sources. Consequently, a HUMINT collector is often the first person to screen them. HUMINT collectors will screen the documents associated with human sources and will extract information of use to them in their immediate collection operation. Any information discovered during this initial screening that might cross-cue another collection effort will be forwarded to the appropriate unit.

1-26. A captured document is usually something that the enemy has written for his own use. For this reason, captured documents are usually truthful and accurate. There are cases in which falsified documents have been permitted to fall into enemy hands as a means of deception but these cases are not the norm. Normal policy of not relying on single-source information should help prevent deceptions of this type from being effective. Documents also do not forget or misinterpret information although it must be remembered that their authors may have. Usually, each document provides a portion of a larger body of information. Each captured document, much like a single piece of a puzzle, contributes to the whole. In addition to tactical intelligence, technical data and political indicators that are important to strategic and national level agencies can sometimes be extracted from captured documents. Captured documents, while not affected by memory loss, are often time sensitive; therefore, they are to be quickly screened for possible exploitation.

CEE OPERATIONS

1-27. CEE includes all types of foreign and non-foreign materiel found on a detainee or on the battlefield that may have a military application or answer a collection requirement. The capturing unit must—

- Recognize certain CEE as having immediate intelligence value, and immediately forward such CEE to the unit's S2. Such items include—
  - All electronic communications equipment with a memory card, including computers, telephones, PDAs, and GPS terminals.
  - All video or photographic equipment.
- Recognize certain CEE as having technical intelligence (TECHINT) value. Such items include—
  - New weapons.
  - All communications equipment not immediately exploitable for HUMINT value.
  - Track vehicles.
  - Equipment manuals.
  - All CEE known or believed to be of TECHINT interest.
- Evacuate the equipment with the detainee.
- Confiscate, tag, and evacuate weapons and other equipment found on the detainee the same as CEDs. (See Appendix D.)
- Secure and report the capture of TECHINT items to the unit's S2 for disposition instructions.

TRAITS OF A HUMINT COLLECTOR

1-28. HUMINT collection is a science and an art. Although many HUMINT collection skills may be taught, the development of a skilled HUMINT collector requires experience in dealing with people in all conditions and under all circumstances. Although there are many intangibles in the definition of a "good" HUMINT collector, certain character traits are invaluable:

- **Alertness.** The HUMINT collector must be alert on several levels while conducting HUMINT collection. He must concentrate on the information being provided by the source and be constantly evaluating the information for both value and veracity based on collection requirements, current intelligence, and other information obtained from the source. Simultaneously, he must be alert not only to what the source says but also to how it is said and the accompanying body language to assess the source's truthfulness, degree of cooperation, and current mood. He needs to know when to give the source a break and when to press the source harder. In addition, the HUMINT collector constantly must be alert to his environment to ensure his personal security and that of his source.

- **Patience and Tact.** The HUMINT collector must have patience and tact in creating and maintaining rapport between himself and the source, thereby enhancing the success of the questioning. Displaying impatience may—
  - Encourage a difficult source to think that if he remains unresponsive for a little longer, the HUMINT collector will stop questioning.
  - Cause the source to lose respect for the HUMINT collector, thereby reducing the HUMINT collector's effectiveness.

- **Credibility.** The HUMINT collector must provide a clear, accurate, and professional product and an accurate assessment of his capabilities. He must be able to clearly articulate complex situations and concepts. The HUMINT collector must also maintain credibility with his source. He must present himself in a believable and consistent manner, and follow through on any promises made as well as never to promise what cannot be delivered.

- **Objectivity and Self-control.** The HUMINT collector must also be totally objective in evaluating the information obtained. The HUMINT collector must maintain an objective and dispassionate attitude regardless of the emotional reactions he may actually experience or simulate during a questioning session. Without objectivity, he may unconsciously distort the information acquired. He may also be unable to vary his questioning and approach techniques effectively. He must have exceptional self-control to avoid displays of genuine anger, irritation, sympathy, or weariness that may cause him to lose the initiative during questioning but be able to fake any of these emotions as necessary. He must not become emotionally involved with the source.

- **Adaptability.** A HUMINT collector must adapt to the many and varied personalities which he will encounter. He must also adapt to all types of locations, operational tempos, and operational environments. He should try to imagine himself in the source's position. By being adaptable, he can smoothly shift his questioning and approach techniques according to the operational environment and the personality of the source.

- **Perseverance.** A tenacity of purpose can be the difference between a HUMINT collector who is merely good and one who is superior. A HUMINT collector who becomes easily discouraged by opposition, non-cooperation, or other difficulties will not aggressively pursue the objective to a successful conclusion or exploit leads to other valuable information.

- **Appearance and Demeanor.** The HUMINT collector's personal appearance may greatly influence the conduct of any HUMINT collection operation and attitude of the source toward the HUMINT collector. Usually an organized and professional appearance will favorably influence the source. If the HUMINT collector's manner reflects fairness, strength, and efficiency, the source may prove more cooperative and more receptive to questioning.

- **Initiative.** Achieving and maintaining the initiative are essential to a successful questioning session just as the offensive is the key to success in combat operations. The HUMINT collector must grasp the initiative and maintain it throughout all questioning phases. This does not mean he has to dominate the source physically; rather, it means that the HUMINT collector knows his requirements and continues to direct the collection toward those requirements.

REQUIRED AREAS OF KNOWLEDGE

1-29. The HUMINT collector must be knowledgeable in a variety of areas in order to question sources effectively. The collector must prepare himself for operations in a particular theater or area of intelligence responsibility (AOIR) by conducting research. The G2 can be a valuable source of information for this preparatory research. The HUMINT collector should consult with order of battle (OB) technicians and analysts and collect information from open sources and from the Secret Internet Protocol Router Network (SIPRNET) to enhance his knowledge of the AOIR. Some of these areas of required knowledge are—

- **The area of operations (AO)** including the social, political, and economic institutions; geography; history; language; and culture of the target area. Collectors must be aware of all ethnic, social, religious, political, criminal, tribal, and economic groups and the interrelationships between these groups.

- **All current and potential threat forces** within the AOIR and their organization, equipment, motivation, capabilities, limitations, and normal operational methodology.

- **Applicable law and policy that might affect HUMINT collection activities. Applicable law and policy include US law; the law of war; relevant international law; relevant directives including DOD Directive 3115.09, "DOD Intelligence Interrogations, Detainee Debriefings, and Tactical Questioning"; DOD Directive 2310.1E, "The Department of Defense Detainee Program"; DOD instructions; and military execute orders including FRAGOs.** HUMINT collectors are subject to applicable law, which includes US law, the law of war (including the Geneva Conventions as applicable), and relevant international law. Additionally, local agreements with HNs or allies and the applicable execute orders and rules of engagement (ROE) may further restrict HUMINT collection activities. However, these documents cannot permit interrogation actions that would be illegal under applicable US or international law.

- **The collection requirements,** including all specific information requirements (SIRs) and indicators that will lead to the answering of the intelligence requirements.

- **Cultural awareness** in the various AOs will have different social and regional considerations that affect communications and can affect the conduct of operations. These may include social taboos, desired behaviors, customs, and courtesies. The staff must include this information in pre-deployment training at all levels to ensure that personnel are properly equipped to interact with the local populace.

1-30. There are other areas of knowledge that help to develop more effective questioning:

- **Proficiency in the target language**. The HUMINT collector can normally use an interpreter (see Chapter 11) and machine translation as they are developed to conduct questioning. Language proficiency is a benefit to the HUMINT collector in a number of ways: He can save time in questioning, be more aware of nuances in the language that might verify or deny truthfulness, and better control and evaluate interpreters.
- **Understanding basic human behavior**. A HUMINT collector can best adapt himself to the source's personality and control of the source's reactions when he understands basic behavioral factors, traits, attitudes, drives, motivations, and inhibitions. He must not only understand basic behavioral principles but also know how these principles are manifested in the area and culture in which he is operating.
- **Neurolinguistics**. Neurolinguistics is a behavioral communication model and a set of procedures that improve communication skills. The HUMINT collector should read and react to nonverbal communications. He must be aware of the specific neurolinguistic clues of the cultural framework in which he is operating.

CAPABILITIES AND LIMITATIONS

CAPABILITIES

1-31. HUMINT collection capabilities include the ability to—

- Collect information and cross-cue from an almost endless variety of potential sources including friendly forces, civilians, detainees, and source-related documents.
- Focus on the collection of detailed information not available by other means. This includes information on threat intentions and local civilian and threat force attitudes and morale. It also includes building interiors and facilities that cannot be collected on by other means due to restrictive terrain.
- Corroborate or refute information collected from other R&S assets.
- Operate with minimal equipment and deploy in all operational environments in support of offensive, defensive, stability and reconstruction operations, or civil support operations. Based on solid planning and preparation, HUMINT collection can provide timely information if deployed forward in support of maneuver elements.

LIMITATIONS

1-32. HUMINT collection limitations include—

- Interpersonal abilities. HUMINT is dependent on the subjective interpersonal capabilities of the individual rather than on the abilities to operate collection equipment. HUMINT collection capability is based on experience within a specific AO that can only be developed over time.

- Identification of knowledgeable sources. There is often a multitude of potential HUMINT sources. Information in response to specific requirements can only be collected if sources are available and identified that have that information.

- Limited numbers. There are never enough HUMINT collectors to meet all requirements. Limited assets must be prioritized in support of units and operations based on their criticality.

- Time limitations. HUMINT collection, particularly source operations, takes time to develop. Collection requirements must be developed with sufficient lead-time for collection.

- Language limitations. Although HUMINT collectors can normally use an interpreter, a lack of language proficiency by the collector can significantly slow collection efforts. Such language proficiency takes time to develop.

- Misunderstanding of the HUMINT mission. HUMINT collectors are frequently used incorrectly and assigned missions that belong to CA, MP, interpreter or translators, CI, or other operational specialties.

- Commanders' risk management. Maneuver commanders, in weighing the risks associated with employing HUMINT collection teams (HCTs), should seriously consider the potential loss of a wealth of information such as enemy activities, locations of high-value personnel, and threats to the force that they will incur if they restrict HCT collection activities. J/G2Xs, operational management teams (OMTs), and HCT leaders must educate maneuver commanders on the benefits of providing security for HCTs and employing them in accordance with their capabilities.

- Legal obligations. Applicable law and policy govern HUMINT collection operations. Applicable law and policy include US law; the law of war; relevant international law; relevant directives including DOD Directive 3115.09, "DOD Intelligence Interrogations, Detainee Debriefings, and Tactical Questioning"; DOD Directive 2310.1E, "The Department of Defense Detainee Program"; DOD instructions; and military execute orders including FRAGOs. HUMINT operations may be further restricted by Status of Forces Agreements (SOFAs) and other agreements, execute orders and ROE, local laws, and an operational umbrella concept. Such documents, however, cannot permit interrogation actions that are illegal under applicable law.

- Connectivity and bandwidth requirements. With the exception of the size, activity, location, unit, time, equipment (SALUTE) report, most HUMINT reporting requires considerable bandwidth. Deployed

HUMINT teams must be able to travel to, and report from, all areas of the battlefield. Digital communication equipment must be able to provide reliable connectivity with teams' reporting channels and sufficient bandwidth for transmission of reports, including digital imagery.

• Timely reporting and immediate access to sources. Except in tactical situations when HUMINT collectors are in immediate support of maneuver units, HUMINT collection and reporting takes time. In stability and reconstruction operations, sources need to be assessed and developed. Once they are developed, they need to be contacted which often takes time and coordination. In offensive and defensive operations, HUMINT collection at detainee holding areas sometimes may still be timely enough to meet tactical and operational requirements. See paragraphs 3-2 and 3-7 for more information on offensive and defensive operations.

Chapter 2

Human Intelligence Structure

ORGANIZATION AND STRUCTURE

2-1. The success of the HUMINT collection effort depends on a complex interrelationship between command and control (C2) elements, requirements, technical control and support, and collection assets. Each echelon of command has its supporting HUMINT elements although no MI organization in the Army is robust enough to conduct sustained HUMINT operations under all operational environments using only its organic HUMINT assets. HUMINT units have specific support requirements to the commander. HUMINT units must be flexible, versatile, and prepared to conduct HUMINT collection and analysis operations in support of any echelon of command. A coherent C2 structure within these HUMINT organizations is necessary in order to ensure successful, disciplined, and legal HUMINT operations. This structure must include experienced commissioned officers, warrant officers, and senior NCOs conscientiously discharging their responsibilities and providing HUMINT collectors with guidance from higher headquarters.

2-2. Regardless of the echelon, there are four basic elements that work together to provide the deployed commander with well-focused, thoroughly planned HUMINT support. The four elements are staff support, analysis, C2, and collection. Each piece of the infrastructure builds on the next and is based on the size, complexity, and type of operation as shown in Figure 2-1.

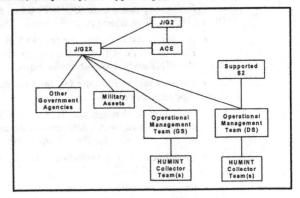

Figure 2-1. Tactical HUMINT Organization.

HUMINT CONTROL ORGANIZATIONS

2-3. HUMINT control organizations are the means by which a commander exercises command of a unit's operations. HUMINT control organizations are vital to the effective use of HUMINT collection assets. HUMINT control organizations consist of the C/J/G/S2X and the HUMINT operations cell (HOC) at the brigade and above level and the OMTs at the battalion and below level.

C/J/G/S2X

2-4. The C/J/G/S2X is a staff element subordinate to the C/J/G/S2, is the primary advisor on HUMINT and CI, and is the focal point for all HUMINT and CI activities within a joint task force (JTF) (J2X), an Army component task force (G2X) or a brigade combat team (BCT) (S2X). The 2X can be organic to the unit staff or can be attached or under operational control (OPCON) to the staff from another organization such as the theater MI brigade. The C/J/G/S2X is part of a coherent architecture that includes organic HUMINT assets and HUMINT resources from national, theater, and non-DOD HUMINT organizations.

2-5. The C/J2X is responsible for controlling Joint Force HUMINT assets, coordinating all HUMINT and CI collection activities, and keeping the joint force C/J/2 informed on all HUMINT and CI activities conducted in the joint force area of responsibility (AOR). The C/J2X is also part of the review and recommendation process concerned with the retention or release of detainees. HUMINT reports maintained at the C/J2X are considered during the review for release process. The C/J2X consists of the 2X Officer, a HOC, a Counterintelligence Coordination Authority (CICA), a HUMINT Analysis Cell (HAC), and a CI Analysis Cell (CIAC). At all echelons, the 2X should also include an Operational Support Cell (OSC) staffed to operate 24 hours a day. The authority and operational responsibilities of a C/J2X in combined or joint contingency operations (CONOP) takes precedence over service-specific CI and HUMINT technical control agencies. Specifically, the C/J/G/S2X—

- Accomplishes technical control and support, and deconfliction of all HUMINT and CI assets through the Army component G2X, the HUMINT and CI operations sections, or the OMTs.

- Participates in planning for deployment of HUMINT and CI assets in support of operations.

- Coordinates, through the HOC and the CICA, all HUMINT and CI activities to support intelligence collection and the intelligence aspects of force protection for the deployed commander.

- Coordinates and deconflicts all HUMINT and CI operations within the operational area.

- Coordinates with the senior US national intelligence representative for specific operational approval when required by standing agreements.

- Is the release authority for HUMINT reporting at his echelon and only releases reports to the all-source system after ensuring all technical control measures for reporting have been met.

- Coordinates with other HUMINT collection agencies not under the control of the command, such as Defense Intelligence Agency (DIA), Central Intelligence Agency (CIA), and Federal Bureau of Investigation (FBI).

- Does not exercise OPCON over HUMINT and CI assets assigned, attached, or reinforcing the unit; however, he is the staff support responsible for creating a cohesive HUMINT and CI effort.

- Coordinates with non-DOD agencies conducting HUMINT collection operations in the joint area of operations (JAO) to ensure deconfliction of sources, informants, or contacts and the HUMINT reporting that is generated by these collection operations.

2-6. The J2X will maintain technical control (see para 4-10) of all CI investigative actions within its AOIR; however, all investigative matters will be in accordance with DOD policies, joint or Military Department doctrine, applicable US law and policy, SOFAs, or other International Standardization Agreements (ISAs). The J2X will advise the responsible Theater CICA (TCICA) of any Army CI element conducting investigative activities that fall under the purview of AR 381-20.

OPERATIONS SUPPORT CELL (OSC)

2-7. The OSC in the C/J/G/S2X staff will maintain the consolidated source registry for all HUMINT and CI activities in the unit's designated AOIR. The OSC will provide management of intelligence property book operations, source incentive programs, and intelligence contingency funds (ICFs) for subordinate HUMINT and CI elements. The OSC responsibilities also include requests for information (RFIs) and/or source-directed requirements (SDRs) management and the release of intelligence information reports (IIRs).

COUNTERINTELLIGENCE COORDINATION AUTHORITY

2-8. The CICA is assigned under the J/G2X and coordinates all CI activities within its designated AOIR. (See FM 34-60 for a detailed explanation of the CI mission.) The CICA—

- Provides technical support to all CI assets and coordinates and deconflicts CI activities in the deployed AOIR.

- Coordinates and supervises CI investigations and collection activities conducted by all services and components in the AOIR.

- Establishes and maintains the theater CI source database.

- Coordinates with the HOC for CI support to detention, interrogation, refugee, and other facilities.

- Manages requirements and taskings for CI collectors in the AO in coordination with the HOC.

- Expedites preparation of CI reports and their distribution to consumers at all levels.

- Coordinates CI activities with senior CI officers from all CI organizations on the battlefield.

- Performs liaison with HN and US national level CI organizations.
- Informs the appropriate TCICA when Army CI elements are conducting CI investigative activities within the purview of AR 381-20.

HUMINT OPERATIONS CELL

2-9. The HOC is assigned under the J/G2X to track all HUMINT activities in the AOIR. The J/G2X uses this information to advise the senior intelligence officer (SIO) on all HUMINT activities conducted within the AOIR. The HOC—

- Provides technical support to all HUMINT collection operations and deconflicts HUMINT collection operations in the designated AOIR.
- Establishes and maintains a consolidated HUMINT source database in coordination with the CICA.
- Coordinates with collection managers and the HAC to identify collection requirements and to ensure requirements are met.
- Coordinates the activities of HUMINT collectors assigned or attached to interrogation, debriefing, refugee, DOCEX, and other facilities.
- Manages requirements and taskings for HUMINT collectors in the AOIR, in coordination with the CICA.
- Expedites preparation of intelligence reports and their distribution to consumers at all levels.
- Performs liaison with HN and US national HUMINT organizations.

OPERATIONAL MANAGEMENT TEAM

2-10. A HUMINT OMT consists of senior individuals in MOS 351M (351E) and MOS 97E. Each OMT can control 2 to 4 HCTs depending upon assigned mission and operational tempo (OPTEMPO). The OMT performs a necessary function when two or more HCTs deploy by assisting the HUMINT element commander in tasking and providing technical support to assigned or attached HCTs. The OMT is optimally collocated with the command post (CP) of the supported unit. However, it must be located where it can provide oversight of team operations and best support the dissemination of tasking, reports, and technical data between the unit and the deployed collection assets. When a higher echelon augments subordinate elements with collection teams, it should include proportional OMT augmentation. When a single collection team is attached in direct support (DS) of a subordinate element, the senior team member exerts mission and technical control over the team. The OMT—

- Provides operational and technical control and guidance to deployed HCTs.
- Normally consists of a WO and noncommissioned officers (NCOs) whose experience and knowledge provide the necessary guidance for effective team collection operations.
- Manages the use of ICFs and incentives for the HCTs.
- Provides the collection focus for HCTs.

- Provides quality control and dissemination of reports for subordinate HCTs.
- Directs the activities of subordinate HCTs and controls their operations.
- Conducts limited single-discipline HUMINT analysis and mission analysis for the supported commander.
- Acts as a conduit between subordinate HCTs, the HOC, and the C/J/G/S2X.
- Reports the HCT mission and equipment status to the HOC and the command element.

HUMINT COLLECTION TEAM

2-11. HCTs are the elements that collect information from human sources. The HUMINT collectors deploy in teams of approximately four personnel in MOS 97E (HUMINT Collector) and MOS 351M (351E) (HUMINT Technician).

2-12. The HCT may be augmented based on factors of mission, enemy, terrain and weather, troops and support available, time available, and civil considerations (METT-TC). Interpreters from the RC or civilian contractors with appropriate security clearances are added when necessary. TECHINT personnel or other specific subject-matter experts (SMEs) may augment the team to meet technical collection requirements. Another example would be pairing HUMINT collectors with dedicated analysts to provide sharper focus to the interrogation effort. In fixed detention facilities, these HUMINT collector or analyst relationships may become more enduring. Commanders are not encouraged to mix HUMINT collectors and CI agents on a single team. Doing so seriously undermines the ability to conduct both the HUMINT collection and CI missions simultaneously. However, commanders may find times when METT-TC factors make it reasonable to augment a CI team with HUMINT support for a mission, or vice versa.

COMMAND DEBRIEFING TEAM

2-13. A command debriefing team is normally not a table of organization and equipment (TOE) organization but may be task organized to meet mission requirements. This task-organized team is normally OPCON to the HOC. Although more prevalent during stability and reconstruction operations, senior personnel will often acquire information of intelligence interest during the normal course of their duties. The HUMINT collection assets, particularly at division echelon or higher, will normally task organize a team of more senior, experienced individuals to debrief these senior unit personnel. In offensive and defensive operations, this same team is prepared to interrogate high-value detainees (including EPWs) or debrief senior civilians. The command debriefing team should not be confused with the G2/S2 debriefing program, which also is critical and is an important conduit of information.

DOCUMENT EXPLOITATION TEAM

2-14. DOCEX teams are normally found at theater and national level organizations. Lower echelon HCTs may also be designated to perform the DOCEX mission based upon mission parameters and linguist availability. However, if organic assets are used, there will be a mission tradeoff. Dependent on the priority of exploitation and volume of documents, HCTs assigned the DOCEX mission may be augmented by military, civilian, or contractor personnel to accomplish their assigned mission. During operations, the DOCEX team will normally screen documents, extract information, and expedite the evacuation of documents to the Joint or Theater Document Exploitation Facility.

HUMINT ANALYSIS AND PRODUCTION ORGANIZATIONS

2-15. HUMINT analysis and production organizations analyze information collected from HUMINT sources, support the requirements management (RM) system, and produce single-discipline intelligence products. HUMINT analysis and production are conducted at all echelons, separate brigades, and higher. (See Chapter 12 for a description of the HUMINT analysis system and methodologies.)

HUMINT ANALYSIS CELL

2-16. The HAC is part of the J/G2X; however, it may be collocated with an analysis and control element (ACE) or Joint Intelligence Support Element (JISE) single-source enclave depending on facilities and operational environment considerations. The HAC works closely with the all-source intelligence elements and the CIAC to ensure that HUMINT reporting is incorporated into the all-source analysis and common operational picture (COP). The HAC is the "fusion point" for all HUMINT reporting and operational analysis in the JISE and ACE. It determines gaps in reporting and coordinates with the RM to cross-cue other intelligence sensor systems. The HAC—

- Produces and disseminates HUMINT products and provides input to intelligence summaries (INTSUMs).
- Uses analytical tools found at the ACE or JISE to develop long-term analyses and provides reporting feedback that supports the HOC, OMTs, and HCTs.
- Provides analytical expertise to the C/J/G/S2X, HOC, and OMTs.
- Produces country and regional studies tailored to HUMINT collection.
- Compiles target folders to assist C/J/G/S2X assets in focusing collection efforts.
- Analyzes and reports on trends and patterns found in HUMINT reporting.
- Analyzes source reliability and credibility as reflected in reporting and communicates that analysis to the collector.
- Develops and maintains databases specific to HUMINT collection activities.
- Produces HUMINT requirements.

- Answers HUMINT-related RFIs.
- Identifies collection gaps and provides context for better collection at their echelon.

JOINT INTERROGATION AND DEBRIEFING CENTER ANALYSIS SECTION

2-17. This section ensures that all members of the Joint Interrogation and Debriefing Center (JIDC) (see para 5-102) are aware of the current situation through the distribution of INTSUMs and products from external agencies. The Analysis Section also supports the JIDC by—

- Providing situation update briefings to all facility personnel every 12 hours.
- Preparing research and background packets and briefings for interrogations and debriefings.
- Developing indicators for each intelligence requirement to support screening operations.
- Conducting single-discipline HUMINT analysis based on collected information to support further collection efforts.
- Correlating reports produced by the JIDC to facilitate analysis at higher levels.
- Answering RFIs from interrogators and formulating RFIs that cannot be answered by the analytical section on behalf of the interrogators.
- Reviewing IIRs and extracting information into analysis tools tailored to support the interrogation process.
- Pursuing products and resources to support the interrogation effort.

HUMINT ANALYSIS TEAM

2-18. The HUMINT analysis team (HAT) is subordinate to the G2 ACE. The HAT supports the G2 in the development of IPB products and in developing and tailoring SIRs to match HUMINT collection capabilities.

Chapter 3

HUMINT in Support of Army Operations

3-1. Army doctrine for full spectrum operations recognizes four types of military operations: offensive, defensive, stability and reconstruction, and civil support. Missions in any environment require the Army to conduct or be prepared to conduct any combination of these operations. HUMINT assets will be called on to provide information in support of all four operations. Simultaneous operations, for example elements of a force conducting offensive operations while other elements are engaged in stability and reconstruction operations, will cause a similar division of the limited HUMINT assets based on METT-TC.

OFFENSIVE OPERATIONS

3-2. Offensive operations aim at destroying or defeating the enemy. Rapid maneuver, constantly changing situations, and a vital need for intelligence support at the point of contact influence HUMINT missions during offensive operations. The guiding principle to the use of HUMINT in support of offensive operations is to minimize the time between when friendly forces encounter potential sources (detainees, refugees, and local civilians) and when a HUMINT collector screens them.

3-3. During offensive operations, at echelons corps and below, HCTs normally operate in the engaged maneuver brigades' AOs and are further deployed in support of maneuver battalions based on advice from the OMTs. These collection assets may be in general support (GS) of the parent brigade or in DS of the maneuver battalions, reconnaissance squadrons, and other forward-deployed maneuver assets. The HCTs and their supporting control structure are deployed in accordance with METT-TC based on three principles:

- The relative importance of that subordinate element's operations to the overall parent unit's scheme of maneuver and the overall ISR plan.

- The potential for that subordinate element to capture detainees, media and materiel, or to encounter civilians on the battlefield.

- The criticality of information that could be obtained from those sources to the success of the parent unit's overall OPLANs.

3-4. HUMINT missions in support of offensive operations include screening and interrogating EPWs and other detainees, questioning and debriefing civilians in the supported unit's AO, and conducting DOCEX, limited to extracting information of immediate tactical value. EAC assets normally support offensive operations through theater interrogation and debriefing facility operations and mobile interrogation teams. These facilities are better equipped to conduct in-depth interrogations and DOCEX, so it is imperative that EPWs and other detainees who will be evacuated to theater facilities be transported there as soon as possible.

HUMINT IN SUPPORT OF FORCED ENTRY OPERATIONS

3-5. Forced entry operations (FEOs) are offensive operations conducted to establish an initial military presence in a target area in the face of expected enemy opposition. HUMINT collection assets may be able to provide vital information to tactical commanders in the critical early stages of the entry operation. Key considerations for HUMINT support to FEOs include:

- HUMINT collectors attached or under OPCON of the initial force package to provide HUMINT collection support for the entry force. Collection teams will normally operate in support of battalion-sized or smaller elements. HUMINT collection assets should be integrated early and should participate in all aspects of planning and training, including rehearsals, to smoothly integrate and execute operations.

- HUMINT assets supporting the entry force must include proportional OMT elements. For example, if 2 to 4 teams are attached to a maneuver brigade, an OMT also needs to be attached. Even if the teams are further attached to maneuver battalions, there must be an OMT at the brigade level to coordinate and control HUMINT collection activities.

- HCTs and OMTs must be as mobile and as survivable as the entry forces. Team leaders should ensure that the supported unit will be able to provide maintenance support to the team vehicles, as appropriate, in accordance with the support relationship.

- Attached or OPCON HUMINT teams must have robust communications connectivity with the supported unit and must have reach connectivity through their OMT.

- HCTs must contain organic or attached language capability in order to conduct HUMINT collection effectively during FEO. It is unlikely that the teams can be augmented with attached civilian interpreters during this type of operation.

HUMINT IN SUPPORT OF EARLY ENTRY OPERATIONS

3-6. Early entry operations differ from FEOs in that early entry operations do not anticipate large-scale armed opposition. Early entry operations establish or enhance US presence, stabilize the situation, and shape the environment for follow-on forces. HUMINT collection provides critical support to defining the operational environment and assessing the threat to US forces. The considerations listed above for FEOs apply equally to early entry operations.

DEFENSIVE OPERATIONS

3-7. Defensive operations defeat an enemy attack, buy time, economize forces, hold the enemy in one area while attacking in another, or develop conditions favorable for offensive operations. Forces conducting defensive operations must be able to identify rapidly the enemy's main effort and rapidly assess the operational conditions to determine the timing of counter-offensive or other operations. HUMINT support to defensive operations centers on the ability to provide the forward-deployed maneuver commander with information and intelligence of immediate tactical value. HUMINT assets should be placed in the AO of the forward elements to minimize the time between when friendly forces encounter potential sources (detainees, refugees, local civilians) and when a HUMINT collector screens them. HUMINT collectors are placed where the potential for HUMINT collection and the criticality of the information are greatest.

3-8. In defensive operations, it may be necessary to divide the HUMINT assets equally among the subordinate elements to provide area coverage until the primary enemy threat is identified. The HUMINT C2 elements (team leader, OMTs, and unit C2) must be prepared to task organize rapidly and shift resources as the situation dictates, based on the changing situation and higher headquarters FRAGO. HUMINT missions in defensive operations normally include interrogation of detainees, refugee debriefings, and assisting in friendly force patrol debriefings.

STABILITY AND RECONSTRUCTION OPERATIONS

3-9. Stability and reconstruction operations sustain and exploit security and control over areas, populations, and resources. They employ military and civilian capabilities to help establish order that advances US interests and values. The immediate goal often is to provide the local populace with security, restore essential services, and meet humanitarian needs. The long-term goal is to help develop indigenous capacity for securing essential services, a viable market economy, rule of law, democratic institutions, and robust civil society. Stability and reconstruction operations involve both coercive and cooperative actions. They may occur before, during, and after offensive and defensive operations; however, they also occur separately, usually at the lower end of the range of military operations. The primary focus of the HCTs during stability and reconstruction operations is to answer the commander's information requirements (IRs) and provide support to force protection. In stability and reconstruction operations, the HUMINT collectors must be able to maintain daily contact with the local population. The nature of the threat in stability operations can range from conventional forces to terrorists and organized crime and civil disturbances. Consequently, intelligence requirements can vary greatly. Examples of HUMINT collection requirements include TECHINT to support arms control; extensive political information and demographic data; order of battle (OB) regarding several different former warring factions during peace operations; or extremely detailed target data. HUMINT collectors also help to ascertain the feelings, attitudes, and activities of the local populace. Stability and reconstruction operations may be conducted in coordination with other US departments and agencies, and in conjunction with other countries and international organizations.

3-10. Centralized management and databasing are key to successful HUMINT operations. The HUMINT assets may operate in GS to the parent unit or operate in the AO of subordinate elements of the parent unit. For example, in a division AO, the HCTs would normally operate in DS to the division but each team would normally have an AOIR that corresponds to the AO of the division's brigades or battalion task forces. There is close coordination between the HUMINT staff officer (C/J/G/S2X) and the OMTs to synchronize HUMINT operations properly, to develop the overall threat awareness, and to deconflict sources. The HCTs screen and debrief contacts to increase the security posture of US forces, to provide information in response to command collection requirements, and to provide early warning of threats to US forces. They may also interrogate detainees if permitted to do so by the mission-specific orders and in accordance with applicable law and policy. Applicable law and policy include US law; the law of war; relevant international law; relevant directives including DOD Directive 3115.09, "DOD Intelligence Interrogations, Detainee Debriefings, and Tactical Questioning"; DOD Directive 2310.1E, "The Department of Defense Detainee Program"; DOD instructions; and military execute orders including FRAGOs.

3-11. Many stability and reconstruction operations are initiated with the establishment of a lodgment or base area. There is a subsequent expansion of operations to encompass the entire AO. The general concept of an HCT's operation is that of a two-phased effort. In the initial phase, the HCT establishes concentric rings of operations around the US forces starting from the supported unit's base of operations and working outward. Each ring is based on the threat environment and the commander's need to develop his knowledge of the tactical situation. The second, or continuation phase, begins once the initial information collection ring is established. The initial ring is not abandoned but rather is added to as the HCT shifts its focus to expand and establish the second and successive rings. The amount of time spent establishing each ring is situationally dependent.

INITIAL PHASE

3-12. The initial phase of stability and reconstruction operations is used to lay the foundation for future team operations. In general, the priority of effort is focused inward on security. The HCT conducts initial and follow-up screenings of locally employed personnel, to establish base data for subsequent source operations. The supported unit S2, with the assistance of the HUMINT team leader, establishes procedures to debrief reconnaissance and surveillance assets operating in the supported unit AO, as well as regular combat patrols or logistics convoys. The HCT lays the groundwork for future collection efforts by establishing liaison with local authorities, as well as developing plans and profiles for HUMINT collection. While establishing the initial and subsequent rings, the HCT actively seeks to collect PIR information, whether it pertains to the current ring or any other geographic location.

CONTINUATION PHASE

3-13. Following the initial phase, the HCT's focus shifts outward. While the HCT continues performing HUMINT collection and analysis functions within the base camp, it also expands its collection effort to outside the base camp to answer the supported unit's requirements. During the continuation phase, the HCT conducts contact operations with local personnel who may be able to provide information of interest to the local commander or to satisfy the requirements of the tasking or request. The HCT also conducts liaison with local authorities, coalition forces (if present), NGOs, and others whose knowledge or activities may affect the success of the US mission. Any time the HCT is outside the base camp, it must be careful to observe the local population and report what it sees. The activities and attitudes of the general population will often have an effect on the commander's decisions on how to conduct US missions in the area.

LEVELS OF EMPLOYMENT

3-14. HCTs may be employed with varying degrees of contact with the local population. As the degree of contact with the population increases, the quantity and diversity of HUMINT collection increases. In many instances, however, there is a risk to the HCT inherent with increased exposure to the local population. The ability of the HCT members to fit in with the local populace can become very important to their safety. Consequently, the commander should consider exceptions to the ROE, as well as relaxed grooming and uniform standards, to help HCT members blend in and provide additional security. Commanders must consider the culture in which the HCT members will be operating. In some cultures, bearded men are more highly respected than clean-shaven men. Relaxing grooming standards for HCTs in these situations will support the team's ability to collect information. The decision regarding what level to employ an HCT is METT-TC dependent. The risk to the collection assets must be balanced with the need to collect information and to protect the force as a whole. The deployment and use of HUMINT collection assets may be limited by legal restrictions, mission-specific orders, directions from higher headquarters, and the overall threat level. The four basic levels of employment for the HCT are discussed below. Figure 3-1 shows these levels as well as their collection potential versus team security.

Base Camp

- Restricting the HCT to operations within the base camp minimizes the risk to the team. This action, however, minimizes the collection potential and maximizes the risk to the force as a whole. While restricted to a base camp, the HCT can maintain an extremely limited level of information collection by—

  - Interviewing walk-in sources and locally employed personnel.

  - Debriefing combat and ISR patrols.

  - Conducting limited local open-source information collection.

- This mode of deployment should be used only when dictated by operational restrictions. These would be at the initial stages of stability and reconstruction operations when the operational environment is being assessed, or as a temporary expedient when the force protection level exceeds the ability to provide reasonable protection for the collectors. A supported unit commander is often tempted to keep the HCT "inside the wire" when the force protection level or threat condition (THREATCON) level increases. The supported unit and parent commanders must compare the gains of the HCT collection effort with the risks posed. This is necessary especially during high THREATCON levels when the supported unit commander needs as complete a picture as possible of the threat arrayed against US or multinational forces.

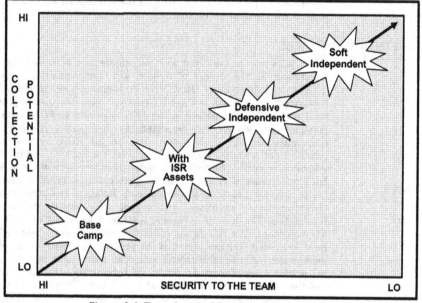

Figure 3-1. Team Level of Employment.

Integrated with Other Operations

- Under some circumstances, when it is not expedient to deploy the HCT independently due to threat levels or other restrictions, it can be integrated into other ongoing operations. The HCT may be employed as part of a combat patrol, ISR patrol, or in support of an MP patrol or stationed at a checkpoint or roadblock. It can also be used to support CA, psychological operations (PSYOP), engineer, or other operations. This method reduces the risk to the team while greatly increasing its collection potential over the confined-to-base-camp method. It has the advantage of placing the team in contact with the local population and allowing it to spot, assess, and interact with potential sources of information.

- The integration into other operations can also facilitate the elicitation of information. However, this deployment method restricts collection by subordinating the team's efforts to the requirements, locations, and timetables of the unit or operation into which it is integrated. Integration can be done at the team or individual collector level. HUMINT collectors should be used only in situations with an intelligence collection potential. It is a waste of a valuable asset to use them in a function that could be performed by a civilian translator.

As an Independent Patrol

- Defensive. One of the key elements of the HCT success is the opportunity to spot, assess, and develop relationships with potential sources of information. Operating as independent patrols, without being tied to ISR or combat assets, enables the HCTs maximum interaction with the local population, thereby maximizing the pool of potential sources of information. The HCT must be integrated into the supported unit's ISR plan and be provided with other command elements as needed to support the collection mission. The team leader will advise the supported unit on the specific capabilities and requirements of the team to maximize mission success. This method also increases the risk to the team. HCT members must carry the necessary firepower for self-protection. They must also have adequate communications equipment to call for help if needed. The team's posture, equipment, and appearance will be dictated by overall force restrictions and posture. When operating as an independent patrol, the HCT should not stand out from overall US forces operations. If US forces are in battle-dress uniforms and operating out of military vehicles, so should the HUMINT collectors.

- Soft. If the threat situation is such that soldiers are authorized to wear civilian clothes when outside base areas, the HUMINT collectors should also move among the civilian population in civilian clothes, so that they do not stand out from others in the area.

CIVIL SUPPORT OPERATIONS

3-15. Army support supplements the efforts and resources of state and local governments and organizations. If a presidential declaration initiates civil support for a major disaster or emergency, involvement of DOD intelligence components would be by exception. Civil support requires extensive coordination and liaison among many organizations—interagency, joint, AC, and RC—as well as with state and local governments, and in any case will require compliance with the Posse Comitatus Act, 18 U.S.C., § 1385, when US forces are employed to assist Federal, state, or local law enforcement agencies (LEAs). The National Response Plan provides a national level architecture to coordinate the actions of all supporting agencies.

MILITARY OPERATIONS IN URBAN ENVIRONMENT

3-16. Units are often task organized with additional ISR units and assets to meet the detailed collection requirements in the urban operations. The complexities of urban terrain cause degradation in the capabilities of many of the sensor systems. HUMINT collectors may have to be placed in DS of lower echelon combat maneuver forces (battalion and lower) to support operations. HUMINT and combat reporting by units in direct contact with threat forces and local inhabitants becomes the means of collection. For successful ISR planning, the S2 must be aware of the capabilities and limitations of the various organic and attached collection systems as they apply to urban operations. As in all environments, commanders must assess the risk involved in the forward deployment of HUMINT assets.

3-17. In urban operations, people (for example, detainees and civilians) are the preeminent source of information. HUMINT collection provides information not otherwise available through signals intelligence (SIGINT) and imagery intelligence (IMINT) such as threat and local population intentions. They collect information on, for example, floor plans, defensive plans, locations of combatants and noncombatants, including civilians in the buildings and surrounding neighborhoods, and other information. The collected information is passed directly to the individuals conducting the combat operation.

3-18. In small-scale contingencies (SSCs) and in peacetime military engagements (PMEs), contact with local officials and populace by the HUMINT collectors can be a prime source of information about the local environment and is a vital component of intelligence support to force protection. During routine patrolling of urban areas it is often expedient to place a HUMINT collector with individual patrols. The key difference between urban and other operations, from major theater war (MTW) to PME, is the number of HUMINT collectors required. The need for HUMINT collectors is a function of population density. Whereas in a rural environment, a HUMINT team may be able to cover an area in excess of 1,200 square kilometers; the same team in a dense urban environment may be able to cover only 10 square blocks or less.

HUMINT COLLECTION ENVIRONMENTS

HUMINT COLLECTION IN A PERMISSIVE ENVIRONMENT

3-19. In a permissive environment, HCTs normally travel throughout their specific AOR as separate teams or as part of a larger reconnaissance team. HUMINT collectors may frequently make direct contact with the individual, view the activity, or visit the area that is the subject of the ISR effort. They normally use debriefing and elicitation to obtain first-hand information from local civilians and officials as their primary collection techniques. Additional information can be obtained from exploitation of open-source material such as newspapers, television, and other media. The priority requirements in this environment are normally linked to force protection. HCTs should establish liaison and casual source contacts throughout their AOIR. Reporting is normally via IIRs, although SALUTE reports are used for critical time-sensitive reporting. Even in a permissive environment, the HUMINT collector conducts the majority of his collection through the debriefing of individuals who have first-hand knowledge of the information they are reporting.

HUMINT COLLECTION IN A SEMI-PERMISSIVE ENVIRONMENT

3-20. In a semi-permissive environment, security considerations increase, but the risk to the collector still must be weighed against the potential intelligence gain. HCTs should still be used throughout their AOIR but will normally be integrated into other ground reconnaissance operations or other planned operations. For example, a HUMINT collector may accompany a CA team or PSYOP team visiting a village. Security for the team and their sources is a prime consideration. The HCTs are careful not to establish a fixed pattern of activity and arrange contacts in a manner that could compromise the source or the collector. Debriefing and elicitation are still the primary collection techniques. Teams are frequently deployed to conduct collection at roadblocks, refugee collection points, and detainee collection points. They may conduct interrogations of EPWs and other detainees within the limits of the mission-specific orders, and applicable law and policy. Applicable law and policy include US law; the law of war; relevant international law; relevant directives including DOD Directive 3115.09, "DOD Intelligence Interrogations, Detainee Debriefings, and Tactical Questioning"; DOD Directive 2310.1E, "The Department of Defense Detainee Program"; DOD instructions; and military execute orders including FRAGOs. DOCEX is also used to accomplish exploitation of threat documents. Reporting is normally via SALUTE report and IIR.

HUMINT COLLECTION IN A HOSTILE ENVIRONMENT

3-21. In a hostile environment, the three concerns for HUMINT collection are access to the sources of information, timeliness of reporting, and security for the HUMINT collectors. Prior to the entry of a force into a hostile AO, HUMINT collectors are used to debrief civilians, particularly refugees, and to interrogate EPWs and other detainees who have been in the AO. HCTs are normally located with the friendly units on the peripheries of the AO to facilitate timely collection and reporting. If a refugee or EPW/detainee population exists prior to this mission, they are screened to determine knowledgability of the AO and are debriefed or interrogated as appropriate. HUMINT collectors accompany the friendly ground reconnaissance elements as they enter the AO. As part of the ground reconnaissance force, they interrogate EPWs and other detainees and debrief refugees, displaced persons, and friendly force patrols. Reporting is normally via oral or written SALUTE reports with more detailed information reported via IIRs. They may also support the S2 through the systematic debriefing of friendly ground reconnaissance assets and the translation of any documents collected by them.

EAC HUMINT

MI BRIGADES AND MI GROUPS SUPPORTING COMPONENT COMMANDS

3-22. Each SCC with an outside continental United States (OCONUS) responsibility has an US Army Intelligence and Security Command (INSCOM) MI brigade or group to provide operational HUMINT support to that command. These MI elements provide peacetime support to the unified command and add a consistent, forward-deployed presence in a particular theater of operations. Theater MI brigade and group assets provide HUMINT support during contingency operations. These HCTs can support a JTF, an army combatant command, or any deployed element that requires augmentation.

JOINT, COMBINED, AND DOD HUMINT ORGANIZATIONS

3-23. The Departments of the Air Force and the Navy have limited HUMINT collection capability. They will normally provide strategic debriefing trained and certified personnel to joint interrogation and debriefing facilities primarily to collect information on areas of particular interest to that Military Department. Within the Department of the Navy, however, the US Marine Corps has a robust tactical HUMINT collection capability that operates primarily in support of engaged Marine Corps forces. Marine expeditionary elements deploy with human exploitation teams (HETs) that provide organic HUMINT and CI support to the deployed Marine force. Marine HETs are rapidly deployable and fully equipped to conduct the full range of tactical HUMINT and CI functions. They can provide support to either the deployed Marine force or as part of JTF HUMINT or CI teams. Each Marine Expeditionary Force (MEF) has organic HETs. HETs can also be attached to a Marine Air-Ground Task Force (MAGTF) for a particular operation.

SUPPORT AGENCIES

3-24. HUMINT agencies from DOD, national level intelligence agencies, and LEAs can support the battlefield commander. In a JTF, a national intelligence support team (NIST) works with the J2X to coordinate national level activities with JTF and component HUMINT and analytical assets. Sometimes liaison officers (LNOs) are assigned directly to the C/J/2X to facilitate collection activities.

- **Defense Intelligence Agency (DIA).** The DIA is a DOD combat support (CS) agency and an important member of the United States Intelligence Community. With more than 7,000 military and civilian employees worldwide, DIA is a major producer and manager of foreign military intelligence. DIA provides military intelligence to warfighters, defense policymakers and force planners in DOD and the Intelligence Community in support of US military planning and operations and weapon systems acquisition.

  - **Defense HUMINT (DH) Service.** The DH Service, a branch of the DIA, is the force provider for strategic HUMINT forces and capabilities. During operations, elements from DH form a partnership within the supported JTF headquarters J2X element for the coordination and deconfliction of HUMINT source-related collection activities. DH support to a joint force is outlined in the classified DIAM 58-11 and DIAM 58-12.

- **Central Intelligence Agency (CIA).** The CIA supports US national security policy by providing accurate, evidence-based, comprehensive, and timely foreign intelligence related to national security. The CIA conducts CI activities, HUMINT collection, special activities, and other functions related to foreign intelligence and national security as directed by the President. Joint Pub 2-01.2 (S//NF) contains details of CIA contributions to the deployed force.

- **Department of State.** The State Department's Bureau of Diplomatic Security provides CI support to diplomatic missions worldwide and gathers extensive information on intelligence capabilities of adversaries within that diplomatic mission's area of concern. The Bureau of Intelligence and Research is the State Department's primary source for interpretive analysis of global developments. It is also the focal point in the State Department for all policy issues and activities involving the Intelligence Community.

- **National Security Agency (NSA).** The NSA is a DOD agency that coordinates, directs, and performs highly specialized activities to protect US information systems and produce foreign intelligence information. It is also one of the most important centers of foreign language analysis and research within the Government.

- **Defense Criminal Investigative Service (DCIS).** The DCIS is the criminal investigative arm of the Inspector General (IG) of DOD. The DCIS's mission is to protect America's warfighters by initiating, conducting, and supervising investigations in support of crucial National Defense priorities.

- **Department of Justice:**

  - Federal Bureau of Investigation. The FBI may provide the deployed commander with national level expertise on criminal and CI issues if currently operating in a task force (TF) AO and liaison is established early.

  - Drug Enforcement Agency (DEA). The DEA provides counterdrug operational expertise to a deployed TF and coordinates its operations with those of a deployed TF.

- **Department of Homeland Security (DHS).** The DHS mission is to prevent terrorist attacks within the United States, reduce the vulnerability of the United States to terrorism, protect the homeland, its citizens, and critical infrastructure and key resources against terrorist attack. DHS provides a lead for Federal incident response, management, and recovery in the event of terrorist attack and natural disasters. The Secretary of Homeland Security is the principal Federal official for domestic incident management. Pursuant to the Homeland Security Act of 2002, the Secretary is responsible for coordinating Federal operations within the United States to prepare for, respond to, and recover from terrorist attacks, major disasters, and other emergencies. DHS operates the Homeland Security Operations Center (HSOC) and the DHS-led Interagency Incident Management Group (IIMG). The DHS AOR is the US and its territories. DHS secures and protects the entry points to the nation, the areas between the entry points, land and water, for people, and cargo or conveyances. DHS enforces immigration, customs, and transportation security laws and regulations, counter-narcotics, counterfeiting, financial crimes, and threats to the President. As legislated in the Homeland Security Act of 2002, DHS is chartered as the primary outreach Federal activity for state, local, and tribal governments, and the private sector. Although DHS has no direct role in support of a "battlefield commander" outside the United States, DHS component organizations have representatives deployed in support of US Government missions in the US Central Command (USCENTCOM) AOR.

- **Department of Energy (DOE).** The DOE can assist with the—

  - Exploitation of weapons of mass destruction (WMD).

  - Protection or elimination of weapons and weapons-useable (dual-use) nuclear material or infrastructure.

  - Redirection of excess foreign weapons expertise to civilian enterprises.

  - Prevention and reversal of the proliferation of WMD.

  - Reduction of the risk of accidents in nuclear fuel cycle facilities worldwide.

  - The capability enhancement of WMD detection including nuclear, biological, and chemical (NBC).

- **National Geospatial Intelligence Agency (NGA).** The NGA is a member of the US Intelligence Community and a DOD Combat Support Agency. NGA provides timely, relevant, and accurate geospatial intelligence in support of national security objectives. Geospatial intelligence is the exploitation and analysis of imagery and geospatial information to describe, assess, and visually depict physical features and geographically referenced activities on the Earth.

- **Counterintelligence Field Agency (CIFA).** The mission of CIFA is to develop and manage DOD CI programs and functions that support the protection of the Department. These programs and functions include CI support to protect DOD personnel, resources, critical information, research and development programs, technology, critical infrastructure, economic security, and US interests against foreign influence and manipulation, as well as to detect and neutralize espionage against the Department.

3-25. Most potential coalition partners have some type of HUMINT capability. Less developed nations may use HUMINT as their primary collection system and may be quite skilled in HUMINT operations. These assets will be present on the battlefield, and US assets are likely to work with them. HCTs should perform regular liaison with coalition HUMINT personnel. It is likely that some coalition partners will be more knowledgeable of the culture in the AO and be able to share insights with US HCTs.

Chapter 4

HUMINT Operations Planning and Management

4-1. HUMINT operations planning and management are supported by a robust structure that includes staff elements such as the C2X when working with non-US forces at the Joint intelligence staff level, G2X at the Division, Corps intelligence staff, the HUMINT operations section in the MI Battalion, and HAT in the Division and Corps ACE. It also includes C2 elements at the MI battalion, company, platoon, and team levels. The OMT provides the first level of staff and C2 functions when two or more HCTs deploy in support of an operation. (See Table 4-1.)

Table 4-1. HUMINT Operations.

ECHELON	ISR PLANNING	TECHNICAL SUPPORT AND DECONFLICTION	MISSION EXECUTION
COMBINED	C2/ACE	C2X/OMT	MI CDR
JOINT	J2/ACE	J2X/OMT	MI CDR (AMIB or MI Battalion)
CORPS/DIVISION	G2/ACE	G2X/OMT	MI CDR/OMT
BRIGADE	S2	MI CDR/OMT	MI CDR/OMT

HUMINT AND THE OPERATIONS PROCESS

4-2. Following the operations process defined in FM 3-0, Chapter 6, there are four components within HUMINT operations: Plan, Prepare, Execute, and Assess.

PLAN

4-3. HUMINT planning defines collection objectives, when to collect it, and which resources will be tasked to do the collection. Commanders with HUMINT collection assets in their units receive collection tasking based on requirements developed during ISR planning. The commander and staff, in concert with their supporting OMTs, assess the requirements and task the team or teams best capable of answering the requirement based on contact placement and access.

4-4. Another aspect to consider carefully during the Plan phase of the operational cycle is technical control. Technical control is ensuring adherence to existing policies and regulations, providing information and guidance of a technical nature, and supervising the MOS-specific TTP required in conducting collection missions. Planning must take into account that technical control does not interfere with or supersede any C2 that a commander has over an asset or unit nor does it interfere with collection of the commander's requirements. For HUMINT collectors, the technical control network includes the C/J/G/S2X, the HOC, and OMTs. Technical control includes the management of source and other sensitive data and databases, the management of intelligence contingency and incentive funds, the liaison with other HUMINT organizations, and the deconfliction of operations. Technical control provides HCTs with specific requirements and data that they need to conduct operations and, in certain circumstances, specific instructions on how to execute missions.

PREPARE

4-5. During this phase, commanders and staff, including HUMINT management sections, review HUMINT mission plans. This review is to ensure all areas of the mission are considered and addressed in the plan and included in rehearsals. Items to cover include but are not limited to—

- Route (primary and alternate).
- Communications.
- Security plan.
- Convoy procedures including actions on contact and rally points.
- Initial requirements to be covered.
- Mission duration.

4-6. The HUMINT collector then researches the topic area addressing the requirement and prepares a questioning plan. The HCTs and OMTs must coordinate all mission requirements. It is important that HUMINT elements are included in all rehearsals conducted by their supported unit. These rehearsals will enable HCTs to carry out essential coordination with other units and ensure that they are included in and familiar with procedures such as resupply, communications, casualty evacuation, fire support, and fratricide avoidance. Rehearsals and briefbacks will allow the supported command to see and correct problems with their support to the HUMINT elements prior to deployment.

EXECUTE

4-7. Mission execution consists of the collection of information in accordance with the integrated ISR plan. The requirements manager validates the requirements based on command guidance. The G3 tasks the requirements to the units and the individual asset managers (that is, OMT) to identify the assets best capable to answer the requirement. When requirements are levied against a specific HCT, the HCT leader decides which of his team's contacts can best answer the requirements. He then turns the requirement into specific team tasks.

ASSESS

4-8. Assessment is the continuous monitoring—throughout planning, preparation, and execution—of the current situation and progress of an operation, and the evaluation of it against criteria of success to make decisions and adjustments. Assessment plays an integral role in all aspects of the intelligence process (see FM 2-0).

HUMINT COMMAND AND CONTROL

4-9. Commanders of organizations that conduct HUMINT operations are responsible for task organization, mission tasking, execution, mission accomplishment, and designation of subordinate AOs (within the guidelines of the OPORD or OPLAN). MI unit commanders who exercise direct control of HUMINT operations, including interrogation operations, at all levels are responsible for and stand accountable to ensure HUMINT collection activities comply with this manual and applicable law and policy. Applicable law and policy include US law; the law of war; relevant international law; relevant directives including DOD Directive 3115.09, "DOD Intelligence Interrogations, Detainee Debriefings, and Tactical Questioning"; DOD Directive 2310.1E, "The Department of Defense Detainee Program"; DOD instructions; and military execute orders including FRAGOs. The MI unit commanders must ensure mission accomplishment by properly allocating resources and logistics in support of all HUMINT collection assets assigned to their units. Commanders must ensure that their HUMINT collection personnel are trained and ready for the mission. There is a need for a partnership between the J/G2X, who exercises technical direction and oversight responsibility and the MI commander, who exercises direct command authority and responsibility. The MI unit commander analyzes the higher headquarters mission, concept of operations, and the specified and implied tasks given to his unit. He restates the unit mission, designs the concept of operations, task organizes his assets, and provides support to subordinate units. Specifically, the MI unit commander—

- Issues mission orders with sufficient details and time for subordinate commanders and leaders to plan and lead their units.
- Must know the threat, his organization, ISR systems, counter-ISR systems, operations, and terrain over which his units will operate and how that terrain enhances or limits HUMINT collection operations.
- Must be aware of the operational and technical limitations of his unit and ensures that all assets are task organized, properly positioned, and fully synchronized to accomplish the mission.
- Oversees the collective and individual training within his unit.
- Coordinates continuously with the higher headquarters staff, the supported maneuver unit staff, and other commanders to ensure integrated R&S operations and support.
- Establishes clear, consistent standards and guidance for current and future operations in order to adhere to policy and the higher headquarters commander's intent without his constant personal supervision.
- Continually assesses his unit's ability to sustain its internal operations and its ability to support assigned missions and keeps the higher headquarters staff informed of unit, equipment, and personnel status that affect collection operations.

- Advises his higher headquarters commander and staff on the capabilities, limitations, and most effective employment of his assets.
- Remains flexible during operations to adjust or execute missions upon receipt of new orders and when the situation changes.
- Ensures personnel are working within legal, regulatory, and policy guidelines.

TECHNICAL CONTROL

4-10. Technical control refers to supervision of the TTP of HUMINT collection. Technical control ensures adherence to existing policies or regulations and provides technical guidance for HUMINT operations. The elements that provide technical control also assist teams in translating collection requirements into executable tasks. Commanders rely on the expertise of intelligence personnel organic to their unit and within higher echelons to plan, execute, and assess the HUMINT collection effort. The OMTs, HATs, and the HOC of the C/J/G/S2X provide technical control. They—

- Define and manage operational coverage and direction.
- Identify critical collection criteria such as indicators associated with targeting.
- Prioritize collection missions in accordance with collection requirements.
- Advise teams on collection techniques and procedures in accordance with policy, regulations, and law.
- Register and deconflict sources.
- Conduct operational reviews.
- Advise commanders.
- Conduct operational coordination with staff elements and other intelligence agencies.
- Manage ICF and incentive usage.

COMMAND AND SUPPORT RELATIONSHIPS

4-11. The activities of HUMINT assets are governed by their command or support relationship. There are subtle differences in the Joint versus the Army description of some of the command and support relationships. Tables 4-2 through 4-4 show these relationships.

4-12. During interrogation operations, close coordination must occur between intelligence personnel and personnel responsible for detainee operations including MP security forces, Master at Arms, and other individuals providing security for detainees. The facility commander is responsible for all actions involving the humane treatment, custody, evacuation, and administration of detainees, and force protection. Whereas, the intelligence commander is responsible for the conduct of interrogation operations.

COMMAND AND SUPPORT RELATIONSHIPS FOR HUMINT OPERATIONS

4-13. Clear command and support relationships are fundamental in organizing for all operations. These relationships identify responsibilities and authorities among subordinate and supporting units. The commander designates command and support relationships within his authority to weight the decisive operation and support his scheme of maneuver. Some forces available to a commander are given command or support relationships that limit his authority to prescribe additional relationships. Command and support relationships carry with them varying responsibilities to the subordinate unit by parent and gaining units. By knowing the inherent responsibilities, a commander may organize his forces to establish clear relationships.

4-14. Command relationships establish the degree of control and responsibility commanders have for forces operating under their tactical control (TACON). When commanders establish command relationships, they determine if the command relationship includes administrative control (ADCON). Table 4-2 shows Army command and support relationships and Table 4-3 shows joint command relationships chart from FM 3-0 (derived from JP 0-2 and JP 3-0).

4-15. Support relationships define the purpose, scope, and effect desired when one capability supports another. Support relationships establish specific responsibilities between supporting and supported units. Table 4-2 shows Army command and support relationships and Table 4-4 shows joint support relationships from FM 3-0 (derived from JP 0-2 and JP 3-0).

HUMINT REQUIREMENTS MANAGEMENT

4-16. The G2/S2 is responsible for RM. He uses the requirements management (RM) process to orchestrate the actions of the unit's organic and supporting ISR capabilities into a unified effort to gain situational understanding and answer the commander's PIRs. Through centralized planning and decentralized execution, RM optimizes the integration of ISR operations into the commander's scheme of maneuver and fire and into the unit's long- and short-range planning. Control mechanisms within the RM structure facilitate the identification of information shortfalls and the redirection of ISR assets to new intelligence production, reconnaissance, or surveillance missions.

Table 4-2. Army Command and Support Relationships.

IF RELATIONSHIP IS:		INHERENT RESPONSIBILITIES ARE:							
		Has Command Relationship with:	May Be Task Organized by:	Receives CSS from:	Assigned Position or AO By:	Provides Liaison To:	Establishes/ Maintains Communications with:	Has Priorities Established by:	Gaining Unit Can Impose Further Command or Support Relationship of:
COMMAND	Attached	Gaining unit	Gaining unit	Gaining unit	Gaining unit	As required by gaining unit	Unit to which attached	Gaining unit	Attached; OPCON; TACON; GS; GSR; R; DS
	OPCON	Gaining unit	Parent unit and gaining unit; gaining unit may pass OPCON to lower HQ. Note 1	Parent unit	Gaining unit	As required by gaining unit	As required by gaining unit and parent unit	Gaining unit	OPCON; TACON; GS; GSR; R; DS
	TACON	Gaining unit	Parent unit	Parent unit	Gaining unit	As required by gaining unit	As required by gaining unit and parent unit	Gaining unit	GS; GSR; R; DS
	Assigned	Parent unit	Parent unit	Parent unit	Gaining unit	As required by parent unit	As required by parent unit	Parent unit	Not Applicable
SUPPORT	Direct Support (DS)	Parent unit	Parent unit	Parent unit	Supported unit	Supported unit	Parent unit; Supported unit	Supported unit	Note 2
	Reinforcing (R)	Parent unit	Parent unit	Parent unit	Reinforced unit	Reinforced unit	Parent unit; reinforced unit	Reinforced unit: then parent unit	Not Applicable
	General Support Reinforcing (GSR)	Parent unit	Parent unit	Parent unit	Parent unit	Reinforced unit and as required by parent unit	Reinforced unit and as required by parent unit	Parent unit; then reinforced unit	Not Applicable
	General Support (GS)	Parent unit	Parent unit	Parent unit	Parent unit	As required by parent unit	As required by parent unit	Parent unit	Not Applicable

NOTE 1. In NATO, the gaining unit may not task organize a multinational unit (see TACON).

NOTE 2. Commanders of units in DS may further assign support relationships between their subordinate units and elements of the supported unit after coordination with the supported commander.

Table 4-3. Joint Command Relationships and Inherent Responsibilities.
(from FM 3-0, derived from JP 0-2 and JP 3-0)

Inherent Responsibilities Are:	If relationship is:		
	COCOM	OPCON	TACON
Has command Relationship with:	Gaining combatant commander; gaining service component commander	Gaining Command	Gaining Command
May be task organized by:	Gaining combatant commander; gaining service component commander	Gaining Command	Parent Unit
Receives logistic support from:	Gaining service component commander	Service component command; parent unit	Parent Unit
Assigned position or AO by:	Gaining component commander	Gaining Command	Gaining Command
Provides liaison to:	As required by gaining component commander	As required by gaining command	As required by gaining command
Establishes and maintains communications with:	As required by gaining component commander	As required by gaining command	As required by gaining command and parent units
Has priorities established by:	Gaining component commander	Gaining Command	Gaining Command
Gaining unit can impose further command relationship/authority of:	OPCON; TACON; direct support; mutual support ; general support; close support	OPCON; TACON; direct support; mutual support; general support; close support	Direct support; mutual support; general support; close support

Table 4-4. Joint Support Categories.

(from FM 3-0, derived from JP 0-2 and JP 3-0)

CATEGORY	DEFINITION
General Support	The action given to the supported force as a whole rather than to a particular subdivision thereof.
Mutual Support	The action that units render each other against an enemy because of their assigned tasks, their position relative to each other and to the enemy, and their inherent capabilities.
Direct Support	A mission requiring a force to support another specific force and authorizing it to answer directly the supported force's request for assistance.
Close Support	The action of the supporting force against targets or objectives that are sufficiently near the supported force as to require detailed integration or coordination of the supporting action with fire, movement, or other actions of the supported force.

DEVELOP HUMINT REQUIREMENTS

4-17. The first step in the RM process is to develop intelligence requirements that accurately identify and prioritize the commander's concerns about the threat and the battlefield environment that must be resolved to accomplish the mission. The G2/S2X, or his representative, normally supports the G2/S2 by identifying HUMINT collection requirements and opportunities and advises the command and staff on HUMINT capabilities. The HUMINT representative must be able to discuss any delays or risks involved in using HUMINT assets. Through participation in the requirements development process, the HUMINT representative has a thorough understanding of the commander's intent and concept of operations and is better able to support the overall ISR effort.

4-18. The analysis of HUMINT requirements is normally a coordinated effort between the HUMINT and CI staff officer (C/J/G/S2X) and the HAT of the supporting analysis element. The C/J/G/S2X team—

- Records all HUMINT requirements whether generated internally (Specific Orders) or received from other echelons or units (Requests).
- Tracks each requirement from receipt to final satisfaction.
- Reviews each requirement for its—
  - **Feasibility.** Feasibility is a determination if a requirement can be answered given available time and resources.
  - **Completeness.** Does the requirement contain all the specifics needed for collection, such as: What the collection requirement is? When the latest time information is of value (LTIOV)? Why it needs to be collected? Who needs the results of the collection?
  - **Necessity.** The C/J/G/S2X team, with the assistance of the HAT, checks available intelligence databases to determine if the required

information has already been collected or is included in an intelligence product.

4-19. The RM team, with the assistance of the C/J/G/S2X team and the HAT, breaks the HUMINT-related PIR into SIRs. Each SIR describes the indicator of threat activity linked to an area or specific location and time. The HOC evaluates—

- Reportable criteria that are linked to the threat activity. The HOC associates these characteristics with a SIR, and compares the characteristics to a particular HUMINT asset's capability to collect.
- Range, which is the distance from the current location of the HUMINT asset or resource to the source. In other words, are there sources available that had or have access to relevant information on the area or activity in question, and can the HUMINT team contact them in a timely manner?
- Timeliness, which is when the information must reach the commander to be of value; that is, the LTIOV.

4-20. The RM team, supported by the C/J/G/S2X and the HAT, attempts to answer the SIRs with intelligence products developed from information available within the existing intelligence databases or pulled from other organizations within the intelligence architecture. If the requirement can be answered in this manner, the intelligence is immediately disseminated. When the required information is neither available nor extractable from archived information or from lower, lateral, or higher echelons, the C/J/G/S2X team develops it into an RFI to higher or an ISR tasking for organic or attached HUMINT assets. The compilation of unanswered requirements and how to answer them form the basis of the ISR plan. The tasking may be in the form of an SDR. An SDR is a specific request or tasking for a collector to question a source on a particular collection requirement. This request involves analysis that results in the conclusion that a specific source possibly has the placement and access to answer a SIR. SDRs are specific; whereas, HUMINT collection requirements (HCRs) are general.

DEVELOP THE HUMINT PORTION OF THE INTEGRATED ISR PLAN

4-21. The HOC within the C/J/G/S2X section assists the G3/G2 in developing the HUMINT portion of the ISR plan in coordination with the HAT and the RM team. The HOC ensures that the HUMINT capabilities and taskings are included in the plan although the plan often will not contain the specifics of HUMINT operations due to the sensitivity of the sources and techniques. The HOC will coordinate with the Office of the SJA to ensure the HUMINT portion of the integrated ISR plan complies with applicable law and policy prior to its implementation. Applicable law and policy include US law; the law of war; relevant international law; relevant directives including DOD Directive 3115.09, "DOD Intelligence Interrogations, Detainee Debriefings, and Tactical Questioning"; DOD Directive 2310.1E, "The Department of Defense Detainee Program"; DOD instructions; and military execute orders including FRAGOs. The HOC coordinates with C/J/G/S2X for mission deconfliction at that echelon to specify the collection capability and current status of the various HUMINT organizations to better enable him to select the "best" organization to collect on various SIRs. HUMINT collection generally requires time to develop the environment and access sources.

4-22. The HUMINT collection environment during an SSC is different from an MTW. During an MTW where the force is moving, a division normally plans 48 hours out; a corps plans 72 hours out. In contrast, the planning focus for units supporting an SSC may be 3 to 6 months out. The longer HCTs are in an area, the better the collector is able to develop leads to answer collection requirements. Requirements may be continuous or may be concerned with specific upcoming events such as national elections. HUMINT is a key asset to determine adversary intentions; however, it is highly dependent on the ability to cultivate or locate sources with the desired information. HUMINT in support of stability and reconstruction operations is not a short-term undertaking. [Example: National level elections are taking place in the AO in 3 months. As a part of integrated ISR planning, an assessment must be conducted to determine the capability to answer post-election collection requirements based upon current contacts and HUMINT leads. If there are no leads or contacts that could answer election-related collection requirements, it is necessary to spot, assess, and contact sources to meet requirements.]

4-23. A second part of the HUMINT portion of the integrated ISR plan is the HUMINT collection focus, which—

- Designates which collection requirements comprise the emphasis for collectors' missions.
- Prioritizes collection requirements based upon the operational environment in the AO and future missions in the AO.
- Includes future operational collection tasks which aid in causing a gap or pause in collection as the unit transitions to the next operational phase or the next operation.

4-24. In addition to specific requirements, a statement of intelligence interest (SII) at the joint level or a collection emphasis message at division or corps is issued to identify the overall collection goals for a time period. As the collection request or requirement is passed down, each echelon performs additional planning for its own specific requirements.

Evaluate HUMINT Resources

4-25. After identifying the SIRs, the HOC and the C/J/G/S2X determine the availability and capability of HUMINT assets and resources that might contribute to requirement satisfaction and which are most suited to collect against each SIR. This does not necessarily imply that the C/J/G/S2X assigns a tasking to a specific team; rather, it develops the requirements or requests for an organization that then executes the mission. The HOC and C/J/G/S2X should also consult the HAT for its analysis of additional potential HUMINT assets and resources which might be available, both on and off the battlefield, to contribute to requirement satisfaction. For example, the HAT may be aware of a group of émigrés now living elsewhere who previously lived near a target site, and who might be able to provide answers to collection requirements if debriefed.

Determine Asset or Resource Capabilities

4-26. The HOC translates the capabilities and limitations of the available HUMINT assets into a set of factors that they can compare to the SIR characteristics. Asset capability factors are technical or performance characteristics, location, and source access. Each HUMINT asset is evaluated for its—

- **Availability.** The HOC reviews the list of viable HUMINT assets for current availability and the addition or deletion of capabilities. This includes considerations such as maintenance time and previous taskings. Coordination with adjacent and higher headquarters and national level agencies by the C/J/G/S2X will determine the availability of higher echelon resources.

- **Survivability.** Survivability must be commensurate with the threats to which the HUMINT assets will be exposed during the course of operations. These assets must be as survivable as, or in certain circumstances more survivable than, the forces they support. The HOC and the commander must weigh the risk versus the gain in using HUMINT assets.

- **Reliability.** Reliability is the ability of the asset to overcome threat deception measures such as misinformation or false information. In HUMINT there are two areas of reliability: source and collector. Source reliability is the determination on the part of the collector if the source is providing accurate information. Collector reliability is a determination on the part of the HOC that the HUMINT collectors within a particular organization have the level of training and experience to collect against a given requirement.

- **Suitability.** Tasking must be based on an asset's capability and on its suitability within the context of the overall plan. For example, HUMINT assets may be capable of collecting against a single target but have unique capabilities against a second target. Intelligence requirements may necessitate tasking these HUMINT assets against the second target if other assets can maintain adequate coverage of the first target.

- **Connectivity.** Connectivity is a critical aspect of any R&S operation. Interoperability, reliability, and robustness of sensors, communications, and supporting automated data processing (ADP) are crucial to the responsiveness, survivability, and overall combat effectiveness of a HUMINT asset. If the automation and communications systems of a HUMINT asset are dissimilar to those of other units in the AO, or if connectivity among assets, supporting systems, and supported systems and elements is too fragile to withstand the stress of operations, commanders will be deprived of important information essential to conducting tactical operations. The HUMINT asset must be able to transmit accurate and timely information to those who must receive it when they need it. Report formats should adhere to established standards in order to ensure that information is easily retrieval at the user desktop through automated queries (push/pull). Planners must look carefully at systems compatibility and the degree of interoperability among the components of the communications architecture. The better the interoperability of assets and the more robust and redundant the communications links, the better the cross-cueing and analytical exchange.

Develop the Scheme of Support

4-27. The scheme of support is the orchestration of HUMINT assets, resources, and requirements to facilitate the collection of information most effectively. It includes all assets that the G3/S3 can task (organic, attached, and DS) and the G2 can request (from higher or adjacent units). By reviewing available HUMINT assets and higher echelon resources, the HOC and the G/S2X determine whether unit assets or higher echelon resources are best able to answer the requirements. If another echelon can answer an SIR, then the J/G/S2, normally through the C/J/G/S2X, requests them to collect the information and deliver the intelligence product. When planning the HUMINT portion of the ISR plan, the HOC should consider the following:

- **Cueing** is using one asset to tip off another to a possible target. The HOC should look for opportunities for HUMINT assets to cue other collection assets and vice versa.

- **Asset redundancy** uses a combination of the same type of assets against a high-priority collection target. This is vital in HUMINT collection since, in dealing with human sources, the information collected is often part of the overall picture or is influenced by the perception and prejudice of the source. The collection on the same target from a number of different assets gives a more accurate intelligence picture and is a method to validate source reporting.

- **Asset mix** uses a combination of different types of assets against a high-priority collection target. When the probability of success of one asset to satisfy the requirement completely is lower than acceptable, the use of multiple capabilities of different assets increases the likelihood of success; for example, using SIGINT assets to intercept voice communications while HUMINT assets observe activities. Neither can collect all the available information, but the information collected by both can be fused into a more complete picture. Like asset redundancy, asset mix places greater demands on the limited assets available, both collection and analysis, and has to be clearly justified by the potential intelligence gain.

- **Integration of new requirements** into ongoing missions may make it possible to reduce timelines, make collection more responsive to the request, and decrease cost and risk. This is critical in HUMINT due to the long time that it takes to develop sources. The use of an existing source to answer new requirements often facilitates collection.

Develop and Prioritize Taskings and Requests for Information

4-28. After the G2/S2X and the G2/S2 approve the HUMINT portion of the ISR plan, the HOC develops specific orders to task assets, develop additional assets, and/or requests to seek higher and lateral support and production. Specific taskings or RFIs are tailored to that specific ISR asset's capabilities and limitations. The G2/S2X supports the requirements manager and the G2/S2 in developing and prioritizing HUMINT taskings. The HOC works with the unit requirements manager to incorporate the HUMINT plan into the overall unit ISR plan and works with the G3/S3 as necessary to help develop OPORDs or FRAGOs to organic or attached ISR units. HUMINT taskings will often include technical data that cannot be passed through normal tasking channels. The HOC will pass that information directly to the applicable HUMINT OMT or unit operations section.

4-29. The HOC and G2/S2X cannot provide operational taskings to a unit for collection. Collection is a stated mission that the commander executes. However, the technical control the HOC can provide as the HUMINT manager affords the J2/G2X the ability to steer and direct collection assets and operations. The MI commander and OMT determine specifically which teams will collect on a given requirement and are responsible for the TTP used. They report on the status and availability of their collection assets. On the HCT level, the team chief determines which sources will be contacted and the details of how the information will be collected from a given source. A specific plan is developed for each source. This plan should—

- Identify the requirement.
- Identify the proposed source.
- Identify questions to be asked during the source meeting.
- Contain an outline of how the meeting should proceed.
- Identify which collector will conduct the source meeting.

4-30. At the HCT level, the senior team member reviews each plan to ensure the proper planning for the collection mission. The plan is a minimum goal for the collection. The collector must be fully aware of the overall collection priorities and be prepared to take advantage of any additional leads.

DIRECT PRODUCTION

4-31. The G2 coordinates intelligence production to provide non-duplicative all-source intelligence products to the commander, staff, and subordinate forces. Some type of production occurs in the intelligence staff or separate analysis element at every echelon from national to battalion level. The HCT of the ACE at echelon's division and higher will support the intelligence production process through the analysis of HUMINT information and the development of single-discipline HUMINT products.

DISSEMINATE INFORMATION

4-32. The 2X element at each level is normally the release authority for HUMINT reporting and products, ensuring that reporting, products, and data are disseminated to the lowest appropriate level. The G/S2X should preplan criteria for the immediate release of combat information on high-value targets, impending attacks, or other time-sensitive requirements. This preplanning will ensure that commanders and other users quickly receive the information in a format that supports situational understanding, strategic responsiveness, and ISR and provides support to effects. Special effort is also made to ensure that information obtained from detainees is passed back down to the unit that detained them. This measure will support the efforts of the commander as well as building trust in the intelligence process.

EVALUATE REPORTING

4-33. The HAT and the HOC provide the requirements manager and the G2/S2 with expertise to support report evaluation. An important part of the evaluation process is providing feedback to the collectors. Feedback is important in HUMINT operations since the same source may be contacted again for additional information. The collector needs feedback on the accuracy, reliability, and appropriateness of the information reported. The G/S2X team tracks reporting to determine how well the HUMINT collection and production efforts are satisfying the PIRs. The G/S2X team supports the RM team's requirements to—

- **Monitor and Maintain Synchronization.** Through coordination with the G2/S2, the G/S2X, and the HAT, the HOC knows when and what critical pieces of information are missing from the commander's estimate of the situation. The HOC uses the HUMINT portion of the ISR plan to ensure synchronization with the overall operation and scheme of maneuver. The other critical tool for the HOC is the decision support template (DST). The HOC must have a complete copy of this document, ensuring the HUMINT assets do not miss a collection requirement.

- **Correlate Reports to Requirements.** The HOC tracks which specific order or group of specific orders originates from which PIR to ensure that the collected information was provided to the original requester. This also allows the HOC to rapidly determine which asset is available for retasking.

- **Screen Reports.** Each report received is screened for accuracy, timeliness, and applicability to the original tasking or request. If the HOC determines that it completely fulfills the tasking or request, the HOC informs the G/S2X and G2/S2 so that the tasking or request can be closed and the information provided to the original requesting unit.

- **Provide Feedback to Collectors and Analysts.** The HOC provides feedback to all the HUMINT R&S assets. This is normally provided through the C2 element of that unit. By doing so, the HOC quickly reinforces if the reporting is answering the original order or request, or the HOC can provide guidance if it is not. This feedback is essential. The RM team may provide additional information on its collection or analysis if the HOC tells the team exactly what is needed or has been missed in the original report.

UPDATE ISR PLAN

4-34. This step aids the G2/G3 in updating the ISR plan by eliminating satisfied collection requirements, redirecting assets to cover non-satisfied requirements, cross-cueing requirements, and adding new collection requirements to the ISR. This process is accomplished by adjusting the HUMINT portion of the overall integrated ISR plan. It maintains intelligence synchronization and optimizes the exploitation of information in response to situation changes in the AO. The updated HUMINT plan is distributed to the G/S2X requirements manager to ensure its incorporation into the overall unit ISR plan. Continuously updating the HUMINT portion of the ISR plan is vital due to the time involved in redirecting HUMINT assets.

HUMINT MISSION PLANNING

4-35. HUMINT mission planning begins when a unit receives a tasking to conduct HUMINT collection in support of a specific mission, operation, or collection plan. The mission analysis portion of the MDMP is explained in FM 5-0. Special factors must be considered when applying the MDMP to HUMINT operations as discussed below.

RECEIVE AND ANALYZE THE HIGHER HEADQUARTERS ORDER

4-36. Attention must be paid to the support relationship (GS or DS) that exists between HUMINT assets and the unit. The operational environment, including applicable law and policy under which the units are operating must be understood, as this affects the ability of the units to perform certain missions. Applicable law and policy include US law; the law of war; relevant international law; relevant directives including DOD Directive 3115.09, "DOD Intelligence Interrogations, Detainee Debriefings, and Tactical Questioning"; DOD Directive 2310.1E, "The Department of Defense Detainee Program"; DOD instructions; and military execute orders including FRAGOs. Because of frequently overlapping AOIRs in HUMINT operations, other unit missions and potential areas of conflict must be identified. Missions of other non-HUMINT units must be understood for coordination and possible integration of HUMINT assets. The availability of assets from higher echelons, requirements to provide support to lower echelons, and the existence of technical control from higher echelons must be identified. Tasking, reporting, and communications channels must be clearly understood.

ISSUE A WARNING ORDER

4-37. After the commander has analyzed his orders and worked out the mission and related tasks, he must quickly pass on this information to his team. This is accomplished through the WARNO. As a minimum, the WARNO must include to whom the order applies, time and nature of the operation, the earliest time of movement, and the time and place where the OPORD will be issued. Unit members should prepare for movement while the leader is performing the remaining preparatory tasks.

MAKE A TENTATIVE PLAN

4-38. When determining how the mission will be carried out, the commander works with the factors of METT-TC. When planning for HUMINT collection missions, focus must be placed on the human beings (threat, friendly, and neutral) as well as the key terrain on the battlefield, including information on—

- The demographics of both the AO and AOI.
- The organization and structure of all opposition in the AO and AOI.
- The history of the AO and AOI pertinent to the current situation.
- The economic and social data of all groups in the AO and AOI.

- All key leaders (political, military, social, religious, tribal), opinion leaders, and other influences on public opinion.
- The media and its influence on the population of both the AO and AOI.
- The primary and secondary languages and dialects spoken in all parts of the AO.

4-39. A target folder, if one is used, provides valuable up-to-date intelligence information about the AO for mission analysis and planning. Once intelligence products identify the contentious areas, trends, capabilities, and latest issues concerning the AO, the commander may request a target folder prepared on specific items, such as a hostile organization with the inclination and potential to cause harm to friendly forces. Target folders may include—

- Imagery of the AO and personalities.
- Terrain models of the AO.
- Latest information reports from the AO.
- Biographical data on key leaders in the AO.

Review Available Assets

4-40. The commander and staff, including the OMTs or HUMINT operations section, must look at organic assets and consider factors such as language capability, experience in various aspects of collection, analysis, and management. If organic assets are inadequate, the commander and staff should consider additional available assets within the organization and resources from higher echelons. The commander and staff must consider the analysis and management structure of a HUMINT operations section in addition to the OMT and HCTs. During this step the mission analysis and planning group should determine, among other things—

- The number of HUMINT collectors available.
- The number of collectors who are qualified linguists.
- The number of linguists available to support the collectors.
- Force protection considerations.
- The optimal number of HCTs, OMTs, and HUMINT operations sections that can be configured from the available assets.
- Whether additional assets such as CI agents, TECHINT personnel, analysts, additional linguists, or other experts need to be added to some or all the HCTs to meet mission requirements.

Determine Constraints

4-41. This is a critical step in HUMINT mission analysis. HUMINT collection operations are affected by applicable law and policy. Applicable law and policy include US law; the law of war; relevant international law; relevant directives including DOD Directive 3115.09, "DOD Intelligence Interrogations, Detainee Debriefings, and Tactical Questioning"; DOD Directive 2310.1E, "The Department of Defense Detainee Program"; DOD instructions; and military execute orders including FRAGOs. The degree of restriction may depend on the type of operation being conducted. Constraints are normally found in the scheme of maneuver, the concept of operations, and coordinating instructions. Specific to intelligence interrogation operations, in accordance with DOD Directive 3115.09, "all captured or detained personnel shall be treated humanely, and all intelligence interrogations or debriefings to gain intelligence from captured or detained personnel shall be conducted humanely, in accordance with applicable law and policy. Acts of physical or mental torture are prohibited."

Identify Critical Facts and Assumptions

4-42. The human factor is preeminent in this step. Assumptions and facts include—

- How HUMINT collectors can interact with the local population.
- What types of sources are available.
- What types of adversary intelligence and unconventional threats are present.

Conduct Risk Assessment

4-43. There are inherent risks involved in HUMINT collection. HUMINT collectors need access to the local population to perform their mission. Rules that restrict all forces to base areas to protect the force may be prudent; however, these restrictions can severely degrade HUMINT collection capabilities, particularly in support of force protection requirements. This measure deprives the collectors of sources needed to anticipate and prevent violent incidents. HUMINT collectors receive cultural training as well as security training to allow them to minimize the dangers of interacting with the local population. Commanders must weigh the risk to collectors against the risk to the force as a whole, and determine whether to provide additional security to the HCT in order to allow the team to perform missions outside the base area to gain needed intelligence. DA Pam 385-1 provides guidance for risk assessment.

Select Courses of Action (COAs)

4-44. During COA development the staff, under the commander's guidance, analyzes various options for deploying and implementing HUMINT assets. Input from HUMINT senior NCOs and WOs is vital to COA development and analysis. Items to consider during COA development include—

- The distribution of the HCTs and OMTs within the AO.
- The support relationship (GS and DS) that exists for the deployed teams.
- The command relationship in effect for the HCTs and OMTs (assigned, attached, or OPCON).
- The manner in which the HUMINT assets are phased into the theater.
- The tactical configuration (personnel and equipment) of the HCT.
- The actual number of the HCTs and OMTs and the size of the supporting HUMINT operations section (if any) deployed.
- The priority of the OMT's efforts.
- The priority of linguist support.

COLLECTION PRIORITY

4-45. During the MDMP, the MI commander advises his higher headquarters on the most efficient use of the HUMINT collectors to meet collection requirements. Depending on the particular higher echelon mission and the capabilities of the specific personnel under his command, the supported S2 must decide whether to concentrate collection efforts on source, debriefing, interrogation, tactical questioning, liaison, or DOCEX operations to answer collection requirements. (See Chapter 5 for a description of these operations.) The MI commander may be required by his operational tasking to support any or all of these operations. He must decide how to task organize his assets to meet these requirements. When faced with limited assets, prioritization of collection is paramount.

4-46. A commander normally must prioritize HUMINT collections and DOCEX. Although the decision is primarily dependent on which type of source (human or document) is most likely to give the priority information, other factors such as phase of operation, ROE, source availability, and collection resource capabilities may influence his decision. At the tactical level, both human sources and documents are screened and the senior HUMINT soldier establishes the priorities. If documents and human sources are determined to be equally likely of containing priority information, human sources are normally exploited first due to—

- The ability of the HUMINT collector to get a human source to elaborate and explain his information, which cannot be done with a document.
- The rate at which people forget detailed information.
- The fact that an individual's resistance is easier to bypass immediately after undergoing a significant traumatic experience (capture). Capture thrusts them into an unfamiliar environment over which they have no control and are vulnerable to various approach techniques. This initial vulnerability passes quickly. An individual's established values begin to assert themselves again within a day or two, and the individual's willingness to cooperate might also decrease.

TASK ORGANIZATION

4-47. Because of the need to place HUMINT collectors in contact with the local population and the need in many cases to integrate the HUMINT collection process into other operations, the planning and analysis staff for HUMINT missions is somewhat expanded from the norm. They should include the C/J/G/S2X, SJA, S1, S2, S3, S4, S5, S6, other staff officers, as necessary, Provost Marshal, MP, and US Army Criminal Investigation Command, CA, unit HUMINT commanders, and senior HUMINT technicians of the deploying unit. If the unit's mission is to replace a currently deployed HUMINT unit, a representative of that unit should be included.

4-48. The challenge to the MI commander is the proper training during operations, task organization, placement, and coordination of movement of HUMINT elements to meet collection requirements. The unit modified table of organization and equipment (MTOE) organization, which is designed for an MTW, may have to be modified to meet the specific requirements of operations in PMEs and SSCs. Augmentation is often needed and must be requested. Task organization must be flexible to adjust to the dynamic mission objectives. Commanders must allow for the augmentation of HCT with other MI specialties and non-MI personnel as mission analysis and planning indicate the need. Mission analysis and planning identify the specific requirements for the HUMINT operations section, HAT, OMTs, and HCTs.

4-49. The composition of the HUMINT elements must be based on METT-TC factors. The number of HCTs and OMTs in the theater depends on the intensity of the collection effort and the geographical coverage of the AO. HCT members should be prepared to support any HUMINT missions they may receive through command channels. They must have the skills to shift easily from one set of functions to another based on the dynamic mission requirements. The number of OMTs in a designated theater will depend on the type and nature of the mission. A single OMT is capable of managing and controlling 2 to 4 HCTs. The size and staffing of the OMT will depend on a number of factors:

- Whether a HUMINT operations section is deployed and how many HCTs are subordinate to it.
- If a single HCT deploys to support a small contingency, there may be no need for an OMT. In this case the team leader must serve as the OMT.
- If three or more OMTs deploy, then a tactical HUMINT operations section should be deployed.
- For every 3 to 4 HCTs and their designated OMT, there should be one headquarters element composed of a platoon leader and a platoon sergeant to handle all administrative and logistical matters.

OPERATIONAL CONSIDERATIONS

RESERVE COMPONENT INTEGRATION

4-50. Given the Army's OPTEMPO and force structure, the integration of RC forces into the AC is highly likely for future operational deployments. Commanders must identify their requirements early and establish proactive coordination (both in garrison and while deployed) with their RC counterparts to fully integrate them during all phases of training and operations. During operations that include significant RC participation, an RC liaison officer normally will be assigned, either temporarily or permanently (at higher echelons), at the appropriate level of command. The commander and staff must ensure that the RC LNO is involved in all aspects of operational planning and execution.

4-51. There are three general categories of RC augmentation:

- Category 1: Formation of specialized units that include a fully integrated AC and RC TOE. The activation of the RC of these units is required for their full operational capability.
- Category 2: Augmentation of active duty units by RC units to fill out unit strength levels or to provide additional functionality. For example, an AC division might require additional HUMINT teams to support it

during a stability operation. If a division required one additional team, it should request a team and not request four HUMINT collectors. If the requirement is for three additional teams, it should request a HUMINT platoon with its organic C2 and OMTs.

- Category 3: The requirement for individual augmentees. This usually occurs when a unit has the C2 structure but needs either additional personnel or additional capability within the command structure. For example, a unit may have a HUMINT platoon but the platoon is at 50 percent strength. Individual augmentation is the easiest method of integration since the individual is integrated in the same manner as any replacement. The augmented unit normally is required to provide all equipment other than initial issue-type equipment.

4-52. There are several items to consider in unit augmentation:

- Accurate Identification of Requirements: During the MDMP, units need to identify those mission-essential capabilities not already present in the unit. The G3/S3, working in conjunction with the G1/S1, considers options that may include RC augmentation of organic units although the final decision to employ RC units is usually determined at Headquarters, Department of Army (HQDA). The requirement for augmentation is forwarded through appropriate personnel channels to US Army Forces Command (FORSCOM) and HQDA, which will identify the appropriate units or personnel. If approved, they will work with the appropriate agencies to establish the timeline in which the units can respond on the Time-Phased Forces Deployment Data List (TPFDDL). When developing requirements, the requesting unit must be sure to articulate its needs accurately, specifying required skills, numbers, and any additional skill identifiers (ASI). [Example: Request augmentation by a HUMINT platoon consisting of at least a platoon headquarters, three HCTs, one OMT, two linguists, and one CI/HUMINT Automated Tool Set (CHATS) proficient operator. The augmenting element will be operating in support of the commander's force protection program in the gaining unit's AOR.]

- Activation Timeline: Units need time to mobilize and conduct any additional collective and individual training that may be specific to the unit's mission or operational environment. The requesting unit needs to be aware of the time required to activate the requested RC and that there may be differences in levels of training or equipment. Timelines should be established by FORSCOM to allow resolution of these problems and should be reflected in the commander's operational planning sequence. Timelines will vary from unit to unit and mission to mission.

- Training: USAR and ARNG units usually cannot train their units or individuals to the same proficiency as the AC. Normally, this is due to the limited amount of training time. Because of this limitation, a certain degree of train-up prior to deployment may be necessary. Commanders should identify available training opportunities and request the participation of personnel identified for augmentation. For an ongoing mission, you should also plan for an extended "right seat ride" mission handover period once the individuals or unit arrives in the theater of operations.

- Command and Control: If the RC augmentation requires activation of an entire unit, it should include their C2 element. If the augmentation is by individuals, then they will fall under the command and control of the gaining units.
- Time on Active Status: USAR and ARNG soldiers are restricted as to the amount of time they can remain on active status. This timeline begins on the date of mobilization and ends on the day the soldier leaves active duty status. Deployed units must take this into account when conducting continuous operations and must identify the requirement to replace RC forces early enough to allow for the required training and handoff procedures.
- Experience: While RC personnel normally lack current military experience, they often perform jobs in the civilian sector that either mitigate this lack of experience or they are able to bring a new and useful capability with them. Care should be taken that reservists who have civilian jobs which are similar to their HUMINT MOS (such as police officers or investigators) recognize the different constraints under which they operate in the military environment. For example, police officers who might normally task informants with minimal oversight cannot do that in their position as a HUMINT collector. Commanders should try to capitalize on these skills, but ensure proper training and understanding of the policies and regulations that govern HUMINT collection operations.

OPERATIONS PLANS, OPERATIONS ORDERS, AND ANNEXES

4-53. An OPLAN is any plan for the conduct of military operations. When a commander issues a directive for the coordinated execution of a military operation, it becomes an OPORD. Although plans are based on specific conditions or assumptions, they are not static. Plans are changed, refined, and updated as a result of continuous estimates and studies. It is critical to include HUMINT plans in the Intelligence Annex to the OPLAN.

4-54. The OPORD gives the HUMINT element approval to execute its mission. OPORDs define the mission, set the parameters of operations, identify who is responsible for what, and how it is to be supported. Additions that are necessary to amplify an OPLAN or OPORD are contained in annexes, appendices, tabs, and enclosures. Tasking for units to conduct HUMINT collection operations is listed in the main body of the OPORD under **Tasks to Subordinate Units.** The HUMINT appendix to Annex B provides the technical guidance for HUMINT collection including the umbrella concept for HUMINT operations.

4-55. The HUMINT appendices provide details on planning, coordinating, approving, and managing HUMINT operations as they relate to the unit's overall mission. These appendices serve as the basic document authorizing most HUMINT operations and programs. They must be reviewed and approved by the appropriate office or commander. The HUMINT appendix to the ISR Annex is necessary to ensure that augmentation of HUMINT assets from other components and agencies are integrated throughout the TF as required to facilitate their specialized collection requirements. Specific tabs may include joint debriefing and interrogation facility operations, source operations, DOCEX, or open-source information.

OPERATIONAL COORDINATION

4-56. HUMINT collection is not conducted in a vacuum. Coordination with MI organizations and non-MI agencies, units, and staff organizations is often critical to expedite and complete HUMINT collection operations. (See Appendix C for predeployment planning.)

MI ORGANIZATIONS

4-57. Elements involved in HUMINT planning, execution, and analysis need to maintain close coordination with their counterparts in the other intelligence disciplines. Coordination includes but is not limited to the disciplines shown below.

Imagery Intelligence:

- Support imagery analysis by using HUMINT sources to identify or confirm the identification of items in imagery. This includes, for example, using human sources to identify the functions of buildings that have been tentatively identified through external imagery.
- Coordinate for current military or civilian imagery to use in the questioning of sources.
- Cue requirements managers and others involved in imagery tasking on locations or activities for imagery collection.
- Coordinate for IMINT information to verify information obtained through HUMINT collection.
- Provide imagery for analysis (through still and video photography and captured imagery).
- Coordinate for technical support as required when questioning personnel on subjects related to imagery.
- Obtain imagery-related collection requirements that can be answered by human sources.

Signals Intelligence:

- Support signals analysis by using HUMINT sources to identify or confirm the information obtained through SIGINT collection.
- Coordinate for current SIGINT information to use in the questioning of sources.
- Cue requirements managers and others involved in SIGINT tasking on locations or activities (including communications types and frequencies) for SIGINT collection.
- Coordinate for information to verify information obtained through HUMINT collection.
- Provide SIGINT-related CEDs for SIGINT analysis.

- Coordinate for technical support as required when questioning personnel on SIGINT-related topics.
- Obtain SIGINT-related collection requirements that can be answered by human sources.

Measurement and Signature Intelligence:

- Support measurement and signature intelligence (MASINT) analysis by using HUMINT sources to identify or confirm the information obtained through MASINT collection.
- Cue requirements managers and others involved in MASINT tasking on locations or activities for the location of MASINT sensors.
- Coordinate for information to verify information obtained through HUMINT collection.
- Provide MASINT-related CEDs for MASINT analysis.
- Coordinate for technical support as required when questioning personnel on MASINT-related topics.
- Obtain MASINT-related collection requirements that can be answered by human sources.

Technical Intelligence:

- Support TECHINT analysis by using HUMINT sources and documents to provide information concerning threat equipment and to support TECHINT materiel analysis. This includes, for example, the interrogation or debriefing of equipment operators of the translation of operators manuals for a piece of equipment being investigated.
- Coordinate for current information on equipment capabilities to use in the questioning of sources.
- Cue requirements managers and others involved in TECHINT tasking on locations or activities for TECHINT collection. This includes forwarding the identification and location of equipment of TECHINT interest obtained during HUMINT collection operations.
- Coordinate for TECHINT information to verify information obtained through HUMINT collection.
- Provide information from CEDs in support of TECHINT.
- Coordinate for technical support as required when questioning personnel on subjects related to areas of TECHINT interest.
- Obtain TECHINT-related collection requirements that can be answered by human sources.

Counterintelligence:

- Support CI analysis by using HUMINT sources to provide information concerning adversary intelligence collection capabilities and operations.
- Identify human and document sources that have information of CI interest.

- Cue requirements managers and others involved in CI tasking individuals or activities of CI interest.
- Coordinate for CI information to verify information obtained through HUMINT collection.
- Provide information from CEDs in support of CI.
- Coordinate for CI support as required when questioning personnel on topics related to areas of CI interest.
- Obtain CI-related collection requirements that can be answered by human sources.
- Integrate CI elements into HUMINT collection operations as applicable.

Open-Source Intelligence:

- Support open-source intelligence (OSINT).
- Provide open source maps, charts, phone directories, business directories, newspapers, video and audio media (including tapes and compact discs) to the appropriate J/G/S2X and Intelligence Community agencies and liaison officers.

OTHER ORGANIZATIONS

4-58. In addition to MI units, HUMINT collection organizations frequently conduct coordination with other military organizations.

- **Military Police Units:** Close coordination between HUMINT collectors and MPs is mutually beneficial. The MPs are responsible for maneuver and mobility support, area security, internment and resettlement, law and order, and police intelligence operations. Both activities (HUMINT collection and MP operations) require close contact with the local civilian, refugee, and detainee populations. HUMINT collection at checkpoints and at EPW and other detainee collection points must be coordinated with the MPs, who are normally responsible for internment and resettlement operations. In return, the HUMINT collectors, because of their screening and questioning of these population groups, can help facilitate the MP's population control missions by providing information about the population's activities and intentions that may be of MP concern. At EPW/detainee collection points, HUMINT collectors should arrange with the MP leadership to be allowed to debrief MPs since MPs are in regular contact with the detainees. This does not constitute tasking. Information collected in this manner may provide valuable insight, which can aid the collector in formulating approach strategies. MPs should be debriefed in such a way so as not to interfere with their mission. Liaison with the MP chain of command is vital to gain their support and assure them that HUMINT collection will not interfere with MP operations. Joint patrols containing MPs and HUMINT collectors can also be mutually beneficial in many situations.
- **Criminal Investigation Division (CID) and Provost Marshal Office (PMO):** The goals of HUMINT collection and those of the MPs (particularly CID) are different. CID and PMO are concerned with

identification and apprehension of criminal elements. The goal of HUMINT collection is the collection of information in response to PIRs that in many situations are centered on force protection. In the situation where the threat includes a criminal element, the HCTs might collect OB type information on the criminal element to ascertain their activities and threat to friendly forces. HUMINT collectors are not trained to conduct criminal investigations and must not be used for this purpose. Criminal investigators and HUMINT collectors must carefully coordinate their activities as necessary. HUMINT collectors are required to report to the proper agency information collected on criminal activities that the HUMINT collectors uncover in the normal course of their activities.

- **Psychological Operations Units:** As with the MP force, HUMINT collectors and PSYOP units are often interested in the same target audience but for different reasons. PSYOP units are interested in modifying the target audience beliefs and actions to be more supportive of US goals. Normally, HUMINT collection elements coordinate with PSYOP elements to obtain information concerning the motivational factors and cultural value systems of the individuals to be questioned. PSYOP units, as a part of their normal operations, develop detailed analysis concerning psychological and cultural factors of friendly and hostile elements in the AO. Such information will help HUMINT collection personnel to understand the source's attitude, value system, and perception; it will also help to obtain information more rapidly. At the same time, PSYOP units often will develop collection requirements to determine local attitudes and for information on the effectiveness of PSYOP campaigns. HUMINT collectors can be tasked to collect on these requirements if they are included as PIRs.

- **Civil Affairs Units:** The CA mission often places CA units in contact with the HUMINT collection target audience. If possible, HUMINT collection missions can be established in coordination with CA missions. If the HUMINT collection mission is viewed as having the potential of interfering with the CA mission and coordinated operations are not possible, CA personnel can still be sensitized to intelligence collection requirements and debriefed by HUMINT collectors as part of a friendly force debriefing operation.

- **Drug and Law Enforcement Agency Operations:** Personnel who are employees of DOD intelligence components may be assigned to assist Federal law enforcement authorities and, when lives are endangered, state and local law enforcement authorities; provided such use is consistent with, and has been approved by an official authorized pursuant to DOD Directive 5525.5, Enclosure 4 (reference (i)). Such official shall ensure that the General Counsel of the providing DOD component concurs in such use. Assistance may be rendered to LEAs and security services of foreign governments or international organizations in accordance with established policy and applicable SOFAs, provided that DOD intelligence components may not request or participate in activities of such agencies undertaken against US persons that would not be permitted activities of such components under the procedures of AR 381-10. HUMINT collectors may assist foreign law enforcement authorities, with prior approval of the J2X. Under no circumstances will HUMINT collectors assist any US or foreign law enforcement authorities in any manner without prior approval by competent authority after a legal review of the proposal.

- **Maneuver Units:** HCTs may be utilized in GS for coverage of an AOIR or in DS to support a specific maneuver unit. The type of coordination needed with maneuver units will vary depending on the type of support relationship the HCT has. HCTs operating in GS should coordinate with maneuver unit commanders when the HCT will be operating in that unit's AO. At a minimum, the HCTs should announce their presence and request information on any conditions or ongoing situations that may affect on the conduct of their mission. An HCT operating in DS of a specific unit will coordinate with the unit for force augmentation to HUMINT patrols as needed in accordance with force protection requirements. The HCT leader should also coordinate with the supported unit's S2 for involvement in debriefings of returning patrol members, checkpoint personnel, convoy leaders and others. HCT leaders may also coordinate to be included in the unit's reconnaissance patrols, as appropriate.

- **Combat Service Support Units:** Current and future combat operations will be conducted in a noncontiguous battlespace. CSS formations and units may be an excellent source for HUMINT collectors. In many situations, DPs and refugees will perceive CSS activities as non-threatening and an activity which can provide them with aid and comfort. CSS operations will naturally draw DPs and refugees hoping to receive support. This could provide opportunities for HUMINT collectors to access this sector of the population. CSS unit S2s should conduct patrol debriefings of returning convoy personnel to capture observations made during convoys, with the goal of cross-cueing the supporting HCT, CI team, or law enforcement element as appropriate.

STAFF COORDINATION

4-59. Successful HUMINT collection operations require support from the staff elements of the supported unit. These elements are collectively responsible for the planning that results in HUMINT tasking. Below is a partial list of the staff responsibilities that affect HUMINT collection:

- G1/S1 HUMINT-related responsibilities include but are not limited to—
  - Supervising the medical support furnished to EPW/detainees.
  - Maintaining a list (by language and proficiency) of qualified linguists within their command.
  - Coordinating with the G4 or G5 for procurement and payment of other interpreters and translators needed to perform intelligence and non-intelligence duties.
  - Ensuring the echelon's OPLAN contains complete provisions for handling and evacuating detainees, refugees, DPs, and local civilians

as required. This plan must satisfy the interests of all other staff officers and provide for—

- Ensuring humane treatment of all personnel.
- Promptly evacuating personnel from the combat zone.
- Integrating procedures for the evacuation, control, and administration of personnel with other combat service (CS) and CSS operations.
- Ensuring delivery of mail to EPWs and other detainees.
- Maintaining detainee (including EPW) statistics.
- Providing administration and control of detainee currency and pay records, including coordinating with appropriate intelligence authorities about investigating large sums of money.

- G2/S2 is responsible for developing intelligence in support of unit operations. The G2/S2 at division and higher and in the interim BCT is supported by a G/S2X and normally a HAT in the performance of his HUMINT-related functions. His HUMINT-related responsibilities include but are not limited to—

  - Obtaining intelligence through intelligence reach to support HUMINT collection.
  - Incorporating HUMINT into the ISR plan.
  - Developing the HUMINT annex to the OPORD and OPLAN.
  - Coordinating to provide technical support for all HUMINT collection operations.
  - Ensuring deconfliction and synchronization for all HUMINT collection assets within the unit's AO. A particular effort must be made to coordinate with all DOD military source operations (MSO), and DOD and other government agencies (OGAs) that may be operating in the AO; with the theater J2X, as part of deconfliction. Failure to deconflict with DOD MSO and OGAs may result in compromise of assets and interruption of collection operations and potentially unintended casualties.
    - Obtaining documents and materials of intelligence interest, including visual and audio media and electronic equipment (such as computers, phones, PDAs) taken from detainees, or seized or loaned, in coordination with the Provost Marshal and other elements.
    - Recording, evaluating, and analyzing collected information and providing feedback to HUMINT collectors.
    - Ensuring adequate HUMINT collection and reporting nets and systems are available.
    - Coordinating with the G3 to ensure plans for HUMINT collection operations are included in unit OPLANs.
    - Coordinating with the G3 to ensure that HUMINT collectors are included in unit training plans, rehearsals, and briefbacks.
    - Drafting instructions for handling, evacuating, and exploiting captured enemy personnel and CEDs. (They coordinate with the G3 to ensure draft instructions are included in the command standing operating procedures (SOPs), OPLANs, and OPORDs.)
    - Projecting capture rates as well as refugee and DP rates.

- Determining the number of interpreters and translators needed to perform intelligence duties.
- Coordinating with other agencies and HUMINT collectors for intelligence sharing.
- Controlling the procedures used to process and grant clearances to the interpreters and translators as required.
- Coordinating with the civil-military operations (CMO) officer for intelligence screening of local nationals, refugees, and DPs.
- Coordinating with SJA for legal review of proposed operations.

- G3/S3 is responsible for operations, plans, organization, and training. His HUMINT collection-related responsibilities include but are not limited to—

  - Ensuring the inclusion of HUMINT collection units in the main body of OPLANs and OPORDs under **Tasks to Subordinate Units** and **Task Organization**.
  - Ensuring instructions for handling, evacuating, and exploiting captured enemy personnel and CEDs in all unit command SOPs, OPLANs, and OPORDs.
  - Incorporating HUMINT collection operations into future plans and operations.
  - Ensuring subordinate units are trained in proper handling and evacuation of captured enemy personnel, materiel, and CEDs.
  - Ensuring that the subordinate elements are trained in OPORDs including ROE and the proper handling of local civilians, foreign nationals, refugees, and DPs.
  - Obtaining, organizing, and supervising employment of additional personnel as guards for EPWs and other detainees where MP assets are not available or insufficient.
  - Tasking the Division/Brigade Engineer Officer in conjunction with the G2/S2 to conduct a site survey for possible EPW/detainee holding area facilities within the operational area. Priority should go to existing facilities needing little or no renovation to meet operational requirements. If suitable facilities cannot be found, the engineer officer should provide detailed facilities design specifications to the G4/S4 for coordination and development of contracted resources.

- G4/S4 responsibilities related to HUMINT collection include but are not limited to—

  - Developing command policy for evacuation and internment of captured enemy personnel, and evacuation and safekeeping of CEE and CEDs.
  - Coordinating contracts for real estate and construction of source-holding facilities if local capabilities are not available. Ideally, existing facilities will be occupied and renovated whenever possible.
  - Collecting and distributing captured enemy supplies. (This is coordinated with the intelligence and operations staffs.)
  - Procuring and distributing rations to personnel holding areas.
  - Transporting EPWs and other detainees in a timely, safe manner to the appropriate facility for processing.

- Determining requirements for use of source labor for the logistical support needed in source-handling operations.
- Providing logistical support to interpreter personnel.
- G5/S5 responsibilities related to HUMINT collection include but are not limited to—
  - Coordinating with local US government, personnel staff representatives, and HN armed forces for procuring native linguists for interpreter support.
  - Coordinating military support of populous.
  - Providing technical advice and assistance in reorientation of sources and enemy defectors.
  - Coordinating MI aspects of CMO activities with the G2.

ADDITIONAL SUPPORT

4-60. In addition to the major staff elements, a HUMINT collection element requires support from several other elements in order to conduct operations. These elements are discussed below.

- The US Army Criminal Investigation Command is the organization with primary responsibility for investigating allegations of criminal acts or reportable incidents committed by or against detainees.
- The SJA can provide legal support and advice on the interpretation and application of applicable law and policy. Applicable law and policy include US law; the law of war; relevant international law; relevant directives including DOD Directive 3115.09, "DOD Intelligence Interrogations, Detainee Debriefings, and Tactical Questioning"; DOD Directive 2310.E, "The Department of Defense Detainee Program"; DOD instructions; and military execute orders including FRAGOS. The SJA is also a channel for reporting known or suspected reportable incidents of abuse or inhumane treatment.
- The Inspector General is a channel for reporting known or suspected reportable incidents of abuse or inhumane treatment.
- The PMO is the channel for reporting criminal activity other than reportable incidents, but also can be used for reporting known or suspected reportable incidents.
- The Chaplain can also receive reports of reportable incidents.
- The G7 provides information on Information Operations and conducts liaison with PSYOP, the Electronic Warfare Officer, the Military Deception Officer, and Operations Security personnel.

PART TWO

HUMINT Collection In Military Source Operations

Part Two discusses HUMINT collection as it pertains to MSO. The Secretary of Defense (SECDEF) has established a DOD-wide HUMINT Enterprise consisting of the following executors: The Office of the Secretary of Defense (OSD), the Combatant Commands (COCOMs), the Military Departments, the Defense Intelligence Agency (DIA). All Defense HUMINT Enterprise executors support and satisfy Defense requirements by employing their available resources and capabilities.

MSO refer to the collection of foreign military and military-related intelligence by humans from humans. MSO are conducted under SECDEF authorities, to satisfy DOD needs in compliance with DOD policy. Within the Army, MSO are conducted by trained personnel under the direction of military commanders. These specially trained personnel may employ the entire range of HUMINT collection operations. MSO sources include one-time, continuous, and formal contacts, from contact operations; and sources from interrogations, debriefings, and liaison activities.

Each type of MSO activity has specific operational requirements, specific legal restrictions, and operational guidelines. HUMINT collection activities in each of these categories require specific approval, coordination, and review. MSO include human source contact operations, debriefing, liaison, and interrogations. This chapter introduces each of these collection operations.

Chapter 5

HUMINT Collection

HUMINT COLLECTION OPERATIONS

5-1. Full spectrum operations require focused MSO with strong capabilities dispersed across the battlefield. In offensive and defensive operations, the HCTs need to be placed in support of the engaged maneuver battalions. In stability and reconstruction operations and civil support operations, the HUMINT teams need to be located in battalion AOs throughout the AOIR.

5-2. The rapid pace of operations, the need to provide near-real time (NRT) support of command decisions and the inherent time delays in moving detainees, including EPWs and civilian refugees to centralized locations, necessitate the dispersion of HUMINT collection assets to forward areas in support of critical operations rather than their retention at detainee and refugee holding facilities at echelons corps and below. This forward deployment gives HUMINT collectors earlier access to sources and is facilitated by enhanced communication and automation capabilities down to the collection team level.

5-3. All operations are different, and deployment of HUMINT assets is METT-TC dependent. Brigades need the capability to provide 24-hour HUMINT collection capability to each battalion AO. The command relationship of the HUMINT collection capability is also METT-TC dependent. The OMT should be located at the echelon that is best able to manage and support the HCTs and to provide the best capability to answer the commander's PIRs.

5-4. The Division and Corps elements should cover their respective areas not covered by their subordinate commands. They also, as needed, reinforce those target areas that are most effective in answering their respective command PIRs already covered by subordinate command capability. EAC HUMINT units normally are responsible for supporting theater or national requirements and providing HUMINT support at theater level facilities such as the JIDC. The EAC units will also augment the echelon below corps units and conduct source operations in the Corps area as required. Operations, particularly in challenging terrain and in stability and reconstruction environments, may require additional HUMINT assets normally obtained from the RC.

HUMAN SOURCE CONTACT OPERATIONS

5-5. HUMINT collection requires the contact between the HUMINT collector, who attempts to gather information through a variety of HUMINT collection techniques, and a human contact, who hopefully has the information that the HUMINT collector wants and who can be convinced to divulge the information. Operations with formal contacts are only conducted by HUMINT collectors and CI agents who are specifically trained and authorized to do so. There are three levels of contacts:

- One-time contact.
- Continuous contact.
- Formal contact.

5-6. The basic goal of all levels of contact is to collect information in response to collection tasking; however, only under certain conditions can HUMINT collectors task contacts to get information for them (see para 5-28). Understanding the types of contacts is key to understanding each type of human source contact operation. The following levels are not all-inclusive nor are the listed categories exclusive. For example, a contact who was initially a one-time contact (such as a walk-in) may later be developed into a continuous contact. A continuous contact may be developed into a formal contact, who can then be tasked, trained, and paid. There is no limit on the number of times a team can meet contacts without recruiting them and making them into a formal contact.

ONE-TIME CONTACT

5-7. The one-time contact is a source of information of value that was, and will be, encountered only once. In all operational environments the HUMINT collector will frequently encounter a source only once, particularly at lower echelons. This may be a local civilian encountered during a patrol, a detainee who is quickly questioned and then evacuated, or a refugee at a checkpoint.

5-8. In addition to the information obtained from a one-time contact, the HUMINT collector must make a reasonable effort to obtain as much basic data as possible about the one-time contact. Complete name, occupation, address, and other basic data of this source are crucial for a thorough analysis of the information provided. The one-time contact and the information he provides cannot be assessed and evaluated independently; however, the information provided by a one-time contact must be reported and corroborated through other HUMINT sources and even other intelligence disciplines.

5-9. Contact reports must be filed with the OMT and source registries maintained in accordance with FM 34-5 (S//NF), AR 381-100 (S//NF), and DIAM 58-11 (S//NF) in order to support analysis of information obtained. If a one-time contact is encountered for a second time and again provides information of value, then the contact may be thereafter treated as a continuous contact.

5-10. A walk-in is a one-time contact who volunteers information of value to US forces on his own initiative. The walk-in source may volunteer information by approaching an HCT, other ISR elements, or US forces or civilian personnel anywhere in the AO. Each unit must have in place a program to identify, safeguard, and direct the walk-in to the appropriate collection asset, to be screened and debriefed as required. For example, a walk-in who wanted to report a crime would be directed to the PMO rather than to a HUMINT collector.

5-11. The collection asset will screen the walk-in to determine the type of information the source has and to determine and evaluate the reliability of the individual. After identifying the type of information, the collector determines if he has the jurisdiction to collect that information. If, for example, the walk-in wishes to report a crime, the collector refers that individual to the proper criminal investigative agency.

5-12. Systematic questioning, deception detection techniques, and cross-checking of information are used extensively in the evaluation process. Concurrently, there are national level directives, DOD directives, and Army regulations that direct specific actions to be taken with a walk-in. When dealing with a walk-in source, HUMINT collectors must guard against adversary intelligence collection. They must also protect legitimate sources of information. The walk-in is thoroughly debriefed on all areas of information relevant to collection requirements, and any information of value is reported.

5-13. On occasion, the HUMINT collector may determine that a one-time contact has the potential to become a continuous contact or a formal contact. This is referred to as a developmental lead. A developmental lead is an individual identified through social and professional status, leads, source profiling, or other techniques, who has knowledge required by the commander. A developmental lead is any person the HUMINT collector expects to see or would like to see again, or a person who indicates that they intend to return in the future.

5-14. When a HUMINT collector identifies a developmental lead, he reports his interest in elevating the source to continuous or formal contact status as soon as possible to the OMT. Although not every developmental lead becomes a source of information, the HUMINT collector should see each developmental lead as a potential source of information and apply the appropriate security measures. The developmental lead is continuously assessed to verify his placement and access to the type of information the HCT is seeking. Additionally, the HUMINT collector continuously assesses the motivation and characteristics of the developmental lead.

5-15. A one-time source cannot be tasked to collect information, but can be sensitized to information in which the HUMINT collector is interested. For example, if a walk-in source provides information on activity in a house in his neighborhood, he might ask if the collector would be interested in more of the same type information in the future. The HUMINT collector cannot tell him to go get more information, but can indicate that he would listen if the walk-in returned with more information on the topic. If the walk-in returns a second time, he must be handled as a continuous contact.

CONTINUOUS CONTACTS

5-16. Continuous contacts are individuals who have been identified as having more information than could be obtained through a one-time contact, and have been met again by HUMINT collection personnel for the purpose of collecting additional information. HUMINT collectors do **not** task continuous contacts, but they can be sensitized in the same way as one-time contacts. Continuous contacts provide their knowledge through informal debriefings and elicitation.

5-17. All contacts who are seen more than once by HUMINT collectors must be tracked by registering them in the Source Registry and reporting the contacts to the OMT. As an example, a one-time contact who reported information to a HCT contacts them again with follow-up information. That person will now be registered as a continuous contact and tracked by the OMT. This registration process helps to prevent the same information from being collected by multiple collectors from the same contact without realizing it. See AR 381-172 (S//NF) and FM 34-5 (S//NF) for further information on source registration and for the required forms. Types of continuous contacts are discussed below.

Local National and Third-Country National Employees

5-18. Local national and third-country national employees are non-US personnel from either the country in which the US forces are operating or a third country who are either employed by US forces directly or through a contractor to provide logistical support and services. One of the purposes of locally employed personnel screening is to assess these individuals as potential sources of information. Local national and third-country national employees can be a prolific source of information about local attitudes and events, particularly in a restrictive environment where US contact with the local population is curtailed. Their information can also be significant in a force protection role. The HUMINT collector must register these individuals with the J/G2X. While the HUMINT collector is assessing the local national employee as an intelligence source, CI agents are assessing the same source pool as potential security risks.

5-19. Coordination between HUMINT collectors and CI elements is essential for deconfliction and to avoid duplication of effort. If the HUMINT collector identifies an employee that may be of CI interest, he should immediately notify the appropriate CI unit.

Displaced Personnel and Refugees

5-20. DPs and refugees are excellent sources of information about denied areas and can be used to help identify threat agents and infiltrators. The degree of access HUMINT collectors have to DPs is dependent on the OPORDs, ROE, and SOFAs in effect. HUMINT collectors can work with CA or other programs dealing with DPs or refugees.

5-21. DPs and refugees are normally considered one-time sources but may be incorporated into other long-term collection programs if their degree of knowledge warrants this. In this case, adherence to the restrictions involving source operations is necessary. Those restrictions can be found in AR 380-10, AR 381-100 (S//NF), DIAM 58-11 (S//NF), DIAM 58-12 (S//NF), and other publications as well as existing ROE and SOFAs.

US Forces

5-22. US forces have many opportunities to interact with the local population in the normal course of their duties in operations. This source perhaps is the most under-utilized HUMINT collection resource. Some US forces, such as combat and reconnaissance patrols, are routinely tasked and debriefed by the appropriate level G2/S2. Others, such as medical teams or engineers who have extensive contact with the local population, should also be debriefed.

5-23. Commanders and staff members who serve as liaison with the local population and local government officials can be fruitful sources of information. CA, PSYOP, MP, and other elements also have legitimate reasons to conduct liaison with local authorities and should be debriefed as appropriate. The friendly force debriefing effort can succeed only with command emphasis.

5-24. HUMINT collection elements need to coordinate with local units to identify those individuals who would be most profitable to debrief and to further coordinate with them for time to conduct the debriefing. Although the S2 and S3 can and should task their soldiers to conduct collection tasks during the course of their normal duties, HUMINT collectors must ensure that their friendly force debriefing effort does not interfere with the primary mission accomplishment of the soldiers being debriefed. HCTs should ensure that the necessary staff S2s and S3s are aware of the HUMINT collection requirements and request that the staffs incorporate these into their respective collection taskings. The results of debriefings by units should also be disseminated to the HCTs for source development, collection targeting, and analysis.

Official Liaison

5-25. Liaison with local military, government, or civilian agency officials provides an opportunity to collect information required by the commander. The HUMINT collector meets with these officials to conduct liaison, coordinate certain operations, collect information, and obtain leads to potential sources of information. Elicitation is the primary technique used with liaison contacts, although in many cases there is a more formal exchange of information. Information obtained by these elements through liaison normally tends to reflect the official positions of their superiors and may not be entirely accurate or complete.

Detainees

5-26. A detainee is any person captured or otherwise detained by an armed force. An EPW is a detainee who meets the criteria of Articles 4 and 5 of the GPW. (See Appendix A.) Detainees may be interrogated. They are frequently excellent sources of information but in many instances the access of the HUMINT collector to the detainees may be curtailed.

5-27. For example, when supporting a counterinsurgency, the supported government may consider all captured insurgents to be criminals and not allow US forces access to them. In these instances, US HUMINT collectors should attempt to sit in during local questioning; they could submit questions or, at a minimum, coordinate to receive the reports from local authority questioning. US HUMINT collectors must remember that regardless of the legal status of the detainees they must be treated in a manner consistent with the Geneva Conventions. (See Appendix A.)

FORMAL CONTACT

5-28. Formal contacts are individuals who have agreed to meet and cooperate with HUMINT collectors for the purpose of providing information. HUMINT collectors who have met with a particular continuous contact three or more times should consider assessing him for use as a formal contact. Formal contacts meet repeatedly with HUMINT collectors, and their operation and tasking must be carried out in accordance with AR 381-172 (S//NF), DIAM 58-11 (S//NF), and DIAM 58-12 (S//NF).

5-29. Formal contacts are generally local nationals or third-country national employees. Knowledge of their meeting with HUMINT collectors is restricted. This can be accomplished by either disguising the fact that the HUMINT collection personnel are indeed HUMINT personnel, or by concealing the purpose of overt meetings with HUMINT personnel. HCTs take extraordinary measures to protect their relationship with these contacts. Depending on METT-TC factors, meetings with formal contacts may range from overt meetings, which are conducted discreetly in order to protect the relationship between the source and HUMINT collectors, to meetings whereby only the collector and the source know the meeting has occurred. When contact operations are conducted using this methodology, the operation must be coordinated in accordance with the Under Secretary of Defense for Intelligence (USD(I)) policy cited in Appendix J. Specific direction regarding documentation required for recruitment, and the designation of approval authority (usually the J/G2X) for recruitment of a formal contact, will be specified in Appendix 5 (HUMINT) of Annex B (Intelligence) to the governing OPLAN or OPORD.

DEBRIEFING OPERATIONS

5-30. Debriefing operations refer to the systematic questioning of individuals not in the custody of the US, to procure information to answer collection tasks by direct and indirect questioning techniques. The primary categories of sources for debriefings are friendly forces and civilians including refugees, DPs, and local inhabitants.

5-31. Debriefing operations are those operations directed towards collecting information from a segment of the target population using primarily debriefing techniques. These debriefing operations are separate from the G2/S2 debriefing program to debrief personnel returning from missions. Debriefing operations often include the debriefing of personnel who may not usually be debriefed as part of their assigned duties.

5-32. Normally Army debriefing operations will be directly related to collection tasks at the operational and tactical levels. Strategic debriefing of high-level personnel in response to theater and national level requirements is often under the purview of the DIA/DH. Army HUMINT collectors frequently participate in this type of collection, which is under the control, rules, regulations, and operational guidance of DH.

PRINCIPLES AND GUIDELINES

5-33. Debriefing operations are conducted under the guidelines of DIAM 58-11 (S//NF) and DIAM 58-12 (S//NF). They are further subject to applicable execute orders and the specific ROE and classified "umbrella concept" that apply to the specific AO.

OPERATIONAL CONSIDERATIONS AND REQUIREMENTS

5-34. Debriefing requires relatively unconstrained access to the target audience. Debriefing operations are frequently constrained by the umbrella concept, overt operational proposal (OVOP), and OPORDs. Debriefing is a time- and resource-demanding operation that often shows limited immediate results. Since the potential target audience is so large, debriefing operations require careful planning and careful screening and selection of specific targets.

DEBRIEFING OPERATIONS AT THE TACTICAL LEVEL

5-35. Debriefing operations at the tactical level include the debriefing of elements of the local and transient civilian population in support of ongoing tactical operations. This is different from but often supportive of tactical SCOs as described in Chapter 1. Although tactical SCOs use specific identified sources to obtain and report information, tactical debriefing operations use one-time and continuous contacts to answer requirements. Tactical debriefing operations are frequently combined with tactical interrogation operations and may identify potential sources for tactical SCOs.

REFUGEE FACILITY AND CHECKPOINT OPERATIONS

5-36. Refugee facility and checkpoint operations involve placing HCTs at points where US forces expect to encounter large numbers of refugees. Deployment of HUMINT collectors at checkpoints is normally preferred due to their ability to collect and report more timely information. As in the questioning of detainees, the debriefing of refugees should not delay their movement out of the danger area.

5-37. Checkpoint debriefing is normally done in coordination with MP or combat forces that are manning the checkpoint. Debriefing at refugee camps is used to obtain longer term and less immediate information. HUMINT collection units established at refugee camps coordinate their activities with the CA, MP, NGO, or other organizations that has responsibility for operating the refugee camp.

5-38. In internment facilities operated by the MPs, HUMINT collectors coordinate with MPs for access to the detainees and for guard support. In facilities operated by NGOs, HUMINT collectors coordinate with NGOs for permission to speak to the refugees. NGOs are civilian agencies and may decide not to permit HUMINT collectors to have access to refugees.

FRIENDLY FORCE DEBRIEFING

5-39. Every member of the friendly force is a potential source for HUMINT collection. Friendly force personnel frequently have contact with the threat, civilian population, or the environment. Although many individuals report their information in the form of combat information, many do not report the information, do not realize its significance, or do not know how to report key information. Frequently a systematic questioning by a trained HUMINT collector will identify key information that can contribute to the intelligence picture and help an individual recall details. It also helps to place his information into a systematic format for the analyst to use.

5-40. HUMINT collectors debrief selected friendly force personnel including combat patrols, aircraft pilots and crew, long-range surveillance teams, deep insert special forces teams, and other high-risk mission personnel. Often the personnel assigned to a sector of responsibility are the first to notice changes in the attitude of the local populace or differences in the mission environment.

5-41. They are also able to provide indicators concerning the mission environment. HUMINT collectors also conduct debriefings of returned prisoners of war (POWs), freed hostages, returned US defectors, and soldiers reported as missing in action. These debriefings help to determine enemy methods of operations, enemy intentions, POW handling and interrogations, enemy weaknesses, information concerning other POWs not returned, and battle damage assessment (BDA).

5-42. HUMINT assets lose access to valuable information if they are not regularly coordinating with the following elements:

- **Cavalry Troops, Unit Patrols, and Scouts.** Unit patrols and scouts have a unique view of the battle area that sensors cannot detect. During operations, units and scouts often patrol villages or populated areas that are contentious and therefore of interest. The unit will gain valuable information on the current status of the AO, potentially answering intelligence requirements, through mission reporting and debriefing by their unit S2 or HUMINT collector.

- **Military Police.** HUMINT collection assets work with the MPs who gain area knowledge through their extensive foot patrols and vehicular convoys. MPs also staff checkpoints and traffic control points (TCPs) where they interact with large numbers of the civilian populace and encounter people and situations that often answer intelligence requirements. MP guards at any internment facility are a valuable source of information on the attitude and behavior of detainees. HUMINT collectors should coordinate with the MP detainee facility commander in order to obtain information on detainees obtained through custodial observation and conversations.

- **Civil Affairs.** CA units have daily interaction with the civilian populace including key members of the civilian community such as politicians, technical personnel, and military leadership.

- **Psychological Operations.** PSYOP teams often interview civilians on the battlefield to determine the effectiveness of friendly and threat PSYOP campaigns. PSYOP elements also gather information on political, social, and other PSYOP requirements. PSYOP elements produce and disseminate intelligence products based partially on their interaction with the civilian populace.

- **Special Operations Forces.** The Special Operations Forces (SOF) team often has greater access to humans and areas on a battlefield than any other collection asset. Their observation of and interaction with the local population provides them access to information that often answers collection requirements. The following are examples of these types of collection missions:
  - Special reconnaissance missions into denied territory to satisfy intelligence gaps or to confirm information from another source.
  - Unconventional warfare (UW) missions normally of a long duration. SOF are inserted into hostile territory to conduct sensitive operations that support US tactical and national objectives. During these missions, SOF units often come in contact with the local population and gather information that meets intelligence requirements.

- **Long-Range Surveillance.** Direct observation and reporting on targets such as activities and facilities may provide timely and accurate intelligence to support a decision or cross-cue other collection capabilities. Long-range surveillance (LRS) is often employed when discreet observation of an activity is necessary over a long period of time or when a collection system that can respond to redirection is necessary.

- **Criminal Intelligence Operations.** CID personnel, in cooperation with MP soldiers, play a key role by linking criminal intelligence to specific groups and events. The criminal intelligence collection effort specifically targets weapons, drugs, organized crime, and identities of smuggling routes. The identification of smuggling routes results in a significant increase in numbers of weapons being confiscated. The timely transfer of criminal intelligence products to tactical units enables a rapid response to serious confrontations, increased confiscation of arms and ammunition, and improved stability in a TF and AO. The Fusion Cell within the ACE develops intelligence products from national, theater, and operational sources. Due to the significant threat that criminal elements pose, CID military agents and CID civilian analysts may be attached to the Fusion Cell to facilitate the police intelligence function.

STRATEGIC DEBRIEFING OPERATIONS

5-43. Strategic debriefing is debriefing activity conducted to collect information or to verify previously collected information in response to national or theater level collection priorities. This avoids surprises of a strategic nature and is used to support long-range strategic planning. Strategic debriefing is conducted in peacetime as well as in wartime. It often fills intelligence gaps on extremely sensitive topics or areas. The sources for strategic debriefing include but are not limited to émigrés, refugees, displaced persons, defectors, and selected US personnel. Strategic debriefing guidance is provided in DIAM 58-11 (S//NF), DIAM 58-12 (S//NF), and DODD 3115.09, "DOD Intelligence, Interrogations, Detainee Debriefings, and Tactical Questioning."

5-44. Strategic debriefing is conducted in a non-hostile, business-like manner. The rapport posture is usually amicable as the source is usually willingly answering national level intelligence needs. Although voluntary sources may not be motivated by a desire for money or other material incentives, it is necessary to ensure that any promised incentives are delivered. The time used in a strategic debriefing can range from days to years. Sources typically have high-level backgrounds in scientific, industrial, political, or military areas.

5-45. Information gathered as strategic intelligence is categorized into eight components. Each of these components can be divided into subcomponents. These components and subcomponents are neither all-encompassing nor mutually exclusive. This approach enhances familiarization with the types of information included in strategic intelligence. An easy way to remember these components is the acronym **"BEST MAPS"**:

Biographic Intelligence
Economic Intelligence
Sociological Intelligence
Transportation and Telecommunications Intelligence

Military Geographic Intelligence
Armed Forces Intelligence
Political Intelligence
Science and Technological Intelligence

- **Biographic intelligence** is the study of individuals of actual or potential importance through knowledge of their personalities and backgrounds. For further guidance on collecting and reporting biographic intelligence, see DIAM 58-12 (S//NF). The subcomponents are—
  - Educational and occupational history—civilian and military backgrounds of individuals.
  - Individual accomplishment—notable accomplishments of an individual's professional or private life.
  - Idiosyncrasies and habits—mannerisms and unusual lifestyles.
  - Position, influence, and potential—present and/or future positions of power or influence.
  - Attitudes and hobbies—significant interests that may affect an individual's accessibility.
- **Economic intelligence** studies economic strengths and weaknesses of a country. The subcomponents are—
  - Economic warfare—information on the diplomatic or financial steps a country may take to induce neutral countries to cease trading with its enemies.
  - Economic vulnerabilities—the degree to which a country's military would be hampered by the loss of materials or facilities.
  - Manufacturing—information on processes, facilities, logistics, and raw materials.
  - Source of economic capability—any means a country has to sustain its economy (for example, black market trade, legitimate business or trades, and imports and exports).
- **Sociological intelligence** deals with people, customs, behaviors, and institutions. The subcomponents are—
  - Population—rates of increase, decrease, or migrations.
  - Social characteristics—customs, morals, and values.
  - Manpower—divisions and distribution within the workforce.
  - Welfare—health and education.
  - Public information—information services within the country.
- **Transportation and telecommunications intelligence** studies systems dedicated to and used during military emergencies and peacetime.

- **Military geographic intelligence** studies all geographic factors (physical and cultural) that may affect military operations. Physical geography is concerned with natural or manmade geophysical features. Cultural geography provides demographics information.
- **Armed forces intelligence** is the integrated study of the ground, sea, and air forces of the country. The subcomponents are—
  - Strategy—military alternatives in terms of position, terrain, economics, and politics.
  - Tactics—military deployments and operations doctrine.
  - OB—location, organization, weapons, strengths.
  - Equipment—analysis of all military materiel.
  - Logistics—procurement, storage, and distribution.
  - Training—as carried out at all echelons to support doctrine.
  - Organization—detailed analysis of command structures.
  - Manpower—available resources and their conditioning.
- **Political intelligence** studies all political aspects which may affect military operations. The subcomponents are—
  - Government structure—organization of departments and ministries.
  - National policies—government actions and decisions.
  - Political dynamics—government views and reactions to events.
  - Propaganda—information and disinformation programs.
  - Policy and intelligence services—organization and functions.
  - Subversion—subversive acts sponsored by the government.
- **Science and technological intelligence** studies the country's potential and capability to support objectives through development of new processes, equipment, and weapons systems. The subcomponents are—
  - Weapons and weapon systems.
  - Missile and space programs.
  - Nuclear energy and weapons technology.
  - NBC developments.
  - Basic applied science.
  - Research and development systems.

LIAISON OPERATIONS

5-46. Liaison is conducted to obtain information and assistance, to coordinate or procure material, and to develop views necessary to understand counterparts. Liaison contacts are normally members of the government, military, law enforcement, or other member of the local or coalition infrastructure. The basic tenet of liaison is *quid pro quo*. An exchange of information, services, material, or other assistance is usually a part of the transaction. The nature of this exchange varies widely depending upon the culture, location, and personalities involved.

5-47. Because the nature of liaison tasks varies widely, the general goals of the liaison operation and the objective of each liaison contact should be clearly defined. The objective should include the type of information to be collected, methods of operations unique to the area, and the command objectives. Additionally, the collector should know limitations on liaison activities. These limitations include—

- Prohibitions against collecting certain types of information or contacting certain types of individuals or organizations.
- Memorandums of understanding with other echelons which delineate each echelon's AOR and AORs for subordinate units.
- Coordination requirements per DCID 5/1 dated 19 December 1984, which are required for selected types of liaison activities.

5-48. Administrative considerations include—

- Type, method, and channels of reporting information obtained from liaison activities.
- Project and contingency fund site numbers to be used.
- Funding and incentive acquisition procedures.
- Limitations on the use of ICFs or incentives.
- Reporting system used.
- Authority under which the specific liaison program is conducted and guidelines for joint and combined operations are set.

5-49. Benefits of liaison include—

- Establishing working relations with various commands, agencies, or governments.
- Arranging for and coordinating joint and combined operations.
- Exchanging operational information and intelligence within legal limits.
- Facilitating access to records and personnel of other agencies not otherwise accessible.
- Acquiring information to satisfy US requirements.
- Accessing a larger pool of information.

INTERROGATION OPERATIONS

5-50. HUMINT interrogation is the systematic process of using approved interrogation approaches to question a captured or detained person to obtain reliable information to satisfy intelligence requirements, consistent with applicable law and policy. Applicable law and policy include US law; the law of war; relevant international law; relevant directives including DOD Directive 3115.09, "DOD Intelligence Interrogations, Detainee Debriefings, and Tactical Questioning"; DOD Directive 2310.1E, "The Department of Defense Detainee Program"; DOD instructions; and military execute orders including FRAGOs. Interrogation is to be conducted by personnel trained and certified to use legal, approved methods of convincing EPWs/detainees to give their cooperation. Interrogation sources are detainees, including EPWs.

5-51. Definitions of EPWs and rules for their treatment are contained in the Geneva Convention Relative to the Treatment of Prisoners of War (GPW). The definition and rules for the treatment of civilians are contained in the

Geneva Conventions Relative to the Protection of Civilian Persons in Time of War (GC). (See Appendix A.) For persons covered by those Conventions, applicable GPW and GC provisions must be adhered to at all times. (Regarding treatment of detained personnel, see also paragraph 5-74.)

5-52. There is an additional protocol to the Geneva Conventions called Protocol I Additional to the Geneva Conventions, 1977, which also contains definitions of who is a civilian and who is an EPW (Articles 50 and 44). The US has not ratified Protocol I nor does it accept the expanded definition of EPWs that it contains. Requirements managers, J/G/S2X personnel, and HUMINT collectors should understand, however, that coalition military personnel with whom they may work may be bound by Protocol I, and those coalition personnel may be required to treat additional personnel as EPWs. Any questions concerning the GPW and Protocol I must be directed to the SJA office for clarification.

5-53. Interrogation operations are specific operations normally conducted at detainee collection facilities directed at the wide-scale collection of information from detainees using interrogation techniques. Although field interrogations are conducted at all echelons and during all operations in which there are detainees, detention facilities where interrogation operations occur are normally located only at theater or JTF level.

5-54. Compliance with laws and regulations, including proper treatment of detainees, is a matter of command responsibility. Commanders have an affirmative duty to ensure their subordinates are not mistreating detainees or their property. HCT leaders must effectively supervise their subordinate collectors during all interrogation operations. Supervisors must ensure that each HUMINT collector has properly completed an interrogation plan and sound collection strategy, and fully understands the intelligence requirements he is seeking to satisfy prior to beginning an interrogation. NCOs and WOs should regularly participate in interrogations with their subordinates to ensure that the highest standards of conduct are maintained. Interrogation supervisors should also monitor interrogations by video, where video monitoring is available. The production, use, and dissemination of interrogation videos must be tightly controlled by HCT leaders. Such videos must not be released for dissemination outside the Intelligence Community without the express permission of the SECDEF or his delegate.

NON-DOD AGENCIES

5-55. Non-DOD agencies may on occasion request permission to conduct interrogations in Army facilities. These requests must be approved by the JTF commander or, if there is no JTF commander, the theater commander or appropriate higher level official. The interrogation activity commander will assign a trained and certified interrogator to escort non-DOD interrogators to observe their interrogation operations. The non-DOD personnel will sign for any detainee they want to question from the MPs, following the same established procedures that DOD personnel must follow. In all instances, interrogations or debriefings conducted by non-DOD agencies will be observed by DOD personnel. In all instances, non-DOD agencies must observe the same standards for the conduct of interrogation operations and treatment of detainees as do Army personnel. All personnel who observe or become aware of violations of Army interrogation operation standards will report the infractions immediately to the commander. The personnel who become aware of mistreatment of detainees will report the infractions immediately and suspend the access of non-DOD personnel to the facility until the matter has been referred to higher headquarters. Non-DOD personnel conducting interrogation operations in an Army facility must sign a statement acknowledging receipt of these rules, and agree to follow them prior to conducting any interrogation operations. Non-DOD personnel working in DOD interrogation facilities have no authority over Army interrogators. Army interrogators (active duty, civilian, or contractor employees) will only use DOD-approved interrogation approaches and techniques.

FOREIGN GOVERNMENT INTERROGATORS

5-56. Foreign governments may request to participate, or may be invited to participate in interrogations in Army facilities. Requests for foreign government access to detainees will be forwarded through the operational chain of command for appropriate action pursuant to DOD policy. Foreign government personnel must comply with US DOD policies and observe the same standards for the conduct of interrogation operations and treatment of detainees as do Army personnel. The interrogation activity commander will assign a trained and certified interrogator to escort foreign government interrogators to observe their interrogation operations. The foreign government personnel will sign for any detainee they want to question from the MPs, following the same established procedures that US DOD personnel must follow. In all instances, interrogations or debriefings conducted by foreign government interrogators will be observed by US DOD personnel. In all instances, foreign government interrogators must observe the same standards for the conduct of interrogation operations and treatment of detainees as do US Army personnel.

MP FUNCTIONS IN ASSOCIATION WITH INTERROGATION OPERATIONS

5-57. MP and MI personnel both have responsibilities with regard to EPW/detainees, but with different goals and responsibilities. (See DOD Directive 3115.09.) Therefore, close coordination must occur between MP and MI personnel in order to facilitate the effective accomplishment of the MP and MI missions. Both MP and MI personnel must ensure that they treat detainees in accordance with the baseline standards of humane treatment.

5-58. MPs are responsible for the humane treatment, evacuation, custody and control (reception, processing, administration, internment, and safety) of detainees; force protection; and the operation of the internment facility, under the supervision of the provost marshal. The MPs do not conduct intelligence interrogations. Intelligence interrogation is strictly a HUMINT function. DOD policy requires that all detainees in its control, whether or not interrogation has commenced, are assigned an internment serial number as soon as possible, normally within 14 days of capture. (See AR 190-8.)

5-59. The standard MP security and internment functions are the only involvement the MPs have in the interrogation process. MPs will not take any actions to set conditions for interrogations (for example, "softening up" a detainee). For purposes of interrogation, military working dogs will not be used.

5-60. MPs may support interrogators as requested for detainee custody, control, escort, and/or additional security (for example, for combative detainees). When interrogators promise an incentive to a detainee, the interrogators must coordinate with the MPs to ensure that the detainee receives the incentive and is allowed to retain it. MPs may provide incentives in support of interrogation operations under the following conditions:

- Using incentives is coordinated with and approved by the MP facility commander.
- Providing and withdrawing incentives does not affect the baseline standards of humane treatment. This means that MPs can provide incentives such as special food items. However, when the incentive is withdrawn, the MPs still must provide the normal rations.
- Using incentives does not violate detainee custody and control or facility security. This means that if a HUMINT collector requests MPs to provide an incentive (for instance, specialty food) but the detainee has been spitting on the guards, then MPs would not provide the incentive because it might reinforce inappropriate behavior.

5-61. MPs exercise the overall responsibility for the safety of detainees, even in those cases in which detainees are in the temporary custody of HUMINT collectors or other agency personnel for the purpose of interrogation. HUMINT collectors should arrange with the MP supervisor to debrief MP guards. Guards who observe and interact with detainees can report the detainees' disposition, activities, mood, and other observable characteristics.

5-62. HUMINT collectors conduct interrogations for intelligence information. They normally work within the confines of the detainee detention facility, but have no involvement in the mission of the security of detainees. MPs follow a strict protocol concerning access to detainees. Accompanied and unaccompanied access to detainees must be coordinated and approved in advance by the MP commander responsible for the detainees or that commander's designated representative.

5-63. When HUMINT collectors coordinate for a detainee interrogation in an internment facility, the MPs escort the detainee to the interrogation site, which is collocated with, or located within the internment facility. MPs verify that the HUMINT collector is authorized access to the detainee. Depending on security concerns, the HUMINT collector may request that the MP remain, or he may request the MP depart until the detainee needs to be returned to the living area. If the MP remains, his functions are to maintain the security, accountability, and safety of the detainee and the safety of the interrogator, interpreter, and others in the interrogation site. The MP will perform no role in the interrogation. When conducting interrogations in a holding area such as a detainee collection point (DCP), MPs may not be available to provide security for interrogation operations. In that case, the

HUMINT collector will need to arrange for security from the unit that has established the holding area.

5-64. If the MP departs the immediate area where the detainee is being questioned (for example, asked to wait outside the interrogation room), the HUMINT collector will assume custody and responsibility for the detainee by signing for the detainee, noting the detainee's physical condition.

5-65. SOPs should be written to comply with a requirement that interrogation operations will always be under observation, whether conducted in fixed sites, holding areas, or in the field. Physical setup and logistical availability will dictate whether observation is conducted directly, from a concealed location, or by video monitoring. HUMINT collectors should never be alone with a detainee without being under observation.

5-66. Once a HUMINT collector has assumed custody of a detainee, he will not turn the detainee over to anyone other than an MP. Specifically, he will not allow another government agency to assume custody from him. The HUMINT collector will instead return the detainee to the custody of the MP, and the agency seeking custody of the detainee will then be required to do so from the MP. Likewise, HUMINT collectors will not assume custody of a detainee directly from another government agency, but will require them to return the detainee directly to the custody of the MP.

LEGAL, REGULATORY, AND POLICY PRINCIPLES AND GUIDELINES

5-67. The GPW (Appendix A, Section I), the GC (Appendix A, Section III), and the UCMJ are relevant documents pertaining to interrogations of detainees.

5-68. The approaches, psychological techniques, and other principles presented in this manual must be conducted in accordance with applicable law and policy. Applicable law and policy include US law; the law of war; relevant international law; relevant directives including DOD Directive 3115.09, "DOD Intelligence Interrogations, Detainee Debriefings, and Tactical Questioning"; DOD Directive 2310.1E, "The Department of Defense Detainee Program"; DOD instructions; and military execute orders including FRAGOs. US policy is to treat all detainees and conduct all interrogations, wherever they may occur, in a manner consistent with this commitment. Authority for conducting interrogations of personnel detained by military forces rests primarily upon the traditional concept that the commander may use all available resources and lawful means to accomplish the mission and to protect and secure the unit.

> *"Prisoners of war do not belong to the power for which they have fought; they are all under the safeguard of honor and generosity of the nation that has disarmed them."*
> —*Napoleon, The Military Maxims of Napoleon*
> *1927, ed. Burnod*

POINT OF CAPTURE THROUGH EVACUATION

MP Functions	HUMINT Functions
• Maneuver and Mobility Support Operations • Area Security • Internment and Resettlement Operations • Law and Order Operations • Police Intelligence Operations • Ensure detainee abuse is avoided and reported	• Screen and question detainees at TCPs and checkpoints • Question contacts, local civilians, refugees, and EPWs • Conduct liaison with military and civilian agencies • Report information obtained • Ensure detainee abuse is avoided and reported • Support DOCEX

DETENTION FACILITY

MP Functions	HUMINT Functions
• Detain and guard EPWs, civilian internees, and other detainees • Conduct reception and processing • Coordinate Classes I, II, and VIII supplies • Coordinate NGOs, PVOs, and interagency visits • Ensure detainee abuse is avoided and reported • Transport detainees within the detention facility to interrogation area • Maintain security during interrogation operations	• Debrief guards • Screen detainees and EPWs for PIR and IR • Provide linguist support when possible • Observe detainees under MP control • Ensure detainee abuse is avoided and reported • Conduct interrogations • Report information obtained • Cross-cue other intelligence disciplines (as needed) • Support DOCEX

Figure 5-1. MP vs HUMINT Responsibilities.

5-69. The Geneva Conventions establish specific standards for humane care and treatment of enemy personnel captured, retained, or detained by US military forces and its allies. All persons who have knowledge of suspected or alleged violations of the Geneva Conventions are obligated by regulation to report such matters through command channels or to designated individuals, such as the SJA or IG. For example, HUMINT collectors who are working with others must ensure that no incidents of detainee abuse occur, whether committed by a fellow HUMINT collector, an interpreter, HN or coalition personnel, MP, representative of another government agency, or anyone else.

5-70. Failure to report a suspected or alleged violation of the law of war may subject the service member to disciplinary actions. Violations of the Geneva Conventions committed by US personnel may constitute violations of the UCMJ. The commander is responsible for ensuring that the forces under his command comply with the Geneva Conventions. If violations occur in the conduct of warfare, the commander bears primary responsibility for investigating and taking appropriate action with respect to the violators.

5-71. Every soldier has the duty to report serious incidents, whether observed or suspected, in accordance with AR 190-40. Such incidents are reported to the chain of command. If the chain of command itself is implicated, the soldier can report the incident to the SJA, IG, chaplain, or provost marshal.

5-72. There are reasons for reporting serious incidents beyond those related to legal requirements. For instance, the publishing of enemy war crimes can be used to influence public opinion against the enemy. Also, reporting war crimes of other countries provides important information that may become relevant, since we would not be able to transfer detainees to any power that we could not rely on to treat them appropriately under the law of war, including the Geneva Conventions.

5-73. Several articles of the GPW apply to HUMINT collectors and interrogation operations. Excerpts from some of the most relevant articles of the Geneva Conventions are listed below. Although the following excerpts are specific to EPWs, service members must treat all detainees captured during armed conflict consistent with the provisions of the GPW unless a determination to the contrary is made. Moreover, US policy requires that US forces apply the principles of the Geneva Conventions, during military operations. (See Appendix A.)

- Article 5 - Should any doubt arise as to whether persons having committed a belligerent act and having fallen into the hands of the enemy, belong to any of the categories enumerated in Article 4, such persons shall enjoy the protection of the present Convention until such time as their status has been determined by a competent tribunal.
- Article 13 - PWs must at all times be treated humanely. Any unlawful act or omission by the Detaining Power causing death or seriously endangering the health of a PW in its custody is prohibited. Likewise, PWs must at all times be protected, particularly against acts of violence or intimidation and against insults and public curiosity.
- Article 14 - PWs are entitled, in all circumstances, to respect for their persons and honor. Women shall be treated with all regard due to their sex, and shall in all cases benefit by treatment as favorable as that granted to men.
- Article 15 - The Power detaining PWs shall be bound to provide, free of charge, for their maintenance and medical attention required by their state of health.
- Article 17 - This article covers several requirements with direct impact on interrogation.
  - Every PW, when questioned on the subject, is bound to give only his surname, first names and rank, date of birth, and army, regimental, personal or serial number, or failing this, equivalent information. If he willfully infringes this rule, he may render himself liable to a restriction of the privileges (emphasis added) accorded to his rank or status.
  - For example, this does not mean if a prisoner fails to give this information he loses status as a prisoner, only special privileges. An example might be an officer who fails to identify himself as such. An officer cannot be compelled to work (Article 49). An officer who fails to identify himself as such could lose this privilege.

- The questioning of PWs shall be carried out in a language they understand.
- No physical or mental torture or any other form of coercion may be inflicted on EPWs to secure from them information of any kind whatever. PWs who refuse to answer may not be threatened, insulted, or exposed to unpleasant or disadvantageous treatment of any kind.

• Article 18 - All effects and articles of personal use, except arms, horses, military equipment and documents, shall remain in the possession of PWs, likewise their metal helmets and protective masks and like articles issued for personal protection. Effects and articles used for their clothing or feeding shall also remain in their possession, even if such effects and articles belong to their regulation military equipment.

- Badges of rank and nationality, decorations and articles having above all a personal or sentimental value may not be taken from PWs.
- Sums of money carried by PWs may not be taken away from them except by order of an officer, and after the amount and particulars of the owner have been recorded in a special register and an itemized receipt has been given, legibly inscribed with the name, rank, and unit of the person issuing said receipt. (Note: Unit SOP should require initial impounding of all sums of money from detainees, properly documented and accounted for, in order to prevent detainees from using money to buy influence of any kind, or participate in black market or other improper activity.)

• Article 19 - PWs shall be evacuated, as soon as possible after their capture, to camps situated in an area far enough from the combat zone for them to be out of danger. Only those PWs, who, owing to wounds and sickness, would run greater risks by being evacuated than by remaining where they are, may be temporarily kept back in a danger zone.

• Article 33 - Medical personnel and chaplains, while retained by the Detaining Power with a view to assisting PWs, shall not be considered as PWs. They shall, however, receive as a minimum, the benefits and protection of the Geneva Convention. They shall continue to exercise their medical and spiritual functions for the benefits of PWs.

5-74. All captured or detained personnel, regardless of status, shall be treated humanely, and in accordance with the Detainee Treatment Act of 2005 and DOD Directive 2310.1E, "Department of Defense Detainee Program," and no person in the custody or under the control of DOD, regardless of nationality or physical location, shall be subject to torture or cruel, inhuman, or degrading treatment or punishment, in accordance with and as defined in US law. All intelligence interrogations, debriefings, or tactical questioning to gain intelligence from captured or detained personnel shall be conducted in accordance with applicable law and policy. Applicable law and policy include US law; the law of war; relevant international law; relevant directives including DOD Directive 3115.09, "DOD Intelligence Interrogations, Detainee Debriefings, and Tactical Questioning"; DOD Directive 2310.1E,

"The Department of Defense Detainee Program"; DOD instructions; and military execute orders including FRAGOs. Use of torture is not only illegal but also it is a poor technique that yields unreliable results, may damage subsequent collection efforts, and can induce the source to say what he thinks the HUMINT collector wants to hear. Use of torture can also have many possible negative consequences at national and international levels.

Cruel, Inhuman or Degrading Treatment Prohibited

All prisoners and detainees, regardless of status, will be treated humanely. Cruel, inhuman and degrading treatment is prohibited. The Detainee Treatment Act of 2005 defines "cruel, inhuman or degrading treatment" as the cruel unusual, and inhumane treatment or punishment prohibited by the Fifth, Eighth, and Fourteenth Amendments to the U.S. Constitution. This definition refers to an extensive body of law developed by the courts of the United States to determine when, under various circumstances, treatment of individuals would be inconsistent with American constitutional standards related to concepts of dignity, civilization, humanity, decency and fundamental fairness. All DOD procedures for treatment of prisoners and detainees have been reviewed and are consistent with these standards, as well as our obligations under international law as interpreted by the United States.[1]

Questions about applications not resolved in the field by reference to DOD publications, must be forwarded to higher headquarters for legal review and specific approval by the appropriate authority before application.

The following actions will not be approved and cannot be condoned in any circumstances: forcing an individual to perform or simulate sexual acts or to pose in a sexual manner; exposing an individual to outrageously lewd and sexually provocative behavior; intentionally damaging or destroying an individual's religious articles.

---

[1] Nothing in this enclosure should be understood to affect the U.S. obligations under the law of war.

5-75. **If used in conjunction with intelligence interrogations, prohibited actions include, but are not limited to—**

- Forcing the detainee to be naked, perform sexual acts, or pose in a sexual manner.
- Placing hoods or sacks over the head of a detainee; using duct tape over the eyes.
- Applying beatings, electric shock, burns, or other forms of physical pain.
- "Waterboarding."
- Using military working dogs.
- Inducing hypothermia or heat injury.
- Conducting mock executions.
- Depriving the detainee of necessary food, water, or medical care.

5-76. While using legitimate interrogation techniques, certain applications of approaches and techniques may approach the line between permissible actions and prohibited actions. It may often be difficult to determine where permissible actions end and prohibited actions begin. In attempting to determine if a contemplated approach or technique should be considered prohibited, and therefore should not be included in an interrogation plan, consider these two tests before submitting the plan for approval:

- If the proposed approach technique were used by the enemy against one of your fellow soldiers, would you believe the soldier had been abused?
- Could your conduct in carrying out the proposed technique violate a law or regulation? Keep in mind that even if you personally would not consider your actions to constitute abuse, the law may be more restrictive.

5-77. If you answer yes to either of these tests, the contemplated action should not be conducted. If the HUMINT collector has any doubt that an interrogation approach contained in an approved interrogation plan is consistent with applicable law, or if he believes that he is being told to use an illegal technique, the HUMINT collector should seek immediate guidance from the chain of command and consult with the SJA to obtain a legal review of the proposed approach or technique. (See paras 5-80 and 5-81 for information on responding to illegal orders.) If the HUMINT collector believes that an interrogation approach or technique is unlawful during the interrogation of a detainee, the HUMINT collector must stop the interrogation immediately and contact the chain of command for additional guidance.

CAUTION: Although no single comprehensive source defines impermissible coercion, certain acts are clearly prohibited. Certain prohibited physical coercion may be obvious, such as physically abusing the subject of the screening or interrogation. Other forms of impermissible coercion may be more subtle, and may include threats to turn the individual over to others to be abused; subjecting the individual to impermissible humiliating or degrading treatment; implying harm to the individual or his property. Other prohibited actions include implying a deprivation of applicable protections guaranteed by law because of a failure to cooperate; threatening to separate parents from their children; or forcing a protected person to guide US forces in a dangerous area. Where there is doubt, you should consult your supervisor or servicing judge advocate.

5-78. Security internees are detainees who are not combatants but who pose a security threat, may be under investigation, or who pose a threat to US forces if released. HUMINT collectors are required to treat all detainees humanely. EPWs are entitled to additional protections guaranteed by the GPW that security internees may not be eligible for. For example, allowing a security internee to communicate with a family member (a right that an EPW has under the Geneva Conventions) could allow him to pass information that would compromise a sensitive investigation and endanger the lives of soldiers and civilians. HUMINT collectors should consult with their SJA for clarification of detainees' status and rights.

5-79. HUMINT collectors are employed below brigade level when the combat situation requires limited tactical interrogation at battalion or lower.

HUMINT collectors should also provide training in the area of tactical questioning to designated S2 personnel. The potential for abuse of the detainee is greatest at initial capture and tactical questioning phase. With the excitement and stress of the battlefield, unskilled personnel may exercise poor judgment or be careless and thus resort to illegal techniques to elicit critical information. Personnel who are not trained HUMINT collectors will not attempt to use approach techniques. Instructions must stress the importance of the proper treatment of detainees. Emphasize that in addition to legal requirements, the abuse of a detainee at the initial stage of contact often renders future interrogation futile. All treatment of detainees must be consistent with the Geneva Conventions. (See ST 2-91.6 for further information on tactical questioning.)

5-80. Orders given to treat detainees in any way that violate the Law of War, including the Geneva Conventions, or that result in detainees being treated in any prohibited manner are unlawful. Every soldier must know how to respond to orders that he perceives to be unlawful. If a soldier receives an order that he knows to be unlawful, or that a person of ordinary sense and understanding would know to be unlawful, or if the order is not clear enough to determine if it is legal or not, he should follow the steps set out below (preferably in the order listed):

- Ask for clarification.
- State that the order is illegal if he knows that it is.
- Use moral arguments against the order.
- State the intent to report the act.
- Ask the senior interrogator to stop the act.
- Report the incident or order if the order is not withdrawn or the act in question is committed.
- If there appears to be no other recourse, refuse to obey the unlawful order.

NOTE: If the order is a lawful order, it should be obeyed. Failure to obey a lawful order is an offense under the UCMJ.

5-81. None of the above actions should be taken in the presence of any detainee. Witnessing actions taken to determine the legality of an order may lead to increased resistance of the detainee and could lead to increased resistance throughout the detainee population if they believe they are being treated unlawfully.

5-82. Illegal orders or incidents must be reported to the chain of command. However, if the chain of command itself is implicated, report the incident or order to the SJA, IG, chaplain, or provost marshal.

OPERATIONAL CONSIDERATIONS AND REQUIREMENTS

EPW Evacuation System

5-83. The MPs are responsible for evacuating detainees, civilian internees, and other detainees, as stipulated in AR 190-8. HUMINT collection assets must be placed to take advantage of the evacuation system the MPs will put into place. The evacuation of detainees and civilian internees normally is a slow and cumbersome process that can severely tax a maneuver unit's resources. Appendix D explains the handling of detainees in detail, including the 5Ss—Search, Silence, Safeguard, Segregate, and Speed to the Rear. The 5Ss are authorized with respect to handling detainees for the purposes of movement of detainees and security. The 5Ss are not authorized for use as interrogation approach techniques.

5-84. The initial evacuation of detainees and civilian internees is the responsibility of the capturing unit. That unit is normally responsible for moving the detainees and civilian internees from the point of capture to the nearest DCP. Under MP doctrine, the MPs are responsible for the detention, security, processing, safety, well-being, accountability, and humane treatment of detainees and civilian internees.

5-85. Normally the MPs assume responsibility for the further evacuation of the detainees and civilian internees; however, under certain circumstances, other units could be charged with this task. The detainees are normally evacuated from a DCP to a short-term collection facility and then finally to a theater internment facility. Once the theater internment facility (joint) is established, dependent on METT-TC factors, the internment facility escort guard units may go forward as far as the initial collection points and escort detainees and civilian internees to a short-term collection facility or straight to a theater internment facility.

5-86. Senior MP commanders coordinate and synchronize transportation and security requirements with MP divisional and BCT leaders. It may take 8 hours for a detainee to reach the DCP; 8 to 16 hours more to reach a short-term collection facility; and 24 additional hours to reach the theater internment facility. Mandatory timelines will be determined in command policy guidance. Critical during this process is that MPs work closely with MI, SJA, and interagency personnel to determine the proper status of individuals detained. Determining whether an individual is an EPW, a criminal insurgent, or in another status is crucial to facilitate the release or transportation, holding, and security requirements. This determination will be used when the individual's biometric data is taken and entered into the Biometric Automated Toolset (BAT).

5-87. The HUMINT collection assets need to be positioned to maximize their collection potential and take advantage of the time available during evacuation. The rapidity of operations and the need to facilitate the commander's situational understanding—coupled with the technological innovations that link the HUMINT collector to databases, analysts, and technical support from anywhere on the battlefield—require placing the HCTs forward into brigade and even maneuver battalion areas to provide immediate access to EPWs/detainees. EPWs/detainees are normally interrogated for tactical information in the maneuver battalion trains areas and then questioned in detail at the theater JIDC.

Security

5-88. When dealing with detainees, the HUMINT collector faces two security considerations: his own physical security and information security. Particularly when operating in support of tactical operations, the HUMINT collector is in close contact with enemy soldiers who could attempt to escape and may attack the HUMINT collector in doing so. Detainees during a stability and reconstruction operation are often people committed to a cause who find themselves in desperate circumstances. Although the detainees are normally under guard, the HUMINT collector must always be alert to any physical threat posed by these individuals. He must also ensure that his own actions do not provide the detainee with the means with which to harm the collector or anyone else.

5-89. The HUMINT collector should also be aware that EPWs and other detainees may attempt to elicit information. Since HUMINT collectors, by virtue of their position, may possess a great deal of classified information, they must be careful not to reveal it unwittingly in the process of questioning a detainee.

PROHIBITION AGAINST USE OF FORCE

Acts of violence or intimidation, including physical or mental torture, or exposure to inhumane treatment as a means of or aid to interrogation are expressly prohibited. Acts in violation of these prohibitions may be a violation of US law and regulation and the law of war, including the Geneva Conventions of 1949, and may be criminal acts punishable under the UCMJ and other US law. Moreover, information obtained by the use of these prohibited means is of questionable value. If there is doubt as to the legality of a proposed form of interrogation, the advice of the SJA must be sought before using the method in question.

Limitations on the use of methods identified herein as expressly prohibited should not be confused with psychological ploys, verbal trickery, or other nonviolent or non-coercive subterfuge used by the trained HUMINT collector in the successful interrogation of hesitant or uncooperative sources. Use of torture by US personnel would bring discredit upon the US and its armed forces while undermining domestic and international support for the war effort. It also could place US and allied personnel in enemy hands at a greater risk of abuse by their captors. Conversely, knowing the enemy has abused US and allied POWs does not justify using methods of interrogation specifically prohibited by law, treaty, agreement, and policy. In conducting intelligence interrogations, the J2/G2/S2 has primary staff responsibility to ensure that these activities are performed in accordance with these laws and regulations. [*The commander bears the responsibility to ensure that these activities are performed in accordance with applicable law, regulations, and policy. The unit must have an internal SOP for execution of the interrogation mission.]

The psychological techniques and principles in this manual should neither be confused with, nor construed to be synonymous with, unauthorized techniques such as brainwashing, physical or mental torture, including drugs that may induce lasting or permanent mental alteration or damage. Physical or mental torture and coercion revolve around eliminating the source's free will, and are expressly prohibited by GWS, Article 13; GPW, Articles 13 and 17; and GC, Articles 31 and 32.

> *Torture is an act committed by a person under the color of law specifically intended to inflict severe physical or mental pain and suffering (other than pain or suffering incidental to lawful sanctions) upon another person within his custody or physical control. (Extracted from Title 18 of the United States Code, Section 2340A).*

*Emphasis added for use in this manual.

Capture Rates

5-90. Anticipating not only overall capture rates but also capture rates linked to specific operations is vital to the correct placement of HUMINT collectors supporting interrogation operations. Defensive and stability and reconstructions operations normally provide a small but steady flow of detainees while successful offensive operations can overwhelm HCTs. To be successful, HUMINT collection support to tactical operations must be carefully planned and prioritized. Available HUMINT collection assets must be balanced against the operations objective, enemy situation estimate, and projected EPW capture rates. The unit S2 is responsible for projecting capture rates.

Interrogating Wounded and Injured Detainees

5-91. Commanders are responsible to ensure that detainees receive adequate health care. Decisions regarding appropriate medical treatment of detainees and the sequence and timing of that treatment are the province of medical personnel. Detainees will be checked periodically in accordance with command health care directives, guidance, and SOPs, applicable to all detainees to ensure they are fit for interrogations. Detainees determined by medical personnel to be medically unfit to undergo interrogation will not be interrogated. Health care personnel will be on call should a medical emergency arise during interrogation. Health care personnel will report detainees' conditions, as appropriate, to the commander. Health care providers shall not be placed in a position to advise on the application or duration of interrogation approach techniques.

5-92. Wounded and otherwise injured detainees can be a valuable source of information. For evacuation purposes, medical personnel may classify detainees as walking wounded or sick or as non-walking wounded or sick. Walking wounded detainees are evacuated through normal evacuation channels. Non-walking wounded are delivered to the nearest medical aid station and evacuated through medical channels.

5-93. HUMINT collectors may interrogate a wounded or injured detainee provided that they obtain permission from a competent medical authority and that the questioning will not delay or hinder medical treatment. Questioning will not delay the administration of medication to reduce pain or the evacuation of the detainee to where they may receive medical treatment, nor will interrogation be allowed if it would cause a worsening of the condition of the detainee. In most cases, this simply requires the HUMINT collector to ask the doctor, medic, or other medical personnel if it is all right to talk to the detainee.

5-94. With the doctor's permission, the HUMINT collector may talk to the detainee before, after, or during medical treatment. The HUMINT collector cannot at any time represent himself as being a doctor or any other type of medical personnel. Nor can he state, imply, or otherwise give the impression that any type of medical treatment is conditional on the detainee's cooperation in answering questions.

TYPES OF INTERROGATION OPERATIONS

5-95. There are two general categories of interrogation operations: **field interrogation operations** and **interrogation facility operations**.

FIELD INTERROGATION OPERATIONS

5-96. Field interrogation operations constitute the vast majority of interrogation operations at echelons corps and below. Field interrogations include all interrogation operations not conducted at a fixed facility. Current doctrine emphasizes the placement of HCTs forward with maneuver units to provide immediate interrogation support while the information is fresh and the detainee may still be susceptible to approaches, due to the shock of capture. The rationale for this method of employment is twofold:

- First, the pace of the modern battlefield no longer allows the luxury of waiting for a detainee to reach a collection point prior to interrogation. Commanders need more timely information, including HUMINT. Also, automated tools and improved communications now permit rapid transmittal of information from forward-deployed HCTs.

- Second, current MP doctrine has the theater level EPW escort companies picking up detainees as far forward as the division forward collection points and bypassing the intervening collection points.

5-97. An added benefit of placing the HCTs with maneuver units is that it allows them to conduct other HUMINT collection activities, such as the debriefing of local civilians and refugees concurrently with interrogation operations. HCTs are allocated to maneuver units based on—

- The relative importance of that subordinate element's operations to the unit's overall scheme of maneuver.
- The potential for that subordinate element to capture detainees, documents, and materiel or encounter civilians on the battlefield.
- The criticality of information obtained from those sources to the success of the parent unit's overall OPLANs.

5-98. As the mission and situation change, the HCTs are redistributed. As MI assets, they should never be kept in reserve.

5-99. During offensive and defensive operations, HCTs normally operate with maneuver brigades and battalions. HUMINT collectors with battalions or brigades should be equipped with vehicles and communications systems that are compatible with the systems organic to the supported unit. HUMINT collectors with brigades and battalions receive their collection priorities from the S2 of the supported unit. In stability and reconstruction operations, the HCTs normally operate in the AOs of battalion and brigade TFs.

INTERROGATION FACILITY OPERATIONS

5-100. Joint interrogation operations are operations conducted at higher echelons, usually at, and in coordination with, EPW and detainee internment facilities. The Joint Forces Commander (JFC) normally tasks the Army component commander to establish, secure, and maintain the EPW internment facility system. The corps may have the mission of establishing an interrogation facility when it is acting as the Army Forces (ARFOR) or Land Component Command (LCC) element.

5-101. An echelon above corps (EAC) MP brigade normally operates the theater internment facility. The subordinate JFC with a J2 staff lead establishes a Joint Interrogation and Debriefing Center as an activity within the theater internment facility. The MI Brigade Commander or other named SIO is normally designated as the JIDC commander. Army interrogation operations are normally carried out in an area of the MP-operated internment facility set aside for that use.

5-102. The JIDC is normally administratively and operationally self-sufficient. A JIDC will function as part of an overall detainee command and control structure as outlined in FM 3-19.40 and/or by policy. Continuous coordination between the JIDC commander and internment facility commander is essential. The JIDC will—

- Normally consist of facility headquarters, operations, analysis, editorial, interrogation, screening, and DOCEX elements.

- Collocate with the theater detainee internment facility.
- Organizationally structure itself to meet METT-TC requirements within the theater.
- Include HUMINT collectors, CI personnel, technical experts, personnel for CEDs and DOCEX, and intelligence analysts, as applicable, from the Army, Air Force, Marine Corps, Navy, and other government agencies.
- Maintain the capability to deploy HCTs forward, as needed, to conduct interrogations or debriefings of sources of interest who cannot be readily evacuated to the JIDC.
- Often establish a combined interrogation facility with allied HUMINT collector or interrogator augmentation if operating as part of a multinational operation.
- Receive collection guidance from the C/J/G2X and send its intelligence reports to the C/J/G2X and to the supported C/J/G/S2.

5-103. The exact size and organizational structure of these elements will vary dependent on METT-TC.

Headquarters Element

5-104. The activity headquarters provides all command, administrative, logistic, and maintenance support to the JIDC. It coordinates with—

- Higher headquarters for personnel, intelligence, and operational and logistical support prior to and after deployment.
- Theater J2 for reporting procedures, operational situation updates, theater and national level intelligence requirements, and collection priorities.
- Provost marshal for location of theater detainee internment facilities and for procedures to be followed by HUMINT collectors and MPs for the processing, interrogating, and internment of EPWs.
- Commanders of theater medical support units and internment facility for procedures to treat, and clear for questioning, wounded EPWs.
- Commanders of supporting CI and TECHINT assets to establish support requirements and procedures.
- The servicing SJA.
- Magistrate for Article 78 issues.
- Commanders of Air Force, Marine, Navy, and national level organizations to arrange administrative and logistic interoperability.

Operations Element

5-105. The operations element controls the daily activities within the JIDC. The JIDC operations element—

- Ensures that work areas are available for all JIDC elements.
- Establishes and maintains JIDC functional files, logs, and journals.
- Makes detainee files available to detainee release boards to assist the board members in their determinations.
- Establishes interrogation priorities.

- Disseminates incoming and outgoing distribution.
- Conducts coordination with local officials, adjacent and subordinate intelligence activities, CI, MP, PSYOP, the Joint Captured Materiel Exploitation Center (JCMEC), Plans and Policy Directorate (J5), and provost marshal.
- Conducts coordination with holding area officer in charge (OIC) for screening site, medical support, access, movement, and evacuation procedures for detainees.
- Conducts operations briefings when required.
- Supervises all JIDC operations and establishes SOPs.
- Supervises all intelligence collection activities within the JIDC.
- Ensures observers are present when OGAs use the JIDC's interrogation rooms.

Analytical Element

5-106. The analytical element normally is directly subordinate to the operations element. The JIDC analytical element ensures that collection requirements are current and validated. It reviews reports to ensure that the information reported is in response to validated collection requirements. In addition, they ensure an up-to-date common operational picture (COP) by maintaining digital mapping of the current tactical situation and with OB updates to help HUMINT collectors maintain their situational awareness. At locations where digital mapping is not possible, paper situation maps (SITMAPs) are maintained. This element also—

- Obtains, updates, and maintains the database.
- Works with interrogators to provide collection focus for interrogations.
- Establishes and maintains OB workbooks and files including data generated by intelligence information which has not been verified.
- Maintains digital or paper SITMAPs, as available, displaying enemy and friendly situations.
- Catalogs, cross-references, and disseminates collection requirements to JIDC collection elements.
- Reviews interrogation reports for inclusion into the database.
- Conducts situation briefings when required.
- Conducts intelligence reach with the J2 analytical cell and other analytical elements, such as INSCOM Information Dominance Center, for relevant information and analysis.

Editorial Element

5-107. The editorial element is normally directly subordinate to the operations element. It reviews all outgoing reports for format, content, and completeness.

DOCEX Element

5-108. At a minimum, the JIDC will contain a small DOCEX element to translate, screen, and extract information from and report on information of intelligence interest from source-associated documents. The theater joint document exploitation facility (JDEF) may be collocated with the JIDC. In this instance, the JDEF will translate, screen, categorize, and exploit all types of CEDs.

Screening Element

5-109. The JIDC normally has a separate screening element to receive and screen all incoming detainees and their personal effects. The screening element will review previous screening reports, which should have been sent along with the detainees; recommend priorities for interrogation; identify individuals of interest to other agencies; and may conduct limited interrogations for PIR information. The exact size of the element will vary based on detainee capture rates and detainee flow. Interrogation elements should use their most experienced interrogators as screeners in order to quickly and effectively select the detainees for interrogation who are most likely to possess useful information.

Interrogation Element

5-110. The interrogation element assigns HUMINT collectors to specific detainees, uses interrogation and other HUMINT collection methods to obtain information in response to intelligence requirements, and produces intelligence reports (IIRs and SALUTE reports) as well as source-related operational reports. The interrogation element may also debrief returning US POWs and other personnel as deemed relevant.

PART THREE

The HUMINT Collection Process

Part Three discusses the logical progression of phases involved in all HUMINT collection. There are five phases and the related task of screening that are critical to HUMINT collection. This remains consistent with previous doctrine as captured in the interrogation process but adds screening as a phase and combines approach and termination. The five phases are screening, planning and preparation, approach and termination strategies, questioning, and reporting.

Chapter 6

Screening

6-1. Available human sources and documents almost always exceed the qualified HUMINT collection assets and resources that can be applied against them. Screening facilitates the efficient application of these limited assets and resources to maximize the collection of relevant information.

HUMAN SOURCE SCREENING

6-2. As it applies to HUMINT operations, screening is the process of evaluating and selecting human sources and documents for the prioritized collection of information based on the collection requirements and mission of the unit conducting the screening or its higher headquarters. Screening categorizes and prioritizes sources based on the probability of a particular source having priority information and the level of cooperation of the source. Screening is also used to determine if a source matches certain criteria that indicate that the source should be referred to another agency. Screening is conducted at all echelons of command and in all operational environments. There are two general categories of screening: human source screening and document screening. Human source screening will be explained in depth in this chapter. Document screening is explained in Appendix I.

6-3. The resources (time and personnel) allocated to screening must be balanced against those required for interrogations, debriefings, and other collection methodologies. Although screening is not in itself an information collection technique, it is vital to the rapid collection of information. Through screening, the effectiveness of limited collection assets can be maximized by targeting those assets against the sources with the highest potential of providing key information. Screening requires experienced individuals with maturity and judgment who are totally knowledgeable of the collection requirements and able to make well-reasoned decisions based on limited information. Collection (interrogation, debriefing, and elicitation) can be integrated into screening activities; however, it slows the screening process and decreases the number of potential sources that can be screened.

6-4. Human source screening is the evaluation of an individual or a group of individuals to determine their potential to answer collection requirements or to identify individuals who match a predetermined source profile. The purpose of screening is to—

- Identify those select individuals among the target audience who have information of potential value and who are willing or can be persuaded to cooperate.
- Identify individuals who match certain criteria that indicate them as being potential subjects for source operations or matching the profile for collection by special interest groups such as TECHINT or CI.

6-5. Screening requires the development of criteria that are indicators of potential information. These might include rank, position, gender, ethnic group, appearance, and location.

6-6. Screening is an integral part to all HUMINT collection operations. While questioning an individual source, a HUMINT collector may switch between screening (finding out general source areas of knowledge) to interrogation, debriefing, or elicitation (finding out detailed information about a specific topic). In operations, such as EPW or refugee operations that involve large numbers of potential sources, screening will normally be conducted as a separate but collocated operation as part of the overall interrogation or debriefing effort. The high number of potential sources being dealt with in most human source screening operations requires a systematic approach be developed and utilized to make the most effective use of the personnel and resources being allocated to the source screening operation.

SCREENING OPERATIONS

6-7. Like all intelligence operations, human source screening operations are focused on certain targets. Although the exact target population group will depend on the requirements of the theater of operations, the target focus of source screening operations is best described as the permanent and transitory population in the AO. This definition includes local indigenous populations, refugees, and travelers in the area, and detainees (including EPWs). Specifically excluded from this definition are members of the HN forces (military and paramilitary), members of allied forces, and members of HN government agencies who are available to US forces through liaison operations. Other personnel not indigenous to the AO (such as legitimate NGOs, humanitarian organizations, UN personnel) are available to US forces for voluntary debriefing and should be excluded from screening operations.

6-8. Screening operations may be conducted in a variety of situations and are dependent on the operational situation and the population. Although every source screening operation has the same basic purpose, each can be directed against different segments of the population in different locations throughout the AO. In order to accommodate the differences in the screening audience and location, different types of source screening operations are employed.

- Tactical Screening. Tactical screening is conducted in support of combat or contingency operations. It can include the screening of EPWs or detainees at the point of capture, the screening of refugees, or the screening of local civilians in cordon and search. At the tactical level, there is no time for elaborate approach techniques so the degree of cooperation becomes a prime concern. Tactical area screening is characterized by rapidly changing requirements, the need to evacuate noncombatants and detainees to a secure area, and the need to collect priority tactical information while operations are in progress. Although the most lucrative type of source is often the detainee, all available sources should be screened for priority tactical information. In tactical screening, the HUMINT collector normally accompanies the maneuver force (OPCON or DS). If the HUMINT collector establishes that the source has information of value during screening, he immediately questions the source. Information collected is passed to the maneuver commander, normally via SALUTE reports. The HUMINT collector may recommend to the commander that individual sources be further detained for additional questioning. Screening must be done accurately in order that a commander can make a decision to detain or release possibly hostile personnel, based on the recommendation of a HUMINT collector.
- Checkpoint Screening. Checkpoints are often established to screen the local populations as they transit through and within the AO or to screen large numbers of individuals such as refugees or DPs as they enter the AO. Screening checkpoints can be static or mobile. HUMINT collectors must pay particular attention to refugees leaving the area ahead of friendly forces (AO or AOI). It is likely that refugees can provide information of tactical value more quickly and easily than detainees. Refugees know the area and may be able to identify for the collector anything that is out of the ordinary, such as insurgent or terrorist activities.
- Local Population Screening. This refers to the screening of the local population within their own neighborhoods. When HUMINT collectors move into a new area, they must observe the local population and determine who may be able and willing to provide the information they have been tasked to collect. Once this determination is made, the collectors must engage those individuals in conversation to assess their level of knowledge.
- Collection Facility Screening. Screening is conducted as a normal part of HUMINT collection operations at collection facilities such as theater interrogation and debriefing facilities and refugee camps. Screening is coordinated with the unit, normally an MP unit that is responsible for the operation of the facility.
- Local Employee Screening. CI personnel periodically screen local employees to determine possible security risks. Concurrently, local employee screening may identify sources who can provide information to answer the CCIRs. Close coordination between HUMINT and CI collection assets is a must in local employee screening.

- Variations and Combinations. All types of screening can be adapted to meet specific circumstances slightly different from those for which they were designed. Additionally, it is possible to use more than one type of screening in an operation if the specific circumstances require it.

6-9. Screening of refugees, EPWs, and other detainees normally occurs at two locations: initially at the point where friendly forces first encounter them and again when they arrive at the theater and other holding areas or refugee camps. The capturing or detaining forces should enforce segregation of EPWs from refugees and other detained civilians; they should be screened in separate operations, one screening for EPWs and one for refugees and other detained civilians. Depending on METT-TC factors, segregation should be conducted as follows:

- Refugees: Refugees, even if of the same nationality as the enemy, are not treated as enemies exclusively based on their nationality and are not automatically subject to control measures. If refugees are encountered on the battlefield, they are segregated from EPWs and screened separately. They are generally not detained further unless some additional reason requires their detention. At a refugee camp, screening will be done in coordination with the NGO operating the refugee camp. If there is a reason to detain refugees for further questioning for intelligence purposes, or because they pose a security threat, they will then be treated as a detainee. Under all circumstances, refugees will be treated humanely. If they are transported to an internment facility, they will be in-processed by MPs and their Geneva Conventions status will be determined. Their status under the Geneva Conventions will afford them certain privileges.

- EPWs: Officers are segregated from enlisted. The enlisted are divided into NCOs and lower enlisted. Males are segregated from females. This segregation facilitates rapid screening for EPWs who may have information to answer PIRs and IRs as well as prohibits officers from influencing enlisted personnel to resist questioning.

- Other Detainees: Civilians should be screened separately from EPWs. As with refugees, if there is a reason to detain civilians for further questioning for intelligence purposes, or because they pose a security threat, they will then be treated as a detainee. Whether or not civilian detainees are released or detained further, screeners should ensure that the civilian detainees are treated humanely. If the civilian detainees are transported to an internment facility, they will be in-processed by MPs and their Geneva Conventions status will be determined. Once detainees are in-processed into an internment facility, they are then considered to be civilian internees and their status as such will afford them certain privileges under the Geneva Conventions.

SCREENING AT FORWARD LOCATIONS

6-10. The initial screening and subsequent questioning should be accomplished as far forward as is operationally expedient. If a HUMINT collector is not available, the unit S2 must ensure initial screening and questioning of sources are completed by qualified personnel. At this level, the individual (military or civilian) is questioned for job, unit (if applicable), mission, PIR and IR, and supporting information (JUMPS). If time allows, the HUMINT collector may collect additional information, such as the source's name, to start a formal source file to preclude duplication at higher echelons. S2s and personnel other than HUMINT collectors should not attempt an approach at this stage.

6-11. HUMINT collectors will only use approach techniques as time and circumstance allow. The prime requirement is to identify the individuals with information of immediate tactical value, to collect that information expediently, and to evacuate the source. In this case, tactical questioning is normally integrated seamlessly into the screening process. This initial screening can also be used to identify individuals for immediate evacuation to a higher echelon facility for detailed questioning. Any screening reports or information reports generated at this level must accompany the EPWs or detainees as they are evacuated. Typically, battlefield screening reports, such as the screening sheet shown in Figure 6-1, will be done on paper in order to allow multiple screeners to work simultaneously. If automation support is available for each screener, an electronic version of the screening report is used, or the "KB Easy" (Figure 10-2), which allows the screener to easily put screening information into a DIA report format and transmit it electronically. (See Chapter 10 for a KB-EZ worksheet.)

6-12. US forces capturing enemy forces or detaining civilians on the battlefield search each individual for weapons, documents, or other material of intelligence interest. Each individual receives a Capture Tag which records basic biographic data such as name, rank, serial number, unit of assignment (military), location of capture, and any special circumstances concerning the capture. (See Appendices E and F.) Each document or item removed from the captive is also "bagged and tagged" to identify from whom it was taken. This initial step is vital, as properly processing captives and their equipment greatly simplifies the screening process. All documents associated with the source and any possessions taken from him must be evacuated with the source, but not on his person. This is to ensure that the next echelon of screeners and interrogators will have the ability to exploit these items for intelligence value, or to support determination of approach strategies.

SCREENING AT REFUGEE CAMPS OR DETENTION FACILITIES

6-13. When a detainee or refugee arrives at an internment facility, refugee camp, or similar facility, a more extensive screening is conducted. The screening sheet is used to facilitate this process. This screening is normally done in conjunction with in-processing into the facility. During in-processing, the MP will assign an Internment Serial Number (ISN) that is registered with the Theater Detainee Reporting Center (TDRC). The ISN will be used to track the detainee throughout the MP detention system. The ISN should not be used in intelligence channels; however, HUMINT collectors should record the ISN on the screening sheet to aid in locating the detainee again. For intelligence reporting purposes, HUMINT collectors will assign the detainee a source reporting number that will be used to identify the detainee and information associated with him, regardless of whether or not the detainee is transported to another facility. The J2 issues source reporting numbers to HUMINT collectors through the OMT.

MP ISN NUMBER:		EVACUATION DATE:

PERSONAL

LNAME(P): _____
LNAME(M): _____
FNAME: _____
MNAME: _____
SVC/ID NO: _____
DOB: _____
LANGUAGES: _____
MARITAL STATUS: M   S   W   D

*** STATUS: M = Military   C = Civilian   P = Paramilitary   ? = Other

CAPTURE DATA

DATE: _____
TIME: _____
PLACE: _____
CAP UNIT: _____
CIRCUMSTANCES: _____
_____
DOCUMENTS: _____
_____
WPNS/EQUIP: _____
_____

MILITARY

BRANCH: AF AR CG MC NY ___
RANK: _____
FULL UNIT DSG: _____
_____
DUTY PPSN: _____
JOB: _____
STATION: _____
SKILLS: _____
EXPERIENCE: _____

ASSESSMENT DATA

PHYSICAL CONDITION: SEX: M F
WOUNDED: Y N _____
REMARKS: _____
_____
MENTAL CONDITION:
   EDUCATION = _____ YRS
INTELLIGENCE: AVG+ AVG AVG-
MENTAL STATE: _____
_____

CIVILIAN

JOB: _____
ORG: _____
DUTIES: _____
_____
SKILLS: _____
_____

SCREENER: _____
DATE: _____ TIME: _____
COOPERATION: 1(High) 2 3(Low)
KNOWLEDGE: A(High) B C(Low)
BGW LIST: Y N BGW CODE: ____
SOURCE CATEGORY: A B C D
APPROACH: _____

SPECIAL HANDLING REQUIREMENT CODES: _____
_____

PIR & IR
_____
_____
_____
_____
_____

REMARKS
_____
_____
_____
_____
_____

Figure 6-1. Screening Sheet.

6-14. When a detainee is in-processed into an internment facility, MPs will assign the detainee's status as an EPW, retained person, protected person, or other status under the Geneva Conventions. Figure 6-2 provides excerpts from FMI 3-19.40 on MP internment and resettlement operations. In an international conflict, individuals entitled to POW status (EPWs) include—

* Members of the regular armed forces.
* Other militias or volunteer corps, and organized resistance movements of a State Party to a conflict, provided they meet each of the following criteria:
  ▪ Commanded by a person responsible for his subordinates.
  ▪ Having a fixed distinctive sign recognizable at a distance.
  ▪ Carrying arms openly.
  ▪ Conducting operations in accordance with the law of war.
* Civilians who accompany the force.
* Crew members of the merchant marine and crews of civilian aircraft of a State Party to the conflict, who do not benefit by more favorable treatment under any other provisions of international law.

6-15. There are other categories specified in Article 4, GPW. Questions with respect to an individual's entitlement to EPW status should be directed to your SJA.

6-16. Retained personnel (see Articles 24 and 26, GWS):
* Official medical personnel of the armed forces exclusively engaged in the search for, or the collection, transport or treatment of wounded or sick, or in the prevention of disease, and staff exclusively engaged in the administration of medical units and facilities.
* Chaplains attached to the armed forces.
* Staff of National Red Cross Societies and that of other Volunteer Aid Societies, duly recognized and authorized by their governments to assist Medical Service personnel of their own armed forces, provided they are exclusively engaged in the search for, or the collection, transport or treatment of wounded or sick, or in the prevention of disease, and provided that the staff of such societies are subject to military laws and regulations.

6-17. Protected persons include civilians entitled to protection under the GC, including those we retain in the course of a conflict, no matter what the reason. A "civilian internee" is a person detained or interned in the United States or in occupied territory for security reasons, or for protection, or because they have committed an offense against the detaining power, and who is entitled to "protected person" status under the GC.

6-18. The term "detainee" may also refer to enemy combatants. In general, an enemy combatant is a person engaged in hostilities against the United States or its coalition partners during an armed conflict. The term "enemy combatant" includes both "lawful enemy combatants" and "unlawful enemy combatants."

* **Lawful enemy combatants:** Lawful enemy combatants, who are entitled to protections under the Geneva Conventions, include members of the regular armed forces of a State Party to the conflict;

militia, volunteer corps, and organized resistance movements belonging to a State Party to the conflict, which are under responsible command, wear a fixed distinctive sign recognizable at a distance, carry their arms openly, and abide by the laws of war; and, members of regular armed forces who profess allegiance to a government or an authority not recognized by the detaining power.

- **Unlawful enemy combatants:** Unlawful enemy combatants are persons not entitled to combatant immunity, who engage in acts against the United States or its coalition partners in violation of the laws and customs of war during armed conflict. For purposes of the war on terrorism, the term "unlawful enemy combatant" is defined to include, but is not limited to, an individual who is or was part of supporting Taliban or al Qaida forces, or associated forces that are engaged in hostilities against the United States or its coalition partners.

Excerpts from FMI 3-19.40, Military Police Internment/Resettlement Operations

ACTIVITIES

The MPs assist MI screeners by identifying captives who may have answers that support PIR and IR. Because MPs are in constant contact with captives, they see how certain captives respond to orders and see the types of requests they make. The MPs ensure that searches requested by MI personnel are conducted out of sight of other captives and that guards conduct same-gender searches.

The MI screeners examine captured documents, equipment and, in some cases, personal papers (journals, diaries, and letters from home). They are looking for information that identifies a captive and his organization, mission, and personal background (family, knowledge, and experience). Knowledge of a captive's physical and emotional status or other information helps screeners determine his willingness to cooperate.

LOCATION

Consider the following when planning an MI screening site:

- The site is located where screeners can observe captives as they are segregated and processed. It is shielded from the direct view of captives and is far enough away that captives cannot overhear screeners' conversations.
- The site has an operation, administrative, and interrogation area. The interrogation area accommodates an interrogator, a captive, a guard, and an interpreter as well as furniture. Lights are available for night operations.
- Procedures are implemented to verify that sick and wounded captives have been treated and released by authorized medical personnel.
- Guards are available and procedures are implemented for escorting captives to the interrogation site.
- Procedures are published to inform screeners who will be moved and when they will be moved.
- Accountability procedures are implemented and required forms are available.

Figure 6-2. MP Support to Screening.

6-19. It may not be immediately evident in a particular theater of operation whether an individual is an unlawful enemy combatant or is associated with or supporting the unlawful enemy combatants of the United States. Consult your J/G/S2 and servicing SJA for information relevant to your theater of operations.

6-20. All captured or detained personnel, regardless of status, shall be treated humanely, and in accordance with the Detainee Treatment Act of 2005 and DOD Directive 2310.1E, "Department of Defense Detainee Program", and no person in the custody or under the control of DOD, regardless of nationality or physical location, shall be subject to torture or cruel, inhuman, or degrading treatment or punishment, in accordance with and as defined in US law. (See Appendix A, GPW Articles 3, 4, 5, 13, and 14.)

6-21. The rights of EPWs are stated in the GPW. They include the right to quarters, rations, clothing, hygiene and medical attention, property, and other rights. EPWs may not renounce their rights to renounce EPW status. (See Appendix A, GPW Article 7, Section I.)

6-22. Retained personnel must receive at least the same benefits as EPWs. They may only be required to perform religious or medical duties, and they may only be retained as long as required for the health and spiritual needs of the EPWs. Retained persons must be returned to their home country when no longer needed.

6-23. Protected persons' rights include protection from physical or moral coercion and from being taken hostage. Protected persons are protected from murder, torture, corporal punishment, mutilation, medical experimentation, and any form of brutality. Protected persons rights are limited, though. They do not have the right to leave captivity and are not immune from prosecution. Protected persons can be screened and identified for intelligence purposes.

SCREENING PROCESS

6-24. At the internment facility, the screening process normally is distinct from the questioning (interrogation or debriefing) process. Dependent on the criticality of the information identified, the source may be questioned immediately for relevant information but will more likely be identified for future questioning. The screening is a more formal process in which the screener attempts to obtain basic biographic data, areas of general knowledge, source cooperation, and vulnerability to select approach techniques in addition to identifying knowledge of critical intelligence tasks. Once the screener has established the basics (source identification, cooperation, and knowledge), he normally passes the source on to the personnel that conduct the questioning. The screener will complete a screening report that will be forwarded in accordance with unit SOPs (see Chapter 10). If a detainee's knowledge is of Joint Forces interest, a knowledgeability brief (KB) should be written and submitted electronically. (A short form KB worksheet is shown at Figure 10-2.) Complete guidance on KBs is contained in DIAM 58-12 (S//NF).

6-25. If the source freely discusses information of PIR value, the screener normally exploits the information fully and completes a SALUTE report. (See Appendices H and I.) If the source's knowledge of PIR information is extensive and he is freely giving the information, the senior screener and the OIC or noncommissioned officer in charge (NCOIC) of the interrogation or debriefing element are notified immediately. They decide if the screener should continue questioning the source or if the source should be handed off to another HUMINT collector. If source-associated documents contain PIR information, the collector will exploit them as fully as possible and write a SALUTE report. (See Appendix H.)

6-26. The source normally is assigned a standard screening code. The screening code is an alphanumeric designation that reflects the level of cooperation expected from the source and the level of knowledgeability the source may possess. Table 6-1 shows the codes for assessing sources. Those sources assigned to the same category are interrogated in any order deemed appropriate by the interrogation or debriefing element.

Table 6-1. Source Screening Codes.

CODE	COOPERATION LEVEL
1	Responds to direct questions.
2	Responds hesitantly to questioning.
3	Does not respond to questioning.
	KNOWLEDGEABILITY LEVEL
A	Very likely to possess PIR information.
B	Might have IR information.
C	Does not appear to have pertinent information.

6-27. Figure 6-3 shows the order in which detainees in the assessed screening categories should be interrogated. Category 1-A sources normally should be the first priority to be questioned. Category 1-B and 2-A would be Priority II. Category 1-C, 2-B and 3-A would be next as Priority III sources, with 2-C, and 3-B being in the fourth group to be interrogated. Category 3-C sources are normally not questioned. This order ensures the highest probability of obtaining the greatest amount of relevant information within the available time. Screening codes may change with the echelon. The higher the echelon, the more time is available to conduct an approach. Appendix B discusses the reliability ratings of information obtained.

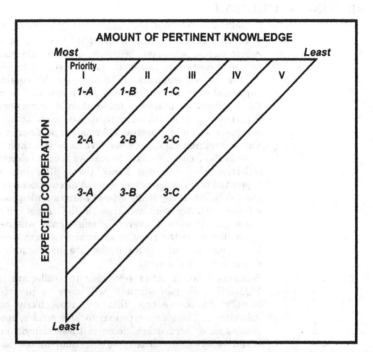

Figure 6-3. Interrogation Priorities by Screening Category.

SCREENING METHODOLOGIES

6-28. Depending on the specific operation or echelon, screening may be a separate operation or may be integrated into a specific collection mission. For example, a HUMINT collector accompanying a patrol encounters a civilian who may have information that is relevant to collection requirements. The HUMINT collector screens the source (that is, asks some general questions to determine the source's level of cooperation and knowledge). Upon receiving a positive response, the HUMINT collector may debrief the civilian on a specific topic or question him on areas of PIR interest. He then reverts to the screening role to determine other relevant knowledge. If the HUMINT collector determines through screening that the source either has no relevant information or cannot be persuaded to cooperate within an operationally expedient timeframe, he is not debriefed as part of the screening process. In detainee or refugee operations, a separate element will normally conduct all screenings. They establish a prioritized list of sources who are then systematically questioned on specific topics by other HUMINT collectors or other technical specialists.

SCREENING REQUIREMENTS

6-29. In addition to potential sources, screening requires several components.

- Collection Requirements. Without a clear list of specific collection requirements, screening becomes virtually impossible. The concept behind screening is to rapidly identify source knowledgeability as it relates to requirements. Screeners should obtain a copy of the supported element's collection requirements and become familiar with the intelligence indicators listed therein. Screeners must use their experience and imagination to devise ways to identify EPWs and detainees who might possess information pertinent to these indicators. Vague requirements (such as "What is the threat doing?") do not provide the focus necessary to make a source selection. The HUMINT collection element must break these SIRs into indicators if the supported intelligence officer has not already done this. The indicators must take into account the type of sources anticipated. For example, a refugee probably will not know if the threat intends to defend a particular ridgeline. However, he might know whether or not there are threat forces on the ridge, if an improvised explosive device (IED) is being employed on a route, if they are digging in, or if engineer type equipment is in the area.

- Selection Criteria. After reviewing the collection requirements, the HUMINT collection element will develop preliminary criteria to identify the source type that will most likely have the required information. The source type may include gender, appearance, military rank, age, or occupation. Some of these criteria are determined by visual observation, thereby saving time in not having to question everyone. Other criteria such as occupation or place of residence may require brief questions.

- Trained Screeners. Screening is possibly the most difficult HUMINT skill. A HUMINT collector must use his experience, questioning skill, cultural knowledge, and knowledge of human nature to decide in a matter of minutes or possibly seconds whether limited HUMINT collection assets and valuable time should be spent talking to an individual based on the way he looks and the answers to a few questions. A wrong decision will mean wasted assets and valuable information missed.

- Language Capability and Cultural Awareness. Screening involves more than asking a series of questions. The HUMINT collector must be able to evaluate the answers, the specific language used, and other clues such as body language to determine the value of an individual to the collection effort. This requires a mature and experienced screener. If the HUMINT collector does not possess the target language, he and his interpreter must be able to work together quickly with mutual trust and confidence.

- Area Conducive to Screening Operations. Effective screening operations must allow the HUMINT collector to speak to the source where the source is not exposed to outside influences or dangers that may inhibit his responses. For that reason, sources should never be screened within the sight or hearing of other potential sources.

HUMINT collectors can use rooms within a building, tents, or other field-expedient methods to isolate the individual being screened. Screening a source within view or hearing of other potential sources may not only pose a danger to the source but also will tend to inhibit the source from freely cooperating.

- Security. The personnel conducting the screening need to be able to concentrate on the individual being screened. Although the collector is ultimately responsible for his own personal security, screening is facilitated by having dedicated personnel present (for example, MPs) who are responsible for security. Screeners coordinate with MP or other security personnel concerning their role in the screening process.

INITIAL DATA AND OBSERVATIONS

6-30. Screening is a filtering process whereby, under ideal circumstances, all potential sources would be questioned to determine if they have information of intelligence interest. In actuality this is often impossible. Screeners often use visual and other aids to "prescreen" the sources in order to eliminate a substantial portion of the target population before conducting detailed screening. For example, if the HUMINT collector at a checkpoint is only interested in information concerning a specific denied geographic area, screeners may ask all refugees if they are from or have been in that denied area recently. A less experienced screener could do this allowing the experienced screener to conduct more extensive screening of the select target audience.

6-31. If time and circumstances permit, screeners should question any friendly personnel who have had extensive contact with the persons to be screened. In the case of detainees, this could include holding area personnel or personnel from the capturing unit. These personnel may be able to help identify sources that might answer the collection requirements or who might match a particular source profile.

6-32. Normally the screener will not have time to conduct any extended observation of the person to be screened; however, the screener should make a quick visual observation of the source prior to questioning him. He should note anything in the source's appearance and behavior that indicates he is willing to cooperate immediately or is unlikely to cooperate. The screener should also note any physical indicators that the source may have the type of information or belong to a certain source profile category.

6-33. Physical indicators include overall appearance such as rank, insignia, and condition of the uniform and type and condition of equipment for military sources and general type and condition of dress (for example, business suit as opposed to work clothes) for civilians. Certain physical indicators (dress, medals such as religious medals, physical type) may be indicators that the source belongs to a specific ethnic or religious group. The source's physical reactions may also indicate a willingness or lack of willingness to cooperate. For example, does the source move forward in the group or attempt to hide within the group; does he intentionally place himself in the wrong segregation group; or does he show any overt signs of nervousness, anxiety, or fright?

6-34. The screeners will also examine all documents and possessions found on the source (if any) and all documents pertaining to the source (if any). **At a minimum, a detainee should have a properly filled out capture tag, which will indicate to the screener where the detainee was captured, by which unit, and under what circumstance.** (See para 6-12 and Appendix D.) Documents such as personal letters, orders, rosters, signal operating instructions (SOIs) and map sections can provide information that identify the source, his organization, mission, and other personal background information (family, knowledge, experience, education). They may in themselves provide information, may identify a source for questioning, and may provide information helpful in assessing a source's susceptibility to an approach.

6-35. Documents pertaining to the source, beginning at the point of capture throughout the evacuation process, such as previous screening and intelligence reports and administrative documents (such as source personnel records prepared by the MPs) help the screener by providing information concerning the source's physical status, emotional status, level of knowledge, level of experience, and other background data. Making timely use of reports from lower echelons can be difficult for the screener, especially when dealing with large numbers of potential sources.

SOURCE ASSESSMENT

6-36. Screeners use standard reporting formats to identify the results of their screening (see Chapter 10). The determination must be made as to whether the source is of any intelligence value to the HUMINT collector. The HUMINT collector will basically place the source within one of four categories.

- Of Immediate Intelligence Interest. This category includes personnel who are assessed, based upon the screening process, who possess information in response to requirements. They are interrogated or debriefed (dependent on their status) to retrieve relevant information. This questioning may be conducted by the same person conducting the screening or by another HUMINT collector.

- Of Interest to Other Agencies. In most cases, the HUMINT collector will be provided with collection requirements by other agencies or disciplines such as TECHINT or CI. In this case the HUMINT collector will question the source on these requirements and report the information appropriately. However, in some instances, particularly in the case of CI, the HUMINT collector may be given a list of topics or a profile of personnel who are of interest to CI. The HUMINT collector will notify the local CI representative when a person matching the "CI profile" is identified. After the HUMINT collector has extracted any relevant intelligence information, he will "pass" the individual off to the CI agents. In many cases, particularly with individuals of TECHINT or other specialized interest, the HUMINT collector will be asked to conduct the questioning with the technical support of the individual from the interested agency. This is coordinated through the HUMINT collector's OMT and the chain of command.

- Of Potential Interest as a Contact Source. On occasion, especially during stability and reconstruction operations, the HUMINT collector may identify an individual who has the potential to provide information in the future, due to his placement or access. Although the individual may not have information of immediate interest, the HUMINT collector will pass his recommendation to the appropriate office, normally the C/J/G/S2X, provided that source operations are authorized (see Chapter 5).
- Of No Interest. This category includes sources who prove (based upon the screening process) to be of no interest to the HUMINT collector or other agencies. Their biographic data is recorded, but they are not questioned further. This category will likely include the bulk of individuals screened. Individuals who have been screened are kept separated from those who have not yet been screened.

OTHER TYPES OF SCREENING OPERATIONS

LOCAL EMPLOYEE SCREENING

6-37. CI personnel conduct local employee screening, primarily to identify individuals who may be a security risk. HUMINT collectors also can use local employee screening as a means to obtain intelligence information or to identify personnel with placement and access to answer information requirements. Employee screening must be conducted in a secure environment and out of the hearing and sight of other employees. Formal written reports of the screening must be maintained.

LOCAL COMMUNITY OR AREA SCREENING

6-38. Local area screening is normally done in coordination with other operations such as a cordon and search operation. The HUMINT collectors accompany the forces conducting the operation and screen the general population to identify individuals of intelligence or CI interest.

SCREENING FOR CI REQUIREMENTS

6-39. Before initiating the screening process, the HUMINT collector establishes liaison with supporting CI agents. The CI element provides CI requirements and provides a profile of personnel of CI interest. CI is normally interested in personnel who—

- Have no identification documents.
- Have excessive or modified identification documents.
- Possess unexplainable large amounts of cash or valuables.
- Are illegal border-crossers.
- Attempt to avoid checkpoints.
- Are on the CI personalities list, which includes members of an intelligence service.
- Request to see CI personnel.
- Have family in the denied area.
- Speak a different language or dialect than is spoken in the area.

6-40. Personnel of CI interest include two general categories of people: The first type of "person of interest" is any individual or group involved in adversary intelligence collection operations or who is attempting to enter the AO to conduct such operations. Examples of these individuals include but are not limited to—

- Known or suspected members and/or supporters of foreign intelligence and security services and known or suspected members and/or supporters of the intelligence activities of non-state entities such as organized crime, terrorist groups, and drug traffickers.
- Known or suspected hostile espionage agents, saboteurs, subversives, or hostile political figures.
- Known or suspected enemy collaborators and sympathizers who may pose a security threat to US forces.
- Personnel known to have engaged in intelligence, CI, security, police, or political indoctrination activities.
- Known or suspected officials of enemy governments whose presence poses a security threat to US forces.
- Political leaders known or suspected to be hostile to the military and political objectives of the US or an allied nation.

6-41. The second type of "person of CI interest" is any individual who possesses information concerning the identification, location, or activities of personnel in the first category.

SCREENING FOR OTHER TECHNICAL COLLECTION REQUIREMENTS

6-42. Other technical areas such as TECHINT, SIGINT, IMINT, MASINT, or other services need to supply the HUMINT collectors with a profile of the individuals with whom they wish to speak. The HUMINT collectors upon identifying such an individual will contact the requesting agency after extracting PIR information.

Chapter 7

Planning and Preparation

7-1. Planning and preparation is one of the five phases of HUMINT collection. HUMINT collection, regardless of the methodology employed, must be a systematic, carefully prepared enterprise. The HUMINT collector engages in general preparation throughout his career. He focuses that preparation to a specific area of the world, specific mission, and specific collection requirements as those become available. Finally, the HUMINT collector focuses his planning and preparation on a specific collection effort with a specific source.

COLLECTION OBJECTIVES

7-2. Each HUMINT collection mission is conducted for a definite purpose. The HUMINT collector must keep this purpose firmly in mind as he proceeds to obtain usable information to satisfy the requirements, and thus contributes to the success of the unit's mission. The HUMINT collector must use the objective as a basis for planning and conducting questioning. The HUMINT collector should not concentrate on the objective to the extent he overlooks or fails to recognize and exploit other valuable information extracted from the source. For example, during HUMINT collection, the HUMINT collector learns of the presence of a heretofore unknown, highly destructive weapon. Although this information may not be in line with his specific objective, the HUMINT collector must develop this important lead to obtain all possible information concerning this weapon.

RESEARCH

7-3. The key to good HUMINT collection is preparation on the part of the collector. The HUMINT collector must understand the environment and particularly its human component, the mission of the supported unit, that unit's intelligence requirements, his source, and the cultural environment. The ultimate success of a questioning session is often decided before the HUMINT collector even meets the source.

GENERAL RESEARCH

7-4. Due to the quickly changing world circumstances, it is impossible to conduct all the specific research required immediately prior to questioning a source. General research should be completed before entering an AO and continues until operation completion. Areas of research include but are not limited to—

- OPLANs and OPORDs. The HUMINT collector must be familiar with the unit OPLAN and that of its higher headquarters. By thoroughly understanding the unit OPLAN and OPORD, the HUMINT collector and HUMINT commanders and leaders can anticipate collection requirements, develop source profiles, recommend deployment strategies, and otherwise integrate HUMINT operations into the overall unit operation. Although the OPORD needs to be read and understood in its entirety, certain areas are of critical importance to the HUMINT collection effort. They include—

- Task organization. This will show where HUMINT C2, staff support, and collection assets will fit into the organizational structure.
- Situation. This gives the friendly and enemy situation.
- Mission. This gives the HUMINT collectors insight into how their operations will integrate into the parent unit's operation.
- Execution. The four execution subparagraphs explain the commander's intent on how the mission is to be carried out:
  - Subparagraph 3a (Concept of Operation) includes how subordinate units' operations will be included in the overall plan.
  - Subparagraph 3a(3) (Reconnaissance and Surveillance) details how HUMINT collection operations will integrate into the overall ISR plan. Additional information on ISR is found in Annex L.
  - Subparagraph 3a(4) (Intelligence), along with Annex A (Task Organization) and Annex B (Intelligence), explains how the Intelligence BOS will support the scheme of maneuver.
  - Subparagraph 3d (Coordinating Instructions) lists the CCIRs and initial PIRs.
- Current events. The HUMINT collector must be knowledgeable about current events in all potential operational areas, especially those events that indicate the populace's feelings or intentions toward the US. This will facilitate a better understanding of the cultural, political, and socio-economic conditions that could influence the attitude and behavior of a source. This knowledge can be obtained and updated through classified periodic intelligence publications and/or military or civilian open sources, including both print and broadcast media, CA and PSYOP databases, and the J/G/S2 analytical elements.
- SOPs. The HUMINT collector must be familiar not only with his own unit's SOP but also with that of any supported unit. The HUMINT collector will be able to obtain specific information about report numbers and formats, as well as information about distribution channels for reports from these SOPs. The SOP will also explain unit policy on source exploitation and evacuation procedures, logistic and maintenance functions, and other C2 and support issues.
- Umbrella concept. The TF commander through the J/G2 and J/G2X issues an umbrella concept for HUMINT operations. When operating under this concept, collection parameters will be established in writing, and it is imperative that the HUMINT collector understands his role. Types of sources will be outlined concerning placement, motivation, and access. The umbrella concept will also specify the types of information against which the HUMINT collector can collect. The umbrella concept is governed by AR 381-100 (S//NF), AR 381-172 (S//NF), DIAM 58-11 (S//NF), and DIAM 58-12 (S//NF).

- Legal guides, SOFAs, operations and execute orders, ROE, and other legal and administrative requirements. The HUMINT collector must be thoroughly familiar with all documents that may set the legal parameters for his collection operations. These are available through the chain of command and from the SJA office. He must know how these requirements apply and to what type of sources each is applied.

- Collection requirements. The HUMINT collector needs not only to know but also to understand the requirements that he will be attempting to answer. These requirements can include CCIRs (PIRs and IRs), essential elements of friendly information (EEFIs), Intelligence Priorities for Strategic Planning (IPSP), specific requests from national level consumers such as HUMINT collection requirements (HCRs), SDRs, or even vocal orders given by the local commander. These all will determine the objective of the questioning plan.

- Databases. Intelligence databases can give the HUMINT collector detailed information about the source's unit, its organization, and its capabilities. They also have information on personalities. The HUMINT collector will use information obtained from databases to control the source and assess his answers for truthfulness. They will also give the HUMINT collector ideas of other areas to research. For example, if the threat is primarily a lightly armed insurgent force, studying similar organizations will provide the HUMINT collector with valuable insights into the possible methods of operation of the current target organization.

- SITMAP and COP. The current situation, both friendly and enemy, is vital for the movement of the HCT and for its collection operations. It reflects enemy unit identification, disposition, and boundaries; major roads or trails for movement of personnel, equipment, weapons; and locations of artillery, minefields, roadblocks, entrenchments, obstacles, staging areas, NBC contaminated areas, and ground surveillance devices. All of this information can be used in source questioning as control questions or in otherwise determining source veracity. The HUMINT collector will be able to identify indicators and predict what should be PIRs and IRs.

- INTSUM. The INTSUM provides a summary of the intelligence situation covering a specific period as dictated by the commander. It is already analyzed intelligence.

- Intelligence estimate. The intelligence estimate is derived from the intelligence preparation of the battlefield (IPB). It is based on all available intelligence and considers everything of operational significance. It will help point out gaps in the intelligence database. It is from these gaps that requirements are derived. It will provide information on the mission, AO, weather, terrain, enemy situation, enemy capabilities, and conclusions. It will cover all of the standard OB topics.

- Weapons and equipment guides. Weapons and equipment guides can assist the HUMINT collector in becoming familiar with the type of equipment employed in the AO. Guides are available in hardcopy and softcopy.

- Area handbooks. These handbooks provide detailed information about a specific area of the world. They provide information on political, economic, sociological, cultural, military, biographic, transportation, and geographic topics. The CIA and other agencies publish area handbooks annually. The US Department of State website also has continuously updated information on trouble spots around the world.

- Previous HUMINT reporting. The HUMINT collector should familiarize himself with all previous relevant reporting from the AO. This will provide him with insight into current operations, the types of information collected, and may help identify information gaps.

- Photographs, maps, and other geospatial products. In conducting general research, the HUMINT collector should become familiar with the AOs. This not only will help identify specific areas of HUMINT collection potential but also will be invaluable in both the questioning of specific sources and the maneuver of the HCT.

- Subject matter experts (SMEs) and technical research. Before deploying to an AO or before supporting on a particular mission, the HUMINT collector may identify particular areas in which he lacks critical knowledge. For example, a HUMINT collector who has previously been operating in an area with a conventional enemy may be deployed to an area with an unconventional threat from irregular forces. Also, intelligence requirements may focus on equipment that is unfamiliar to the HUMINT collector. In order to prepare himself, the HUMINT collector contacts SMEs or analysts or uses technical materials to gain background information.

- Other reports. Intelligence agencies publish numerous reports and summaries that are readily available to the HUMINT collector.

CLOSED AND OPEN-SOURCE INFORMATION (USE OF REACH)

7-5. Reach is a process by which deployed military forces rapidly access information from, receive support from, and conduct collaboration and information sharing with other units and organizations (deployed in theater and from outside the theater) unconstrained by geographic proximity, echelon, or command. Intelligence support is established based on requirements that will help the commanders (regardless of echelon) make decisions. Reach can be accomplished in various ways. There is no requirement for all intelligence functional areas or echelons to use the same approach; hence, there is no common standard for all units to use. Each organization or section should develop its strategy on using the various intelligence reach components. Standard enabling tools will provide for easier access than ever before (for example, access to the INSCOM Information Dominance Center).

INTELLIGENCE REACH COMPONENTS

7-6. Intelligence reach requires the G2/S2 to develop a strategy on how best to support the unit's mission with intelligence reach capabilities. There are eight basic elements of the strategy:

- Push: Push occurs when the producers of intelligence or information are knowledgeable of the customer's requirements and are able to send the desired intelligence to the customer without further requests. Push is accomplished through the Joint Dissemination System (JDS) and/or the Automated Message Handling System (AMHS).

- Pull: Pull occurs when the customer is familiar enough with existing databases to be able to anticipate the location of the desired information. Pull is greatly enhanced through the use of portals and homepages with hyperlinks to the various categories of information available to the user. This requires the establishment of such a homepage at each echelon, thus enabling higher echelons to research and pull from lower databases and homepages.

- Database Access: Access to local, theater, DOD, non-DOD, and commercial databases allows analysts to leverage stored knowledge on topics ranging from basic demographics to OB information. A validated DIA Customer Number (acquired by the J2/G2/S2) in combination with SIPRNET and Joint Worldwide Intelligence Communications System (JWICS) connectivity establishes access to most of the databases online.

- Integrated Broadcast Services (IBS): IBS is an integrated, interactive dissemination system, focusing on tactical user's information requirements using a common message Data Element Dictionary (DED) and J-series family of message formats. The goal of IBS is to resolve the uncoordinated proliferation of "stovepiped" intelligence or information broadcasts by providing the tactical commander with integrated time-sensitive tactical information.

- Collaborative Tools: Collaborative tools are computer-based tools (groupware) that help individuals work together and share information. They allow for virtual on-line meetings and data sharing. As much as possible, collaborative tools should be emplaced with all necessary echelons and centers prior to deployment.

- Request for Information: Reach includes the ability of an intelligence officer at any level to request information that is beyond what is available at his location, using the Community On-Line Intelligence System for End Users and Managers (COLISEUM) System. Once an RFI is entered into the system every other user of that system can see it. Hence, an echelon several echelons above the actual requester can and often does become aware of the request and may, in fact, answer it. Reach is also provided through INSCOM's Information Dominance Center and other nodes at J2 and G2.

- Leveraging Collection Management: The collection and ISR management system is established to provide a mechanism for tasking and managing collection assets for required information. Analysts who are trained and familiar with the system and the various tasking procedures can leverage the system for refined information.

- Distributed Common Ground System-Army (DCGS-A): DCGS-A is the ISR fusion and processing system for the future, as part of the overarching DOD-directed DCGS-A surface system family of systems. It will bring national and joint ISR capabilities down to JTF level, corps and division levels and BCT level to provide leaders with NRT information and visualization of threat, weather, and terrain information and intelligence. DCGS-A consolidates the capabilities of the following current-force ground processing systems:
  - All-Source Analysis System (ASAS).
  - Counterintelligence and Human Intelligence (CI/HUMINT) Single-Source Workstation.
  - Tactical Exploitation System (TES).
  - Guardrail Information Node (GRIFN).
  - Guardrail Common Sensor (GRCS) Intelligence Processing Facility (IPF).
  - Prophet Control.
  - Joint STARS Common Ground Sensor (CGS).

7-7. For more information on Intelligence Reach, see FM 2-33.5/ST.

SOURCE-SPECIFIC RESEARCH

7-8. Source-specific research is done immediately prior to questioning the source. The HUMINT collector may have to respond spontaneously in the case of a walk-in source in tactical operations, or if the HUMINT collector has advanced warning as in the case of a planned meeting with a source, a long-term debriefing, or an invitational source. Areas of research include but are not limited to—

- Screening Reports, KBs, Other Reports: Reports about the source not only can provide specific information about the type of information the source can provide to answer specific collection requirements but also can give the HUMINT collector extensive background information about the source. This background information can give clues to information the source might possess and to possible approach techniques. Information contained in screening reports and KBs may provide insight into—
  - Geographic Area: This area may show information about the source's ethnic background, political affiliation, religion, and customs. Information can be obtained from databases, locally registered vital statistics, and residence registries.
  - Languages: Determining the languages and dialects spoken, written, and understood by a source can provide valuable insights into that source's geographic and ethnic or tribal background, education, and social status. This determination of languages and dialects can be facilitated by the use of "flash cards" specific to the battlefield.
  - Other Reports: This can include other reports collected from this source at other echelons or reports from other sources from the same unit or location as the source. It can also include reports or documents published by the ACE at your request.

- Political Group: This area can provide information on the source's beliefs as well as provide information on political leaders and goals. Additionally, political affiliation can sometimes provide information about subversive groups and paramilitary ties. Knowing the goals of the political organization can also assist the HUMINT collector in choosing an approach or establishing rapport.
- Religious Affiliation: The source's religious affiliation may provide insight into his motivation, moral strengths and weaknesses, and other motivational factors.
- Technical Field: Having knowledge about the source's technical field can assist the HUMINT collector in deciding upon which questions to ask. It will also assist the HUMINT collector in verifying the source's truthfulness because the HUMINT collector will have an understanding of the source's specialty.
- Employment: By researching the source's employment history, the HUMINT collector can discover other areas of information that the source may be able to provide.
- Education: The source's education level and educational history can not only give the HUMINT collector insight into the possible information the source can provide but also provide insight into possible approach strategies.
- Social Status: Knowledge of the source's social status may provide a clue to a good approach strategy because the source may be accustomed to a certain type of treatment. It may also provide a clue to biographical information that the source may be able to provide.
- Criminal Records: Criminal records may also indicate possible approach strategies. Additionally, they may indicate which groups or organizations the source may have knowledge about.

- Documents and Other Media Captured on or in Immediate Association with a Detainee or Brought in by a Debriefing Source: Documents captured with or otherwise pertaining to the source may give the HUMINT collector information about the source, his unit, or his role within that unit. They may answer requirements or indicate knowledge of PIRs. Personal letters, for example, could be used during the approach phase. If a source comes in voluntarily and provides documents, they should be reviewed prior to debriefing the source.
- Photographs, Maps, and Other Geospatial Products: Maps and photographs of the area about which the source is being questioned can give the HUMINT collector an idea of where the source has been and in what kind of terrain he operated, which might indicate knowledge or use of certain tactics. If the HUMINT collector is not familiar with the area the source was in, the HUMINT collector should take some time to look over the map so he can more readily relate when the source mentions locations or dispositions. Aerial photographs show more detailed up-to-date information than maps. They will not normally be as readily available as maps. Maps and other geospatial products will also be needed for use in the map-tracking portion of an interrogation. The HUMINT collector should work with the ACE of the supported unit to obtain them for the AO.

- SMEs: There will be occasions when the HUMINT collector will talk to sources about subjects of which the HUMINT collector has no knowledge. In that case, the HUMINT collector will want to talk to personnel who are SMEs. Depending on the depth of knowledge that the source is expected to have and the time available to prepare, the HUMINT collector may arrange for a technical expert to support the questioning (see Chapter 9).
- Technical Manuals: There are various weapon and equipment identification guides available in hardcopy, softcopy, and off the Internet that can assist the HUMINT collector in identifying any equipment mentioned by the source.
- Source Physical and Mental Condition: HUMINT collectors should observe the source prior to questioning if possible and also talk to anyone available who has relevant information concerning the source. MP guards can be an especially valuable source of information based on source observation and should be debriefed periodically. This can prevent surprises at the onset of the questioning session and can help the HUMINT collector assess the source's physical and mental condition as well as provide insights to possible approaches.
- Databases: Collectors should review source information and reports contained in the various databases available to them. The CHATS system, BAT database, and other databases can provide collectors with source information and previous reporting.

HUMINT COLLECTION PLAN

7-9. After conducting appropriate research, the HUMINT collector working with an analyst, if available, develops a source-based collection plan. This is geared to the specific source that is going to be questioned. The amount of time spent in preparing this plan depends on the operational circumstances. This may range from a quick mental review by an experienced HUMINT collector in a tactical environment to a formal written plan submitted by a subordinate to a team leader. The source collection plan will vary from source to source. It will also vary with the conditions under which the source is questioned. It serves as a checklist to ensure that all steps necessary to prepare for questioning are conducted. Whether written or oral, the HUMINT collection plan should contain at least the following items:

- HUMINT collection requirements.
- Serial number of EPW/detainee to be questioned.
- Location and time for the questioning.
- Primary and alternate approaches.
- Questioning plan including topics to be covered and the planned sequence of these topics.
- Prepared questions for unfamiliar or highly technical topics.
- Method of recording and reporting information obtained.

OBJECTIVE

7-10. The HUMINT collector will first determine the objective of his questioning. The objective is the set of collection requirements that the HUMINT collector will attempt to satisfy during the questioning session. A number of circumstances including the intelligence requirements, the time available, and the source will set the objective. Determining the objective consists of three parts:

- Identify the intelligence requirements. The primary objective of any questioning session is to answer or confirm PIR or other collection requirements.
- Identify the subject: The HUMINT collector will want to consider the source; for example, who he is, what he may know. The HUMINT collector will also want to consider the legal and other restrictions based on the type of source (contact source, EPW, refugee, strategic). For a military source (EPW) this includes rank, position specialty, and unit of assignment. For a civilian source it includes job, placement and access, associations, area of residence, and employment.
- Identify the intelligence requirements that the source may be able to answer. The HUMINT collector cannot normally waste time "fishing" for information. He must determine based on screening, what collection requirements the source can answer. The HUMINT collector compares the information that he gathered through his general and source-specific research and compares it to his list of collection requirements. He compares that list to the identity of the source and refines the list including all requirements that the source can be expected to be able to answer. The HUMINT collector will approach those areas first while staying aware of leads into other collection topics.

LOCATION

7-11. In most cases, the location for the questioning will be determined by operational requirements. However, the HUMINT collector should ensure some basic requirements are met:

- Each questioning session should be conducted outside the hearing and view of third parties. Even in the case of a source meeting in a public place, the HUMINT collector should choose a location where they cannot be overheard and where their meeting will not arouse suspicion.
- The location should be in a place that has reasonable security for the HUMINT collector and the source. In contact operations, the risk cannot always be eliminated but the acceptable risk levels should be based on the expected intelligence gain. In combat operations, most questioning (interrogation, debriefing of civilians on the battlefield) will take place in forward combat areas, but it cannot be done if it increases the risk to the source. Safe evacuation of the sources has priority over questioning.
- The location should provide ready access to the chosen method of recording and reporting the information.

7-12. When conducting military source operations, the location of the questioning will have psychological effects on the source. The questioning location should be chosen and set up to correspond to the effect that the HUMINT collector wants to project and his planned approach techniques. For example, meeting in a social type situation such as a restaurant may place the source at ease. Meeting in an apartment projects informality while meeting in an office projects more formality. Meeting at the source's home normally places him at a psychological advantage, while meeting in the HUMINT collector's work area gives the collector a psychological edge. The HUMINT collector should consider the status and level of the source, security, the workspace available, furnishings, the amount of lighting provided, and the ability to heat or cool the room as needed.

TIME

7-13. Time to conduct questioning should be estimated based on the source, the type of information that the HUMINT collector expects to get, and the complexity of that information. Other considerations include expected evacuation times for sources in tactical situations, the number of other sources that need to be spoken to; and in contact operations, the estimated time that the HUMINT collector can meet with the source without increasing the risk.

7-14. The HUMINT collector must also consider the physical conditions of the source and himself. After extended operations, there may be a limit on how long either the HUMINT collector or source can concentrate on a given subject. Even if the HUMINT collector has an unlimited time period (such as at a joint interrogation and debriefing facility), he must break his questioning down into topical sessions to maximize effectiveness. Time is only an estimate and should be modified based on the circumstances. It may be extended, for example, if the source has a greater than expected amount of information, or critical information in unforeseen areas. The time may be curtailed if the HUMINT collector has met his requirements, the source does not possess the expected information, or a more valuable source is identified.

PRIMARY AND ALTERNATE APPROACHES

7-15. In most circumstances, if the HUMINT collector is meeting with the source for the first time, he should select at least two alternate approaches to use if the direct approach is unsuccessful (see Chapter 8). These approaches need to be based on the HUMINT collector's source-specific research, his general area research, knowledge of the current situation, and knowledge of human nature. There are four primary factors that must be considered when selecting tentative approaches:

- The source's mental and physical state. Is the source injured, angry, crying, arrogant, cocky, or frightened?
- The source's background. What is the source's age and level of military or civilian experience? Consider cultural, ethnic, and religious factors.
- The objective of the HUMINT collection. How valuable is the source's potential information? Is it beneficial to spend more effort convincing this source to talk?

- The HUMINT collector himself. What abilities does he have that can be brought into play? What weaknesses does he have that may interfere with the HUMINT collection? Are there social or ethnic barriers to communication? Can his personality adapt to the personality of the source?

7-16. If the HUMINT collector has a screening sheet or KB, he can use it to help select his approaches. After reviewing the information, the HUMINT collector will analyze the information for indicators of psychological and/or physical weakness that would make a source susceptible to a specific approach. The HUMINT collector also needs to consider his particular strengths and weaknesses in conducting specific approaches. He must consider what immediate incentives he may possibly need and ensure that they are available. Also, if incentives had been previously offered or promised, the collector needs to know if they were in fact provided. If the HUMINT collector has previously questioned the source, he must evaluate the approaches he used and decide if they need to be modified or if additional approach techniques will be needed (see Chapter 8.)

ADDITIONAL SUPPORT REQUIRED

7-17. The HUMINT collector must decide if he will need technical support to include interpreter support.

- Technical support. The HUMINT collector must decide if he will need additional support including analytical, technical, or interpreter support.
- Analytical or technical support. The HUMINT collector must decide if he has the analytical or technical capability to question a specific source. If not, he must decide what degree of support from advice to participation is required of the analyst or technical expert. Any request for analytical or technical support must be coordinated with the 2X. On rare occasions, it may be desirable for the HUMINT collector to seek polygraph support or support from a Behavioral Science Consultant (BSC). BSCs are authorized to make psychological assessments of the character, personality, social interactions, and other behavioral characteristics of interrogation subjects and advise HUMINT collectors of their assessments, as needed.
- Interpreter support. If the HUMINT collector does not speak the needed language or does not speak the needed language well enough to conduct questioning, an interpreter will be required. If the HUMINT collector will need an interpreter, the HUMINT collector will also have to consider the clearance needed to complete the questioning and the availability of the interpreter, as well as the extra time necessary to complete the questioning session. The HUMINT collector will also have to brief the interpreter on the method of interpretation and the HUMINT exploitation plan. Also, he should determine whether there are any cultural aspects associated with the interpreter that may enhance or detract from the success of the meet. (See Chapter 11 for detailed information on HUMINT collection using an interpreter.)

DEVELOP A QUESTIONING PLAN

7-18. The HUMINT collector must develop a plan that will guide his questioning of the source. This includes general topics to be exploited and the sequence in which they will be covered.

7-19. There are two general sequences used in questioning: topical and chronological.

- Topical questioning is used when time is a prime concern, when the source is believed to possess key information in a limited area, when the questioning is concerning a technical topic, or when the source has been talked to previously and this is a subsequent questioning to expand on earlier topics.
- Chronological questioning normally is used when the HUMINT collector is uncertain of the areas of source knowledge, when time is not a factor in questioning, during initial questioning when the source is believed to have knowledge on a large number of topics, and in friendly force mission debriefing.

7-20. A topical sequence is an outline of topics to be questioned in a selected sequence and is based on intelligence requirements or HCRs, as well as a specific source's potential to provide information pertinent to those requirements. The plan serves as a checklist for the HUMINT collector to ensure that all subjects pertinent to the collection objective are questioned in an efficient and organized manner. The HUMINT collector uses his estimate of the type and extent of knowledge possessed by the source to modify the basic topical sequence of questioning. He selects only those topics in which he believes the source has pertinent knowledge. In this way, the HUMINT collector refines his element's overall objective into a set of specific HUMINT collection subjects. In OB factors questioning in either a tactical or strategic setting, and across the full spectrum of operations, the topics covered include missions and the nine major OB factors:

- Composition.
- Strength.
- Dispositions.
- Tactics.
- Training.
- Combat effectiveness.
- Logistics.
- Electronic technical data.
- Miscellaneous.

7-21. See Appendix G for questioning quick reference examples of topics covered under the nine OB factors.

7-22. In strategic and operational debriefing operations the relevant HCR or SDR will guide the HUMINT collector. Regardless of which tasking document is referenced, the topical sequence is established by collection requirements, modified or sequenced, based on source knowledge and time.

7-23. The nine OB factors are not the only guideline that may be used by the HUMINT collector. If the collection objective is something other than a military unit, many of the OB factors will not fit the collection plan. A helpful memory aid, in this case, is mission, identification, location, and organization (MILO). MILO gives a short, easily remembered structure for questioning nonmilitary or strategic topics. The MILO factors can be questioned in any order, but often the most logical sequence of MILO questioning is identification, organization, location, and mission. Many of the nine OB factors can also fit into the MILO format.

IDENTIFY MEANS OF RECORDING AND REPORTING

7-24. The HUMINT collector will want to decide upon a means of recording the information obtained through source questioning. If the HUMINT collector is planning to use a sound or video recorder, he will also have to consider the availability of the equipment and its positioning (see Chapter 9). Along with the method of recording the information, the HUMINT collector will have to decide on the means of reporting the information (see Chapter 10). Tapes of interrogations must be safeguarded in accordance with DOD Regulation 5200.1-R.

FINAL PREPARATIONS

7-25. After the source-specific questioning plan is developed, the HUMINT collector takes some final preparatory steps.

- Review plan. The HUMINT collector should always go over his collection plan with his supervisor. This review can be written or oral. In addition to the obvious requirements to keep the chain of command informed, this review helps identify any weaknesses in the plan and is a means to effect required coordination and support.
- Collect questioning support materials. The HUMINT collector will want to collect the various references and other guides that he will use to support his questioning. These materials may include source documents, maps, aerial photographs, imagery, OB data, extra lights, extra tables, drawing templates, graph paper, questioning guides, technical reference manuals, city plans and handbooks, and recording devices.
- Conduct required coordination. The HUMINT collector coordinates any support requirements including analytical, technical, or interpreter support, questioning location, ICFs, recording equipment, security, and transportation.
- Organize. The HUMINT collector organizes his materials in a logical manner that will complement his topical sequence. By being organized, the HUMINT collector will not waste time trying to locate the correct manual or guide. Additionally, the HUMINT collector will present a professional appearance to his source.
- Reconnoiter the questioning location. If the questioning location is to be somewhere other than the HUMINT collector's normal AO, such as a public restaurant, the HUMINT collector should conduct an unobtrusive reconnaissance of the site. If at all possible, this should be at the same time and day of the week as the planned meeting. This allows the HUMINT collector to assess the possible security problems

of the location, judge the traffic flow, and identify any other items that might affect the questioning. He can also judge where within the meeting site he can set up for maximum security and psychological advantage. He must be careful that in doing so he does not set up patterns of operation that will increase rather than decrease security problems.

- Set up questioning site. If the HUMINT collector has control over the site where the collection is being conducted, the last step in preparing is the actual setup of the questioning site. The HUMINT collector will want to decide on the placement of the furniture and lighting and where everyone will be seated and decide where he will place his technical support materials.

- Question guards. If the person to be questioned is a detainee, the HUMINT collector should arrange to question MP guards who have been in contact with the detainee to ascertain source behavior, attitude, and other useful information that guards may be able to provide.

- Check with medical personnel. If the detainee was injured or ill, ensure that he was treated by medical authorities and released for questioning.

7-26. The supervisor reviews each plan for legal considerations, appropriate goals in accordance with the collection objectives of the supported unit, and makes any changes he thinks are necessary. The supervisor ensures that contract interrogators are utilized in accordance with the scope of their contract and current policy. (See Appendix K.) After the plan is approved, the collection operation is executed. Prior to execution, the supervisor ensures mission brief back, rehearsal, and pre-combat inspections are conducted.

Chapter 8

Approach Techniques and Termination Strategies

8-1. Regardless of the type of operation, the initial impression that the HUMINT collector makes on the source and the approach he takes to gain the source's cooperation will have a lasting effect on the continuing relationship and the degree of success in collecting information. The approach used will vary based on the type of operation; the operational environment; the status of the source; the personality, position, and identity of the source; and the personality and experience level of the HUMINT collector and the time available.

8-2. The MPs will not take any actions to set conditions for interrogations (for example, "softening up" a detainee). Additionally, in accordance with DOD Directive 3115.09, military working dogs, contracted dogs, or any other dog in use by a government agency shall not be used as a part of an interrogation approach nor to harass, intimidate, threaten, or coerce a detainee for interrogation purposes. Leadership throughout the chain of command is responsible to ensure that HUMINT operations are in compliance with these governing regulations and guidelines, whether the HUMINT collection is to take place as part of HCT operations or in an internment facility.

8-3. The only authorized interrogation approaches and techniques are those authorized by and listed in this manual, in accordance with the Detainee Treatment Act of 2005. Two approaches, Mutt and Jeff and False Flag, require approval by the first O-6 in the interrogator's chain of command. The restricted interrogation technique "Separation" requires COCOM commander approval for use, and approval of each interrogation plan using "Separation" by the first General Officer/Flag Officer (GO/FO) in the chain of command. Coordination may also be required with the C/J/G2X, security, legal, or other personnel. Regardless of the coordination efforts required, use of all techniques at all locations must carefully comply with this manual and additional instructions contained in the latest DOD and COCOM policies.

NOTE: The word "source" will be used in this chapter to mean any person who is the objective of the HUMINT collector's approach, and is applicable in any collection situation unless otherwise noted in the text. This use of the term "source" is consistent with US Army Intelligence Center HUMINT collector training.

APPROACH PHASE

8-4. During the approach phase, the HUMINT collector establishes the conditions of control and rapport to facilitate information collection. The approach begins with initial contact between the source and the HUMINT collector. Extreme care is required since the success of the collection effort hinges, to a large degree, on the early development of the source's willingness to communicate. Interrogators must have a deep understanding of the cultural norms, anomalies, and emotional triggers of the person being interrogated in order to select appropriate approach strategies and to interrogate effectively.

8-5. The HUMINT collector's objective during this phase is to establish a relationship with the source that results in the source providing accurate and reliable information in response to the HUMINT collector's questions. The HUMINT collector adopts an appropriate persona based on his appraisal of the source but remains alert for verbal and non-verbal clues that indicate the need for a change in the approach techniques. The amount of time spent on this phase will depend mostly on the probable quantity and value of information the source possesses, the availability of other sources with knowledge on the same topics, and available time. At the initial contact, a businesslike relationship should be maintained. As the source assumes a cooperative attitude, a more relaxed atmosphere may be advantageous. The HUMINT collector must carefully determine which of the various approach techniques to employ.

8-6. Sources will cooperate with the HUMINT collector for various reasons ranging from patriotic duty to personal gain, such as material gifts or money. They may also respond to emotion or logic. Regardless of the type of source and his outward personality, every source possesses exploitable characteristics that, if recognized by the HUMINT collector, can be used to facilitate the collection process. These characteristics may be readily apparent or may have to be extrapolated from the source's speech, mannerisms, facial expressions, physical movements, involuntary responses (perspiration, changes in breathing, eye movement), and other overt indications that vary from source to source. From a psychological standpoint, the HUMINT collector must be cognizant of the following behaviors. People tend to—

- Want to talk when they are under stress and respond to kindness and understanding during trying circumstances. For example, enemy soldiers who have just been captured have experienced a significant stress-producing episode. The natural inclination is for people to want to talk about this sort of experience. If the EPW has been properly segregated and silenced, the HUMINT collector will be the first person the EPW has a chance to talk to. This is a powerful tool for the collector to use to get the subject talking. The desire to talk may also be manifested in refugees, DPs, and even local civilians when confronted by an unsettled situation.
- Show deference when confronted by superior authority. This is culturally dependent but in most areas of the world people are used to responding to questions from a variety of government and quasi-government officials.
- Operate within a framework of personal and culturally derived values. People tend to respond positively to individuals who display the same value system and negatively when their core values are challenged.
- Respond to physical and, more importantly, emotional self-interest. This may be as simple as responding to material rewards such as extra

food or luxury items for their personal comfort or as complex as responding to support in rationalizing guilt.

- Fail to apply or remember lessons they may have been taught regarding security if confronted with a disorganized or strange situation.
- Be more willing to discuss a topic about which the HUMINT collector demonstrates identical or related experience or knowledge.
- Appreciate flattery and exoneration from guilt.
- Attach less importance to a topic if it is treated routinely by the HUMINT collector.
- Resent having someone or something they respect belittled, especially by someone they dislike.

8-7. HUMINT collectors do not "run" an approach by following a set pattern or routine. Each approach is different, but all approaches have the following in common. They—

- Establish and maintain control over the source and collection effort. This does not necessarily equate to physical control. Rather it means that the HUMINT collector directs the conversation to cover the topics that are of interest to him. This may be overt in a debriefing or an interrogation or subtle in an elicitation. In a very basic sense, the HUMINT collector is in control if he is asking questions and receiving answers. If the source is asking questions, refusing to answer questions, or directing or attempting to direct the exchange, he is challenging for control. If the source challenges this control, the HUMINT collector must act quickly and firmly to reestablish control.

- Establish and maintain a rapport between the HUMINT collector and the source. Rapport is a condition established by the HUMINT collector that is characterized by source confidence in the HUMINT collector and a willingness to cooperate with him. This does not necessarily equate to a friendly atmosphere. It means that a relationship is established and maintained that facilitates the collection of information by the HUMINT collector. The HUMINT collector may establish a relationship as superior, equal, or even inferior to the source. The relationship may be based on friendship, mutual gain, or even fear.

- Identify the source's primary emotions, values, traditions, and characteristics and use them to gain the source's willing cooperation.

8-8. The successful application of approach techniques, coupled with measures to ensure source veracity, results in the source providing accurate information in response to the HUMINT collector's requirements. The source may or may not be aware that he is providing the HUMINT collector with needed information. The approach does not end when the source begins providing information but is reinforced as necessary throughout the questioning.

DEVELOPING RAPPORT

8-9. The basis of rapport is source confidence in the HUMINT collector, which leads to a willingness to cooperate. Rapport does not necessarily mean a friendly relationship, although that may be the case. It means an establishment of a relationship in which the HUMINT collector presents a realistic persona designed to evoke cooperation from the source. The source responds with relevant, truthful information. Rapport is established during the approach and must be maintained throughout the questioning of the source. If the HUMINT collector has established good rapport initially and then abandons the effort, the source would rightfully begin to question the HUMINT collector's sincerity and may cease answering questions.

BUILDING RAPPORT

8-10. Building rapport is an integral part of the approach phase. The establishment of rapport begins when the HUMINT collector first encounters the source. Depending on the situation, the HUMINT collector may introduce himself to the source. In debriefing and liaison operations, this will normally be the collector's true name and affiliation. In elicitation, the requirement and type of introduction depends on the operation. In interrogation operations, the HUMINT collector normally will not introduce himself unless he is laying the groundwork for an approach. If he does introduce himself, normally he will adopt a duty position and rank supportive of the approach strategy selected during the planning and preparation phase. The HUMINT collector must select a rank and duty position that is believable based on the HUMINT collector's age, appearance, and experience. A HUMINT collector may, according to international law, use ruses of war to build rapport with interrogation sources, and this may include posing or "passing himself off" as someone other than a military interrogator. However, the collector must not pose as—

- A doctor, medic, or any other type of medical personnel.
- Any member of the International Committee of the Red Cross (ICRC) or its affiliates. Such a ruse is a violation of US treaty obligations.
- A chaplain or clergyman.
- A journalist.
- A member of the US Congress.

8-11. The HUMINT collector should seek advice from his SJA concerning representing himself as holding any other sensitive position.

8-12. A good source assessment is the basis for the approach and vital to the success of the collection effort. The HUMINT collector continually assesses the source to see if the approaches—and later the questioning techniques—chosen in the planning and preparation phase will indeed work. Approaches chosen in planning and preparation are tentative and based on the limited information available from documents, guards, and personal observation. This may lead the HUMINT collector to select approaches that may be totally incorrect for obtaining this source's willing cooperation. Thus, careful assessment of the source is critical to avoid wasting valuable time in the approach phase. Whether the HUMINT collector is using reasoned argument or emotion to get the source to cooperate, he must be convincing and believable and appear sincere.

RAPPORT POSTURE

8-13. Unless there is rationale for acting otherwise, the HUMINT collector will begin his interaction with the source in a businesslike manner. He will be neither hostile nor overly friendly. Based on the tentative approaches developed during planning and preparation and the verbal and physical clues from the source, the HUMINT collector will modify this posture to facilitate collection.

8-14. Based on planning and preparation, the HUMINT collector may decide to adopt a stern posture. He presents himself as a person in a superior position to the interrogation source and demands proper deference and obedience by the interrogation source. In the case of an EPW this is manifested by having the source remain at attention and address the HUMINT collector as "Sir." This can be effective in dealing with lower ranking military personnel or members of oppressed ethnic, tribal, or religious groups who are conditioned to respond to authority or civilians in lower economic or social positions who are used to responding to directions from various bureaucrats and civilian superiors. This posture can have negative results since many persons in the positions mentioned above have developed mechanisms for dealing with superiors that mostly involve giving minimal information and agreeing with whatever the authority figure says.

8-15. In most cases, either initially or after the interrogation source has begun answering questions, the HUMINT collector will adopt a more relaxed or even sympathetic posture. The HUMINT collector addresses the interrogation source in a friendly fashion, striving to put him at ease. Regardless of the posture selected by the HUMINT collector, he must stay detached emotionally while maintaining the appearance of total involvement and stay within his adopted persona. The HUMINT collector must control his temper at all times. He must not show distaste, disgust, or unease at anything the source says unless that reaction is a planned part of the approach strategy. He should not show surprise at anything that the interrogation source says since it might undermine source confidence in the HUMINT collector and their relationship.

8-16. The HUMINT collector must support his verbal approaches with appropriate body language. Just as the HUMINT collector is observing the source to identify non-verbal clues that support or contradict the verbal message, the HUMINT collector is being scrutinized by the source to identify the same clues. The techniques used in an approach are a totality of effort, not just verbal conversation between the HUMINT collector and the source. Body language is in many instances culturally dependent. Standing at a given distance from an individual may be perceived as comforting in some societies and hostile in others. The HUMINT collector must adapt his body language to the culture in which he is working rather than expect the source to adapt to his.

APPROACH TECHNIQUES

8-17. The approaches listed are not guaranteed solutions for every situation. Some individual approaches that may be suitable for one operating environment, such as when conducting HUMINT contact operations, may be ineffective in another, such as interrogation. Some will be successful with one source and ineffective with another. In any case, everything the HUMINT collector says and does must be in compliance with the applicable law and policy under which the HUMINT collector is operating. Applicable law and policy include US law; the law of war; relevant international law; relevant directives including DOD Directive 3115.09, "DOD Intelligence Interrogations, Detainee Debriefings, and Tactical Questioning"; DOD Directive 2310.1E, "The Department of Defense Detainee Program"; DOD instructions; and military execute orders including FRAGOs.

8-18. There are 18 approach techniques that can be employed on any detainee regardless of status or characterization, including EPWs. Additionally, there is one restricted interrogation technique called separation (see Appendix M). Separation cannot be employed on EPWs. With the exception of the direct approach, which may be effective by itself, approach techniques are used in combination with other approaches and techniques. Transitions from one approach to another must be smooth, logical, and convincing.

DIRECT APPROACH

8-19. (Interrogation and Other MSO) Almost all HUMINT collection begins with the direct approach. The exception to this is during elicitation operations that by their very nature are indirect. In using the direct approach, the HUMINT collector asks direct questions (see Chapter 9). The initial questions may be administrative or nonpertinent but the HUMINT collector quickly begins asking pertinent questions. The HUMINT collector will continue to use direct questions as long as the source is answering the questions in a truthful manner. When the source refuses to answer, avoids answering, or falsely answers a pertinent question, the HUMINT collector will begin an alternate approach strategy. The fact that the source is answering questions does not preclude the HUMINT collector from providing an incentive to reward the source and continue his cooperation as long as that incentive does not slow down the collection. For example, a HUMINT collector might offer the source coffee or cigarettes to reward his cooperation. See Chapter 9 for the use of Repeat and Control questions in detecting deception.

8-20. Statistics from interrogation operations in World War II show that the direct approach was effective 90 percent of the time. In Vietnam and in Operations URGENT FURY (Grenada, 1983), JUST CAUSE (Panama, 1989), and DESERT STORM (Kuwait and Iraq, 1991), the direct approach was 95 percent effective. The effectiveness of the direct approach in Operations ENDURING FREEDOM (Afghanistan, 2001-2002) and IRAQI FREEDOM (Iraq, 2003) are still being studied; however, unofficial studies indicate that in these operations, the direct approach has been dramatically less successful. The direct approach is frequently employed at lower echelons when the tactical situation precludes selecting other techniques, and where the EPW's or detainee's mental state is one of confusion or extreme shock. However, the HUMINT collector must remember that just because a source is answering a direct question does not mean he is being truthful.

INCENTIVE APPROACH

8-21. (Interrogation and Other MSO) The incentive approach is trading something that the source wants for information. The thing that you give up may be a material reward, an emotional reward, or the removal of a real or perceived negative stimulus. The exchange of the incentive may be blatant or subtle. On one extreme, the exchange may be a formal cash payment for information during some contact operations while on the other extreme it may be as subtle as offering the source a cigarette. Even when the direct approach is successful, the HUMINT collector may use incentives to enhance rapport and to reward the source for cooperation and truthfulness. The HUMINT collector must be extremely careful in selecting the options offered to a detainee source. He cannot deny the detainee anything that he is entitled to by law.

8-22. The HUMINT collector also should not offer anything that is not in his power to give. Although this might be expedient in the short term, in the long run it will eliminate source cooperation. When asked to provide something beyond his authority, the HUMINT collector can agree to help, check into, or otherwise support the request without committing himself to its successful accomplishment. HUMINT collectors must be cautious in the use of incentives for the following reasons:

- There is an inherent suspicion of the truthfulness of "bought" information. Sources may manufacture information in order to receive or maintain an incentive. Sources may also "hold back" information in the hopes of trading it at a later date for greater incentives. They may also hold back information if the incentive is not immediately available or guaranteed.

- The incentive must be believable and attainable. The incentive must be within the capability of the HUMINT collector's assumed persona to achieve. For example, if the detainee was captured after killing a US soldier, an incentive of release would not be realistic or believable. Likewise, if the interrogator is presenting himself as being a "harmless clerk" at the detention center, it would be unrealistic to expect a detainee to believe that a clerk could arrange to have the detainee's girlfriend brought to visit him. Such a visit might be possible, but the interrogator's assumed persona would not seemingly provide him with the authority to make it happen.

- The HUMINT collector must provide any promised incentive. A simple promise of an incentive may be sufficient to obtain immediate cooperation. If, however, the HUMINT collector does not follow through on providing the incentive, he will lose credibility and rapport with his source. This may end the cooperation of not only that source but also possibly any potential source who has contact with that source.

- The HUMINT collector may not state or even imply that the basic human rights guaranteed by applicable national and international laws, regulations, and agreements will be contingent on a detained source's cooperation. An incentive for cooperation is viable only if the HUMINT collector has or is perceived to have the authority to withhold the incentive if the source is not cooperative. A HUMINT collector cannot promise an EPW that he will be treated in accordance with the GPW if he cooperates. This statement implies that the EPW will not be treated properly if he does not cooperate. Since the EPW must be treated in accordance with the GPW whether he cooperates or not, the HUMINT collector will rapidly lose credibility.

EMOTIONAL APPROACHES

8-23. (Interrogation and Other MSO) Emotional approaches are centered on how the source views himself and his interrelationships with others. Through source observation and initial questioning, the HUMINT collector can often identify dominant emotions that motivate the EPW/detainee. The motivating emotion may be greed, love, hate, revenge, or others. The emotion may be directed inward (feelings of pride or helplessness) or outward (love of family). The HUMINT collector employs verbal and emotional ruses in applying pressure to the source's dominant emotions. He then links the satisfaction of these emotions to the source's cooperation. Often, the presentation of like experiences and presenting the source with an opportunity to express his emotions is sufficient to result in cooperation. However, sometimes the source must be presented with a specific action or tangible manifestation of support.

8-24. Although the emotion is the key factor, an emotional approach is normally worthless without an attached incentive. The incentive must meet the criteria listed above for the incentive approach to ensure that the incentive is believable and attainable. For example, this technique can be used on the EPW/detainee who has a great love for his unit and fellow soldiers. Simply having the source express this emotion is not enough. After the source expresses this emotion, the HUMINT collector can take advantage of this by telling the EPW/detainee that by providing pertinent information, he may shorten the war or battle in progress and save many of his comrades' lives, but his refusal to talk may cause their deaths. This gives the source the alternatives of facing the status quo or expressing love of comrades through cooperating with the HUMINT collector.

8-25. Religion is an especially difficult topic to use in any emotional approach. An approach using religion may encourage the source to be further motivated by love, remorse, futility, or even pride to cooperate with the interrogator. On the other hand, an approach using religion may also encourage the source to end any rapport and cooperation with the interrogator. Although it is acceptable to use religion in all interrogation approaches, even to express doubts about a religion, an interrogator is not permitted to denigrate a religion's symbols (for example, a Koran, prayer rug, icon, or religious statue) or violate a religion's tenets, except where appropriate for health, safety, and security reasons. Supervisors should carefully consider the experience level of their subordinates before permitting the use of religion in any interrogation approach.

8-26. Similarly, supervisors should question the appropriateness of demeaning any racial group, including the source's, to elicit an emotional response during an interrogation approach.

8-27. One common danger to the use of emotional approaches is the development of an emotional attachment on the part of the HUMINT collector. It is natural that a source will develop an emotional attachment to the HUMINT collector. The HUMINT collector will often foster this attachment. However, it is vital the HUMINT collector not develop a corresponding emotional attachment to the source. This problem normally develops when a HUMINT collector has contact with one source or a group of similar sources over an extended period of time. There is transference of the source's problems to the HUMINT collector. For example, HUMINT collectors working in a refugee camp frequently begin to view the welfare of the refugees as a greater concern than HUMINT collection. The HUMINT collector, while developing emotion within the source, must act believably but at the same time he must remain detached. He must remember that the emotion is a means to an end (that is, information collection). Supervisors must carefully observe HUMINT collectors for signs of this emotional attachment to the source and take appropriate action ranging from counseling to reassignment.

8-28. The following are types of emotional approaches.

Emotional Love Approach

8-29. (Interrogation and Other MSO) Love in its many forms (friendship, comradeship, patriotism, love of family) is a dominant emotion for most people. The HUMINT collector focuses on the anxiety felt by the source about the circumstances in which he finds himself, his isolation from those he loves, and his feelings of helplessness. The HUMINT collector directs the love the source feels toward the appropriate object: family, homeland, or comrades. If the HUMINT collector can show the source what the source himself can do to alter or improve his situation or the situation of the object of his emotion, the approach has a chance of success.

8-30. The key to the successful use of this approach is to identify an action that can realistically evoke this emotion (an incentive) that can be tied to a detained source's cooperation. For example, if the source cooperates, he can see his family sooner, end the war, protect his comrades, help his country, help his ethnic group. A good HUMINT collector will usually orchestrate some futility with an emotional love approach to hasten the source's reaching the breaking point. In other words if the source does not cooperate, these things may never happen or be delayed in happening. Sincerity and conviction are critical in a successful attempt at an emotional love approach as the HUMINT collector must show genuine concern for the source, and for the object at which the HUMINT collector is directing the source's emotion. The emotional love approach may be used in any MSO where the source's state of mind indicates that the approach may be effective.

Emotional Hate Approach

8-31. (Interrogation and Other MSO) The emotional hate approach focuses on any genuine hate, or possibly a desire for revenge, the source may feel. The HUMINT collector must clearly identify the object of the source's hate and, if necessary, build on those feelings so the emotion overrides the source's rational side. The source may have negative feelings about his country's regime, immediate superiors, officers in general, or fellow soldiers. The emotional hate approach may be used in any MSO where the source's state of mind indicates that the approach may be effective.

8-32. The emotional hate approach may be effective on members of racial or religious minorities who have or feel that they have faced discrimination in military and civilian life. The "hate" may be very specific. For example, a source may have great love for his country, but may hate the regime in control. The HUMINT collector must be sure to correctly identify the specific object of the hate. The emotional hate approach is most effective with the immature or timid source who may have had no opportunity up to this point for revenge, or never had the courage to voice his feelings.

8-33. As in the emotional love approach, the key to the successful application is the linking of the emotion with a tangible manifestation of that emotion. The HUMINT collector must be extremely careful that he does not promise anything that would be contrary to national or international law or US interests or goals. For example, if an EPW feels he has been treated unfairly in his unit, the HUMINT collector can point out that, if the source cooperates and divulges the location of that unit, the unit can be destroyed, thus affording the source revenge. But he cannot promise that the unit if attacked would not be allowed to surrender or that the unit if it surrenders will be treated badly.

8-34. The HUMINT collector must be careful that he does not assume that casual negative comments equate to a strong hate. Many soldiers will make negative comments against their army but will support and defend their army against any "outsider." The HUMINT collector should also not assume generalities; for example, assuming that a member of an ethnic minority hates the ethnic majority just because most ethnic minorities hate those in the ethnic majority.

Emotional Fear-Up Approach

8-35. (Interrogation and Other MSO) Fear is another dominant emotion that can be exploited by the HUMINT collector. In the fear-up approach, the HUMINT collector identifies a preexisting fear or creates a fear within the source. He then links the elimination or reduction of the fear to cooperation on the part of the source. The HUMINT collector must be extremely careful that he does not threaten or coerce a source. Conveying a threat may be a violation of the UCMJ. The HUMINT collector should also be extremely careful that he does not create so much fear that the source becomes unresponsive. The HUMINT collector should never act as if he is out of control or set himself up as the object or focal point of the source's fear. If the HUMINT collector acts in this manner, it is extremely difficult to then act as the outlet for the fear. Supervisors should consider the experience level of their subordinates before approving their use of this approach.

8-36. If there is a justifiable fear, the HUMINT collector should present it and present a plan to mitigate it if the source cooperates (combination of emotional and incentive approaches). For example, an EPW source says that he will not cooperate because if he does his fellow prisoners will kill him or, if a contact source says that if people find out he is cooperating, his family will suffer. In these cases, the HUMINT collector can point out that the source has already placed himself at risk and he or his family may suffer whether he cooperates or not (justified fear). But if he cooperates, the HUMINT collector will do his best to ensure that either no one will find out or that he will be protected (incentive).

8-37. If there is no justified fear, the HUMINT collector can make use of non-specific fears. "You know what can happen to you here?" A fear-up approach is normally presented in a level, unemotional tone of voice. For example, "We have heard many allegations of atrocities committed in your area and anyone that was involved will be severely punished" (non-specific fear). "If you cooperate with me and answer all of my questions truthfully, I can make sure you are not falsely accused" (incentive). The source should demonstrate some indication of fear, whether verbal or non-verbal, prior to using this approach. If a fear is pre-existing, the approach will work and is legal. If there is no indication of fear, another approach should be considered.

8-38. It is often very effective to use the detainee's own imagination against him. The detainee can often visualize exactly what he is afraid of better than the HUMINT collector can express it.

8-39. The "fear-up" approach is frequently used in conjunction with the emotional love or hate approaches. For example, the HUMINT collector has already established that a detainee source has a strong love of family but is now separated from them. He may state, "I wonder how your family is getting along without you?" (fear of the unknown). He then promises to allow the detainee more than the minimum two letters a month required by the GPW.

Emotional Fear-Down Approach

8-40. (Interrogation and Other MSO) The emotion of fear may dominate the source to the point where he is unable to respond rationally to questioning, especially in interrogation sources. However, the fear-down approach may be used in any MSO where the source's state of mind indicates that it would be an appropriate approach to use. In the fear-down approach the HUMINT collector mitigates existing fear in exchange for cooperation on the part of the source. This is not normally a formal or even voiced agreement. Instead, the HUMINT collector through verbal and physical actions calms the source. Psychologically, the source then views the HUMINT collector as the protector or the one who is providing the calm and wishes to help the HUMINT collector in gratitude and in order to maintain the HUMINT collector as the protector. When used with a soothing, calm tone of voice and appropriate body language, a fear-down approach often creates rapport and nothing else may be needed to get the source to cooperate. At times, however, the

HUMINT collector must describe concrete actions that he will take in order to remove the source's fear.

8-41. Frequently the object of the fear is too traumatic for the source to face directly. While calming the source, the HUMINT collector may initially ask nonpertinent questions and avoid the subject that has caused the source's fear. This develops rapport and establishes communication. The HUMINT collector must remember that his goal is collecting information, not concern with the psychological well being of the source. He will be concerned with the latter only insofar as it helps him obtain the former. This approach technique may backfire if allowed to go too far. After convincing the source he has nothing to fear, the source may cease to be afraid and may feel secure enough to resist the HUMINT collector's pertinent question.

Emotional-Pride and Ego-Up Approach

8-42. (Interrogation and Other MSO) The emotional-pride and ego-up approach may be used in any MSO. It exploits a source's low self-esteem. Many HUMINT sources including EPWs and other detainees, retained persons, civilian internees, or refugees may suffer from low self-esteem and feelings of helplessness due to their immediate circumstances. Others, such as individuals or members of social or ethnic groups that have been discriminated against or low-ranking members of organizations (including the military), may also show low self-worth. In this technique, the source is flattered into providing certain information in order to gain credit and build his ego. The HUMINT collector must take care to use a flattering somewhat-in-awe tone of voice, and speak highly of the source throughout this approach while remaining believable. This should produce positive feelings on the source's part as he receives desired recognition. The source will eventually reveal pertinent information to solicit more favorable comments from the HUMINT collector.

8-43. This technique can also be employed in another manner—by flattering the source into admitting certain information in order to gain credit. For example, while interrogating a suspected saboteur, the HUMINT collector states: "This was a smooth operation. I have seen many previous attempts fail. I bet you planned this. Who else but a clever person like you would have planned it? When did you first decide to do the job?"

8-44. A variation of this approach can also be used on individuals with strong egos. It is based on the premise that everyone likes to talk about what they do best. The HUMINT collector shows interest in and asks the source to explain an aspect of his job. The questioning begins with nonpertinent aspects of the source's job. The HUMINT collector displays interest and asks increasingly technical and pertinent questions. For example, if the source is an EPW who was a pilot, the HUMINT collector might begin by asking him what it is like to fly. As the source talks about this, the collector demonstrates interest and gradually uses questions to lead the conversation to capabilities of specific aircraft, specific missions that the pilot has flown, tactics, or whatever topic is a priority for collection.

Emotional-Pride and Ego-Down Approach

8-45. (Interrogation) The emotional-pride and ego-down approach is based on attacking the source's ego or self-image. The source, in defending his ego, reveals information to justify or rationalize his actions. This information may be valuable in answering collection requirements or may give the HUMINT collector insight into the viability of other approaches. This approach is effective with sources who have displayed weakness or feelings of inferiority. A real or imaginary deficiency voiced about the source, loyalty to his organization, or any other feature can provide a basis for this technique.

8-46. The HUMINT collector accuses the source of weakness or implies he is unable to do a certain thing. This type of source is also prone to excuses and rationalizations, often shifting the blame to others. An example of this technique is opening the collection effort with the question, "Why did you surrender so easily when you could have escaped by crossing the nearby ford in the river?" The source is likely to provide a basis for further questions or to reveal significant information if he attempts to explain his surrender in order to vindicate himself. He may give an answer such as, "No one could cross the ford because it is mined."

8-47. The objective is for the HUMINT collector to use the source's sense of pride by attacking his loyalty, intelligence, abilities, leadership qualities, slovenly appearance, or any other perceived weakness. This will usually goad the source into becoming defensive, and he will try to convince the HUMINT collector he is wrong. In his attempt to redeem his pride and explain his actions, the source may provide pertinent information. Possible targets for the emotional-pride and ego-down approach are the source's—

- Loyalty.
- Technical competence.
- Leadership abilities.
- Soldierly qualities.
- Appearance.

8-48. There is a risk associated with this approach. If the emotional-pride and ego-down approach fails, it is difficult for the HUMINT collector to recover and move to another approach without losing his credibility. Also, there is potential for application of the pride and ego approach to cross the line into humiliating and degrading treatment of the detainee. Supervisors should consider the experience level of their subordinates and determine specifically how the interrogator intends to apply the approach technique before approving the interrogation plan.

Emotional-Futility

8-49. (Interrogation and Other MSO) The emotional-futility approach is generally used in an interrogation setting, but may also be used for other MSO, if indicated by the source's state of mind. In the emotional-futility approach, the HUMINT collector convinces the source that resistance to questioning is futile. This engenders a feeling of hopelessness and helplessness on the part of the source. Again as with the other emotional approaches, the HUMINT collector gives the source a "way out" of the helpless situation. For example "it is hopeless for your forces to continue fighting because they can no longer get supplies, but you can help end the war and their suffering." When employing this technique, the HUMINT collector must have factual information. The HUMINT collector presents these facts in a persuasive, logical manner. He should be aware of and able to exploit the source's psychological and moral weaknesses, as well as weaknesses inherent in his society.

8-50. The futility approach is effective when the HUMINT collector can play on doubts that already exist in the source's mind. Factual or seemingly factual information must be presented in a persuasive, logical manner, and in a matter-of-fact tone of voice. Making the situation appear hopeless allows the source to rationalize his actions, especially if that action is cooperating with the HUMINT collector. When employing this technique, the HUMINT collector must not only have factual information but also be aware of and exploit the source's psychological, moral, and sociological weaknesses. Another way of using the futility approach is to blow things out of proportion. If the source's unit was low on, or had exhausted, all food supplies, he can be easily led to believe all of his forces had run out of food. If the source is verging on cooperating, it may aid the collection effort if he is told all the other sources have cooperated.

8-51. The futility approach must be orchestrated with other approach techniques (for example, love of comrades). A source who may want to help save his comrades' lives may be convinced the battlefield situation is hopeless and they will die without his assistance. The futility approach is used to paint a bleak picture for the prisoner, but it is not normally effective in and of itself in gaining the source's cooperation.

Other Approaches

8-52. There are numerous other approaches but most require considerable time and resources. Most are more appropriate for use with sources who are detainees, but some, such as change of scenery, may have application for elicitation or MSO.

8-53. **We Know All**. (Interrogation) In the "we know all" approach technique, the HUMINT collector subtly convinces the source that his questioning of the source is perfunctory because any information that the source has is already known. This approach may be employed in conjunction with the "file and dossier" technique or by itself. If used alone, the HUMINT collector must first become thoroughly familiar with available data concerning the source and the current situation. To begin the collection effort, the HUMINT collector asks questions based on this known data.

8-54. When the source hesitates, refuses to answer, or provides an incorrect or incomplete reply, the HUMINT collector provides the detailed answer himself. The HUMINT collector may even complete a source's answer, as if he is bored and just "going through the motions." When the source begins to give accurate and complete information, the HUMINT collector interjects pertinent questions. Questions to which answers are already known are also asked periodically to test the source's truthfulness and to maintain the deception that the information is already known. There are some inherent problems with the use of the "we know all" approach. The HUMINT collector is required to prepare everything in detail, which is time consuming. He must commit much of the information to memory, as working from notes may show the limits of the information actually known. It is also only usable when sufficient prior information exists to convince the source that "we know all."

8-55. **File and Dossier.** (Interrogation) The file and dossier approach is a variation of the "we know all" approach. The HUMINT collector prepares a dossier containing all available information concerning the source or his organization. The information is carefully arranged within a file to give the illusion that it contains more data than actually there. The file may be padded with extra paper if necessary. Index tabs with titles such as education, employment, criminal record, military service, and others are particularly effective. It is also effective if the HUMINT collector is reviewing the dossier when the source enters the room and the source is able to read his name on the dossier and sees the numerous topics and supposed extent of the files.

8-56. The HUMINT collector proceeds as in the "we know all" approach. He refers to the particular labeled segment of the dossier before, during, or after asking a question. In the early stages of questioning, the HUMINT collector asks questions to which he has the answer. He may answer along with the source, complete the information for the source, or even show the source where the information is entered in the dossier. He never lets the source physically handle the dossier. As the source becomes convinced that all the information that he knows is contained within the dossier, the HUMINT collector proceeds to topics on which he has no or little information. In doing so, he still refers to the appropriate section of the dossier and may even nod his head knowingly or tell the source that the information the source is providing still matches what is in the dossier.

8-57. This technique has several limitations and drawbacks. The preparation time in developing the dossier is extensive. The success of this technique is largely dependent on the naiveté of the source, volume of data on the subject, and skill of the HUMINT collector in convincing the source that the dossier is more complete than it actually is. There is also the risk that a less naïve source will refuse to cooperate, claiming that, if the collector already knows everything, there is no need for him to talk. Also with this technique, the HUMINT collector is limited in the method he may use to record new information. If the HUMINT collector writes down information, it destroys the illusion that all the information has already been obtained. The HUMINT collector is normally limited to using electronic recording devices or his memory. The HUMINT collector can also arrange ahead of time for another interrogator or analyst to take notes for him, undetected by the source. This could be especially effective in a situation where a separate monitoring area (for oversight) is used by the analyst.

8-58. **Establish Your Identity.** (Interrogation) In using this approach, the HUMINT collector insists the detained source has been correctly identified as an infamous individual wanted by higher authorities on serious charges, and he is not the person he purports to be. In an effort to clear himself of this allegation, the source makes a genuine and detailed effort to establish or substantiate his true identity. In so doing, he may provide the HUMINT collector with information and leads for further development. The HUMINT collector should initially refuse to believe the source and insist he is the individual wanted by the ambiguous higher authorities. This will force the source to give even more detailed information in order to convince the HUMINT collector he is who he says he is.

8-59. **Repetition**. (Interrogation) The repetition approach is used to induce cooperation from a hostile source. In one variation of this approach, the HUMINT collector listens carefully to a source's answer to a question, and then repeats the question and answer several times. He does this with each succeeding question until the source becomes so thoroughly bored with the procedure, he answers questions fully and candidly to satisfy the HUMINT collector and gain relief from the monotony of this method. The repetition technique must be judiciously used, as it will generally be ineffective when employed against introverted sources or those having great self-control. It may also provide an opportunity for a source to regain his composure and delay the collection effort. In this approach, the use of more than one HUMINT collector or a tape recorder has proven effective.

8-60. **Rapid Fire**. (Interrogation) The rapid-fire approach is based upon the principles that—

- Everyone likes to be heard when he speaks.
- It is confusing to be interrupted in mid-sentence with an unrelated question.

8-61. This approach may be used by one, two, or more HUMINT collectors to question the source. In employing this technique, the HUMINT collectors ask a series of questions in such a manner that the source does not have time to answer a question completely before the next one is asked. This confuses the source, and he will tend to contradict himself as he has little time to formulate his answers. The HUMINT collectors then confront the source with the inconsistencies causing further contradictions. In many instances, the source will begin to talk freely in an attempt to explain himself and deny the HUMINT collector's claims of inconsistencies. In this attempt, the source is likely to reveal more than he intends, thus creating additional leads for further exploitation. This approach may be orchestrated with the emotional-pride and ego-down or fear-up approaches. Besides extensive preparation, this approach requires experienced and competent HUMINT collectors, with comprehensive case knowledge and fluency in the source's language.

8-62. **Silent**. (Interrogation) The silent approach may be successful when used against either a nervous or confident source. When employing this technique, the HUMINT collector says nothing to the source, but looks him squarely in the eye, preferably with a slight smile on his face. It is important not to look away from the source but force him to break eye contact first. The source may become nervous, begin to shift in his chair, cross and re-cross his legs, and look away. He may ask questions, but the HUMINT collector should not answer until he is ready to break the silence. The source may blurt out questions such as, "Come on now, what do you want with me?" When the HUMINT collector is ready to break silence, he may do so with questions such as, "You planned this operation for a long time, didn't you? Was it your idea?" The HUMINT collector must be patient when using this technique. It may appear the technique is not succeeding, but usually will when given a reasonable chance.

8-63. **Change of Scenery.** (Interrogation and Other MSO) The change-of-scenery approach may be used in any type of MSO to remove the source from an intimidating atmosphere such as an "interrogation" room type of setting and to place him in a setting where he feels more comfortable speaking. Bringing a source into a formal setting to conduct an interrogation or debriefing has psychological implications. On the positive side, it places the HUMINT collector in a superior position since he is operating on his "home turf" and has set the conditions for the meeting. It allows the HUMINT collector control over the immediate environment including the positioning of the participants, to establish the desired atmosphere for the approach.

8-64. However, there are potential negative factors in the conduct of questioning in an "Interrogation Room" environment. The source may be intimidated and more guarded; he may consider the formal setting in terms of an adversarial relationship; and he may limit his answers as a mode of self-protection. In some circumstances, the HUMINT collector may be able to invite the source to a different setting for coffee and pleasant conversation. When removed from the formal environment, the source may experience a feeling of leaving the interrogation behind. The perceived reduced pressure may lower his guard and allow him to attach less significance to conversation that occurs outside the formal setting, even though pertinent information is still being discussed. During the conversation in this more relaxed environment, the HUMINT collector steers the conversation to the topic of interest. Through this somewhat indirect method, he attempts to elicit the desired information. The source may never realize he is still being questioned.

8-65. **Mutt and Jeff. (Interrogation)** The goal of this technique is to make the source identify with one of the interrogators and thereby establish rapport and cooperation. This technique involves a psychological ploy that takes advantage of the natural uncertainty and guilt that a source has as a result of being detained and questioned. Use of this technique requires two experienced HUMINT collectors who are convincing actors. The two HUMINT collectors will display opposing personalities and attitudes toward the source. For example, the first HUMINT collector is very formal and displays an unsympathetic attitude toward the source. He may, for instance, be very strict and order the source to follow all military courtesies during questioning. Although he conveys an unfeeling attitude, the HUMINT collector is careful not to threaten or coerce the source. Conveying a threat of violence is a violation of the UCMJ.

8-66. At the point when the interrogator senses the source is vulnerable, the second HUMINT collector appears (having received his cue by a signal, hidden from the source, or by listening and observing out of view of the source), and scolds the first HUMINT collector for his uncaring behavior and orders him from the room. The second HUMINT collector then apologizes to soothe the source, perhaps offering him a beverage and a cigarette. He explains that the actions of the first HUMINT collector were largely the result of an inferior intellect and lack of sensitivity. The inference is that the second HUMINT collector and the source share a high degree of intelligence and sensitivity.

8-67. The source is normally inclined to have a feeling of gratitude towards the second HUMINT collector, who continues to show sympathy in an effort to increase rapport and control for the questioning that will follow. If the source's cooperation begins to fade, the second HUMINT collector can hint that he is a busy person of high rank, and therefore cannot afford to waste time on an uncooperative source. He can broadly imply that the first HUMINT collector might return to continue the questioning. The Mutt and Jeff approach may be effective when orchestrated with Pride and Ego Up and Down, Fear Up and Down, Futility, or Emotional Love or Hate.

8-68. **Oversight Considerations:** Planned use of the Mutt and Jeff approach must be approved by the first O-6 in the interrogator's chain of command. The HUMINT collector must include as a part of the interrogation plan—

- No violence, threats, or impermissible or unlawful physical contact.
- No threatening the removal of protections afforded by law.
- Regular monitoring of the interrogation shall be performed by interrogation personnel.

8-69. **False Flag.** (Interrogation) The goal of this technique is to convince the detainee that individuals from a country other than the United States are interrogating him, and trick the detainee into cooperating with US forces. For example, using an interrogator who speaks with a particular accent, making the detainee believe that he is actually talking to representatives from a different country, such as a country that is friendly to the detainee's country or organization. The False Flag approach may be effectively orchestrated with the Fear Down approach and the Pride and Ego Up.

8-70. **Oversight Considerations:** The interrogation chain of command must coordinate an interrogation plan that uses the False Flag approach with the legal representative and the 2X, and receive approval from the first O-6 in the interrogator's chain of command for each specific use of the False Flag approach.

- The use of the False Flag approach must complement the overall interrogation strategy and other approach techniques listed in the interrogation plan.
- When a HUMINT collector intends to pose as a national of a third-party country, that country must be identified in the interrogation plan.
- No implied or explicit threats that non-cooperation will result in harsh interrogation by non-US entities.
- HUMINT collectors will not pose or portray themselves as any person prohibited by this manual, paragraphs 8-10 and 8-11 (for example, an ICRC representative).

8-71. **Separation.** See Appendix M, Restricted Interrogation Technique – Separation.

Selecting an Approach

8-72. There often is insufficient information available to determine an approach other than the direct approach. In this case where the source answers questions but will not discuss pertinent issues, the HUMINT collector may ask direct but nonpertinent questions to obtain sufficient information to develop an approach strategy. This technique is also useful in debriefing to establish rapport. Nonpertinent questions may include—

- Asking about immediate past events. This includes asking an EPW about the circumstances of his capture or asking a refugee about the circumstances concerning his arrival at the refugee point or checkpoint. By doing this, the HUMINT collector can gain insight into the source's current state of mind and, more importantly, he can ascertain his possible approach techniques.
- Asking background questions. This includes asking about the source's family, work, friends, likes, and dislikes. These types of questions can develop rapport and provide clues as to the source's areas of knowledge or reveal possibilities for incentives or emotional approaches.
- Considering what are culturally and socially acceptable topics of discussion. For example, asking an Arab male about his wife could be considered extremely rude, whereas not asking an American the same question might be seen as insensitive.

Making Smooth Transitions

8-73. With the exception of the direct approach, no other approach is effective by itself. HUMINT collectors use different approach techniques or combine them into a cohesive, logical technique. Smooth transitions, sincerity, logic, and conviction are needed to make a strategy work. HUMINT collectors must carefully assess the source's verbal or nonverbal clues to determine when a change in approach strategy is required. The HUMINT collector must guide the conversation smoothly and logically, especially when moving from one approach technique to another. Using transitional phrases can make logical and smooth tie-ins to another approach. By using nonpertinent questions, the HUMINT collector can move the conversation in the desired direction and, as previously stated, sometimes can obtain leads and hints about the source's stresses or weaknesses or other approach strategies that may be more successful.

Recognizing Source Cooperation

8-74. Each source has a point where he will begin to cooperate and answer questions. Some sources will begin answering questions completely and truthfully with no preparation; others might require hours or even days of work. The amount of time that a HUMINT collector spends on an approach depends on a variety of factors. These include—

- The quality and criticality of the information believed to be possessed by the source.
- The presence or absence of other sources that probably possess that information.
- The number of HUMINT collectors and sources available.
- The LTIOV that the HUMINT collector is attempting to obtain.

8-75. The HUMINT collector needs to identify the signs that the source is approaching or has reached the point of cooperation. For example, if during the approach the source leans forward with his facial expression indicating an interest in the proposal or is more hesitant in his argument, he is probably nearing the point where he will cooperate. The HUMINT collector must also be aware of the fact that a source can begin to cooperate in certain areas while continuing to resist strongly in other areas. The HUMINT collector should recognize the reason for refusal, overcome the objection, and stress the benefit of cooperating (reinforce the approach). Once the HUMINT collector determines the source is cooperating, he should interject pertinent questions. If the source does not answer the question, the HUMINT collector should continue with his approach or switch to an alternate approach technique and continue to work until he again believes the source will cooperate. If the source answers the pertinent question, the HUMINT collector continues asking relevant questions until the questioning session is completed.

8-76. If a cooperative source balks at answering a specific line of questions, the HUMINT collector must assess the reason for the refusal. The HUMINT collector may have arrived at a topic that the source finds particularly sensitive. Other reasons that might cause a source to stop answering questions are fatigue or unfamiliarity with the new topic. If this topic is critical, the HUMINT collector may have to reinforce the previously successful approach or may have to use a different approach.

APPROACH STRATEGIES FOR INTERROGATION

8-77. Interrogation does not mean a hostile relationship between the HUMINT collector and the source. In fact, most interrogation sources (90 percent or more) cooperate in response to the direct approach. Unfortunately, those sources who have the placement and access to make them high priority sources are also the ones with the highest degree of security awareness. A source who uses counter-interrogation techniques such as delaying, trying to control the conversation, or interrogating the HUMINT collector himself may—

- Be an intelligence trained soldier.
- Be survival, evasion, resistance, and escape (SERE) trained.
- Be a terrorist.
- Have been a detainee or previously incarcerated.

8-78. In stability and reconstruction operations and civil support operations, detainees are often politically motivated and resistant to most approaches.

8-79. EPWs are normally vulnerable to basic incentive and emotional approach techniques. Most EPWs are traumatized to various degrees by the events preceding or surrounding their capture. They tend to be disoriented and exhibit high degrees of fear and anxiety. This vulnerable state fades over time, and it is vital for HUMINT collectors to interrogate EPWs as soon as and as close to the point of capture as possible. The earlier that an EPW is questioned the more likely he is to cooperate. And the earlier that he begins to cooperate, the more likely he is to continue to cooperate. It is also vital that the HUMINT collector be the first person that the EPW has a chance to talk to. This means that proper silencing and segregation of the sources by whoever is transporting them is an important part of a successful approach.

8-80. The vulnerability of civilian detainees to approach techniques available to the HUMINT collector may be dependent on the exact nature of the conflict. US HUMINT collectors are obligated to treat all detainees in accordance with applicable law and policy. Applicable law and policy include US law; the law of war; relevant international law; relevant directives including DOD Directive 3115.09, "DOD Intelligence Interrogations, Detainee Debriefings, and Tactical Questioning"; DOD Directive 2310.1E, "The Department of Defense Detainee Program"; DOD instructions; and military execute orders including FRAGOs. Detainees and, in particular, EPWs are guaranteed certain rights and privileges. The HUMINT collector may not take any action to remove, state that he will remove, or imply that he will remove any guaranteed right if a detainee fails to cooperate. Under the GPW, EPWs cannot be denied their rights or their privileges accorded them by rank as guaranteed by the GPW. Privileges afforded to them, however, which are not guaranteed by the Geneva Conventions or other applicable law or agreements, may be withheld. (See Appendix A, Section I.) Consult your SJA for questions concerning rights and privileges.

8-81. The HUMINT collector is frequently under a great deal of pressure to "produce results." This situation, coupled with the facts that the HUMINT collector is dealing with threat personnel who may have been attempting to kill US personnel just minutes before questioning and the fact that the source is in a vulnerable state, leads to a tendency to use fear-up techniques. This may, in some circumstances, be the proper approach; however, the HUMINT collector must ensure that in doing so he neither loses control of his own emotions nor uses physical or mental coercion.

APPROACH STRATEGIES FOR DEBRIEFING

8-82. Sources who are debriefed vary even more widely than those who are interrogated. Since debriefing is the systematic questioning of individuals not in the custody of the questioning forces, the HUMINT collector needs to engender an atmosphere of cooperation and mutual benefit. Some sources for debriefing include members of the friendly forces and local personnel. HUMINT collectors often believe that approach techniques are not required for friendly forces and that friendly forces should view debriefing as part of their duties and in their own best interest. However, this is not necessarily the case.

8-83. Many people see debriefing as an interruption in their normal duties and a waste of their time. HUMINT collectors must be sure to stay focused on the purpose and goals of the debriefing. They should be businesslike and must maintain the proper relationship with the source based on his rank and position. The HUMINT collector should allow senior sources more latitude to interpose their opinions and evaluations. A change of scene often facilitates the debriefing of a high-level source since it removes him from his normal distractions, such as the telephone, and allows him to concentrate on the topics being discussed.

8-84. Refugees and DPs are subject to many of the same anxieties and trauma that are experienced by EPWs or other detainees, with the added benefit to the HUMINT collector that they normally have an obvious vested interest in cooperating. Basic incentives usually are sufficient to induce their willing cooperation. The emotional support that can be provided by the HUMINT collector by simply listening and commiserating with their hardship is often sufficient to gain cooperation. The emotional approaches such as love of family and hate toward those who made them refugees are strong motivators toward cooperation.

8-85. The approach techniques used in the questioning of local civilians are probably the most difficult. The approach techniques chosen must take into consideration the attitude of the local population toward the US and its presence and cultural considerations. The local population must see their cooperation as self-beneficial.

APPROACH STRATEGIES FOR ELICITATION

8-86. Elicitation is a sophisticated technique used when conventional collection techniques cannot be used effectively. Of all the collection methods, this one is the least obvious. However, it is important to note that elicitation is a planned, systematic process that requires careful preparation. It is always applied with a specific purpose in mind. This objective is the key factor in determining the subject (which source to question), the elicitor, and the setting. The subject will be selected based on access to or knowledge of the desired information.

8-87. Before approaching the subject, it is necessary to review all available intelligence files and records, personality dossiers, and knowledge possessed by others who have previously dealt with the subject. This will help determine the subject's background, motivation, emotions, and psychological nature. It also may require unobtrusive observation of the subject to establish such things as patterns of activity and likes and dislikes. The setting can be any number of social or official areas. It is important to note that the source should be approached in his natural surroundings, as this will diminish suspicion.

8-88. The key to elicitation is the establishment of a rapport between the elicitor and the source, normally based on shared interests. In the initial stages of an elicitation, the collector confines his conversations to innocuous subjects such as sports and social commentary. Dependent on the value of the source, the collection environment, and the security consciousness of the source the initial stage could last from a few minutes to numerous seemingly accidental meetings over a period of weeks or months. The HUMINT collector will gradually shift the conversation to topics of collection interest but will be prepared to return to more unthreatening topics based on negative reactions on the part of the subject. Once a topic of interest has been introduced, the HUMINT collector keeps the conversation going by asking for clarification (for example, "I agree, however, what did you mean by....?") or expressing a hypothetical situation.

8-89. There are two basic elicitation approaches: mild flattery and provocation.

- Mild Flattery: Most people like talking about their interests and like talking to those who are knowledgeable and interested in the same topics. People also like to speak to someone who values their opinion on shared interests. The HUMINT collector takes advantage of this. The HUMINT collector leads the conversation into areas that he wishes to collect but does it in such a way that it appears to the source that the source is leading the conversation. Above all in elicitation, the HUMINT collector plays the role of the rapt, attentive, and inquisitive listener.
- Provocation: This is a more dangerous approach and, if used too early in an operation, can alienate the source. Once the HUMINT collector has established shared interests with the source, he can selectively challenge some of the source's statements, encouraging the source to provide more information in support of his view. The HUMINT collector can also insert bits of actual information into the conversation to cause the source to confirm and expound on the topic. Care must be taken so as not to give away more information than is gained.

TERMINATION PHASE

8-90. When it is necessary or prudent, the HUMINT collector will terminate the questioning of a particular source. Whatever the reason for terminating, the HUMINT collector must remember there is a possibility that someone may want to question the source at a later date. There are many reasons why a HUMINT collector may want or need to terminate questioning:

- The source remains uncooperative during the approach phase.
- The collection objective cannot be met in one questioning session.
- The HUMINT collector fails to maintain rapport and loses control of the questioning.
- The collection objectives have been satisfied.
- The HUMINT collector or the source becomes physically or mentally unable to continue.
- Information possessed by the source is of such value that his immediate evacuation to the next echelon is required.
- The HUMINT collector's presence is required elsewhere.

8-91. There are many ways to conduct a termination, but the following points must be conveyed to the source:

- The HUMINT collector should sincerely and convincingly reinforce successful approaches. All promised incentives should be rendered.
- The source must be told the information he gave will be checked for truthfulness and accuracy. His reaction to this statement should be closely monitored. The exact form of this statement will be dependent on the situation. It should not be done in a manner to alienate a cooperative source.
- The source must be told that the same or another individual may speak to him again. This sets the stage for future contacts.
- Any identification must be returned to the source. If the HUMINT collector has other documents or belongings of the detainee (such as letters or photographs), he will either return them to the detainee, if appropriate, or will turn them over to the MP guard. Depending on the circumstances and the legal status of the detainee, the MPs will retain the detainee's property and return the property to him at the end of his internment.
- In a debriefing, the HUMINT collector will normally ask the source not to discuss the subject of the questioning for his own protection. In interrogation operations, the HUMINT collector normally coordinates with the holding area guards to have the detainees who have been interrogated kept separate from sources who have not yet been interrogated if the situation allows.

Chapter 9

Questioning

9-1. Questioning is one of the five phases of HUMINT collection. Developing and using good questioning techniques enable the HUMINT collector to obtain accurate and pertinent information and to extract the maximum amount of information in the minimum amount of time. The HUMINT collector must know when to use different types of questions.

GENERAL QUESTIONING PRINCIPLES

9-2. Questions should be presented in a logical sequence to avoid neglecting significant topics. The HUMINT collector begins the questioning phase with the first topic in the sequence he tentatively established as part of his questioning plan. He obtains all of the source's pertinent knowledge in this topical area before moving on to the next topic in his sequence. The only exception is exploiting a hot lead, which is discussed in paragraph 9-21.

9-3. The HUMINT collector must at all times remember that his mission is the rapid collection and dissemination of accurate information. He must not allow himself to be sidetracked into nonpertinent discussions or debates nor should he express distaste or value judgments on the information being supplied unless that is a planned part of his approach technique. The HUMINT collector uses vocabulary that is clear, unambiguous, and understandable by the source. The source may not be on the same intellectual level or have the same degree of education as the HUMINT collector, so the HUMINT collector must adapt his questioning to the level of the source. The source may also have specific technical knowledge, more education and/or a higher intellectual level than the HUMINT collector. In this case, the HUMINT collector normally relies on prepared questions or technical support for his questioning. Without good systematic questioning techniques, even the most cooperative source may provide only minimal usable information.

DIRECT QUESTIONS

9-4. Direct questions are basic questions normally beginning with an interrogative (who, what, where, when, how, or why) and requiring a narrative answer. They are brief, precise, and simply worded to avoid confusion. The HUMINT collector must consider the probable response of the source to a particular question or line of questioning and should not, if at all possible, ask direct questions likely to evoke a refusal to answer or to antagonize the source.

TYPES OF DIRECT QUESTIONS

9-5. The HUMINT collector must be able to use the following types of direct questions:

- Initial, topical.
- Follow-up.
- Nonpertinent.
- Repeat.
- Control.
- Prepared.

Initial Questions

9-6. The HUMINT collector begins his questioning with the first topic in his collection plan and asks all the basic questions necessary to cover the topic. The answers to the basic questions will determine the requirements for follow-up questioning. The initial questions are directed toward obtaining the basic information on the topic. In other words, they are the "who, what, where, when, how, and why" of each topic.

Follow-up Questions

9-7. Follow-up questions are used to expand on and complete the information obtained from the initial questions. Often even if the initial question is a well-constructed direct question, it will elicit only a partial answer. For example, when asked, "Who is going to attack?" The source might say, "My unit." Follow-up questions are used to determine precisely what the source means by "my unit" and what other units may also attack. The answer to follow-up questions may lead to more follow-ups until the source's knowledge on a given topic is exhausted. At a minimum, upon receiving a positive answer to an initial question, the HUMINT collector needs to ask "Who (what, where, when, why, how) else?" For example, if the HUMINT collector asks the source, "Who, in the local government is collaborating with the insurgents?" and is told a name in response, he will ask follow-up questions to determine all the required information about this individual and then will ask, "Who else, in the local government is collaborating with the insurgents?" This will continue until the source's knowledge in this area is exhausted.

Nonpertinent Questions

9-8. Nonpertinent questions are questions that do not pertain to the collection objectives. They are used to conceal the collection objectives or to strengthen rapport with the source. They are essential when the collector is using the elicitation technique. Nonpertinent questions may be used to gain time for the HUMINT collector to formulate pertinent questions and may also be used to break the source's concentration, particularly, if the HUMINT collector suspects the source is lying. It is hard for a source to be a convincing liar if his concentration is frequently interrupted.

Repeat Questions

9-9. Repeat questions ask the source for the same information obtained in response to earlier questions. They are a method to confirm accuracy of important details such as place names, dates, and component parts of technical equipment and to test truthfulness. Repeat questions should not be exact repetitions of an earlier question. The HUMINT collector must rephrase or otherwise disguise the previous question. The repeat question also needs to be separated in time from the original question so that the source cannot easily remember what he said. Repeat questions may also be used to develop a topic the source had refused to talk about earlier.

Control Questions

9-10. Control questions are developed from recently confirmed information from other sources that is not likely to have changed. They are used to check the truthfulness of the source's responses and should be mixed in with other questions throughout the questioning. If a source fails to answer a control question as expected, it may be an indicator that he is lying. However, there are other possible explanations. The source—

- Could have misunderstood the question.
- Could be making up information in order to please the questioner and/or receive a promised incentive.
- Could have answered the question truthfully to the best of his ability, but his information could be wrong or outdated.
- May be correct and the information that the control question was based on is no longer true.

9-11. It is the responsibility of the HUMINT collector to determine, through follow-up questions, which of the possibilities is the case. The HUMINT collector should also consult with the HAT for assistance in verifying the source reporting through all-source analysis.

Prepared Questions

9-12. Prepared questions are questions developed by the HUMINT collector, normally in writing, prior to the questioning. Prepared questions are used primarily when dealing with information of a technical nature or specific topic, which requires the HUMINT collector to formulate precise and detailed questions beforehand. The HUMINT collector may have to research analytical or technical material or contact SMEs to assist him in preparing questions. HUMINT collectors must not allow the use of prepared questions or any limitations to their education or training to restrict the scope and flexibility of their questioning. In many instances, the HUMINT collector should have an analyst or technical expert "sit in" on the questioning as well.

9-13. The HUMINT collector must be able to use the different types of questions effectively. Active listening and maximum eye-to-eye contact with the source will provide excellent indicators for when to use follow-up, repeat, control, and nonpertinent questions. The HUMINT collector must use direct and follow-up questions to fully exploit subjects pertinent to his interrogation objectives. He should periodically include control, repeat, and nonpertinent questions in order to check the truthfulness and consistency of the source's responses and to strengthen rapport.

TYPES OF QUESTIONS TO AVOID

9-14. When using the questioning methodologies of interrogation, HUMINT collectors should avoid using negative, compound, or vague questions. Leading questions are usually to be avoided, but some special questioning techniques, such as use of a polygraph, require the use of leading questions.

Leading Questions

9-15. Leading questions are questions that are constructed so as to require a yes or no answer rather than a narrative response. They generally begin with a form of the verb "to be" (such as "is," "was," "were," "will," "are"). For example, "Is the mayor working with the insurgents?" Leading questions should generally be avoided for the following reasons:

- They make it easier for the source to lie since the source only provides minimal information.
- It takes longer to acquire information.
- A source, particularly one that is frightened or trying to get an incentive, will tend to answer in the way that he thinks the HUMINT collector wants him to answer.

9-16. Although normally avoided during questioning, an experienced HUMINT collector may use leading questions when the technical nature of the subject matter or the specific information needed leaves no alternatives. Leading questions can be used to—

- Verify specific facts.
- Pinpoint map locations.
- Confirm information obtained during map tracking.
- Transition from one topic area to another.

Negative Questions

9-17. Negative questions are questions that contain a negative word in the question itself such as, "Didn't you go to the pick-up point?" If the source says "yes," the HUMINT collector is faced with the question of whether he means "yes, I went to the pick-up point" or "yes, I didn't go to the pick-up point." When the source answers, the HUMINT collector cannot be sure what the answer means; therefore, he must ask additional questions. This can be particularly confusing when working with an interpreter. Other cultures may interpret a negative question in a way other than what the HUMINT collector meant. Negative questions should never be used during questioning unless they are being used deliberately during the approach to make the source appear to contradict himself. In other instances, the insertion of negative words within the question makes them impossibly open-ended. For example, "Who didn't attend the meeting?"

Compound Questions

9-18. Compound questions consist of two questions asked at the same time; for example, "Before you were captured today, were you traveling north or south?" Or "Where were you going after work and who were you to meet there?" They are easily misunderstood and may confuse the source or force him to give an ambiguous answer. Compound questions allow the source to evade a part of the question or to give an incomplete answer.

Vague Questions

9-19. Vague questions do not have enough information for the source to understand exactly what the HUMINT collector is asking. They may be incomplete, general, or otherwise nonspecific and create doubt in the source's mind. Vague questions confuse the source, waste time, and are easily evaded. They result in answers that may confuse or mislead the HUMINT collector and require further follow-up questions.

ELICITATION

9-20. Elicitation is the gaining of information through direct interaction with a human source where the source is not aware of the specific purpose for the conversation. Elicitation is a sophisticated technique used when conventional questioning techniques cannot be used effectively. Of all the collection methods, this one is the least obvious. However, it is important to note that elicitation is a planned, systematic process that requires careful preparation. Elicitation is always applied with a specific objective in mind and normally directed toward a specific source.

LEADS

9-20. A lead is a statement made by a source spontaneously or in response to questioning that leads the questioner to believe that the source has information on a topic other than the one currently under discussion. Documents captured with or on the source may also be exploited as sources of leads. Leads are referred to as either "hot" or "cold."

HOT LEADS

9-21. A hot lead is a statement made by a source either spontaneously or in response to questioning that indicates he has information that could answer intelligence requirements on a topic other than the one currently under discussion. The lead could also be on a topic that although not listed as a requirement is, based on the HUMINT collector's experience, of critical importance. Information on WMD and information on US personnel being held by threat forces are normally considered hot leads even if not listed as requirements. The HUMINT collector will normally question the source immediately on a hot lead, unless he is already asking questions on another topic. In this case, he completes questioning and reports the information on the priority topic, as appropriate, and then immediately questions on the hot lead. As soon as the HUMINT collector is sure he has obtained and recorded all the details known to the source, he reports the hot lead information by the most expedient means available, normally in SALUTE report format. The HUMINT collector then resumes his questioning of the source at the point where the hot lead was obtained.

COLD LEADS

9-22. A cold lead is a statement made by a source either spontaneously or in response to questioning that indicates he has information on a topic of interest other than the one currently under discussion but that would not answer PIRs. The HUMINT collector makes note of the cold lead and exploits it after the planned questioning objectives have been satisfied or at the appropriate time during the questioning sequence.

DETECTING DECEIT

9-23. HUMINT information often has the capability to be more accurate and reliable than other disciplines. SIGINT information, for example, is not always able to return to the original source of the information to determine the reliability of the information, and interpretation of IMINT information may be uncertain. However, while HUMINT can be reviewed for reliability, determining the reliability of human sources is a continuous process accomplished by carefully assessing not only the sources of information but also assessing the information itself.

9-24. Detection of deception is not a simple process, and it normally takes years of experience before a HUMINT collector can readily identify deliberate deceit. Inconsistencies in the source's actions or words do not necessarily indicate a lie, just as consistency is not necessarily a guarantee of the truth. However, a pattern of inconsistencies or unexplainable inconsistencies normally indicate deceit.

TECHNIQUES FOR IDENTIFYING DECEIT

9-25. Techniques for identifying deceit include but are not limited to the following:

- Repeat and control questions (see paras 9-9 and 9-10).
- Internal inconsistencies. Frequently when a source is lying, the HUMINT collector will be able to identify inconsistencies in the timeline, the circumstances surrounding key events, or other areas within the questioning. For example, the source may spend a long time explaining something that took a short time to happen, or a short time telling of an event that took a relatively long time to happen. These internal inconsistencies often indicate deception.
- Body language does not match verbal message. An extreme example of this would be the source relating a harrowing experience while sitting back in a relaxed position. The HUMINT collector must be careful in using this clue since body language is culturally dependent. Failing to make eye contact in the US is considered a sign of deceit while in some Asian countries it is considered polite.
- Knowledge does not match duty position or access. Based on the source's job, duty position, or access the HUMINT collector should have

developed a basic idea of the type and degree of information that an individual source should know. When the source's answers show that he does not have the expected level of information (too much or too little or different information than expected), this may be an indicator of deceit. The HUMINT collector needs to determine the source of unexpected information.

- Information is self-serving. Reporting of information that is self-serving to an individual or his group should be suspect. For example, a member of one ethnic group reporting generic atrocities by an opposing ethnic group or a source reporting exactly the right information needed to receive a promised incentive should be suspect. That is not to say that the information is necessarily false, just that the HUMINT collector needs to be sure to verify the information.

- Lack of extraneous detail. Often false information will lack the detail of truthful information, especially when the lie is spontaneous. The HUMINT collector needs to ask follow-up questions to obtain the detail. When the source is unable to provide the details that they should know, it is an indicator of deceit. If the source does provide this additional information, it needs to be checked for internal inconsistencies and verified by repeat questions.

- Repeated answers with exact wording and details. Often if a source plans on lying about a topic, he will memorize what he is going to say. If the source always relates an incident using exactly the same wording or answers repeat questions identically (word for word) to the original question, it may be an indicator of deceit. In an extreme case, if the source is interrupted in the middle of a statement on a given topic, he will have to start at the beginning in order to "get his story straight."

- Source appearance does not match story. If the source's physical appearance does not match his story, it may be an indication of deceit. Examples of this include the source who says he is a farmer but lacks calluses on his hands or the supposed private who has a tailored uniform.

- Source's language usage does not match story. If the type of language, including sentence structure and vocabulary, does not match the source's story, this may be an indicator of deceit. Examples of this include a farmer using university level language or a civilian using military slang.

- Lack of technical vocabulary. Every occupation has its own jargon and technical vocabulary. If the source does not use the proper technical vocabulary to match his story, this may be an indictor of deceit. The HUMINT collector may require the support of an analyst or technical expert to identify this type of deceit.

- Physical cues. The source may display physical signs of nervousness such as sweating or nervous movement. These signs may be indicators of deceit. The fact that an individual is being questioned may in itself be cause for some individuals to display nervousness. The HUMINT collector must be able to distinguish between this type of activity and nervous activity related to a particular topic. Physical reaction to a particular topic may simply indicate a strong emotional response rather than lying, but it should key the HUMINT collector to look for other indicators of deceit.

- Failure to answer the question asked. When a source wishes to evade a topic, he will often provide an answer that is evasive and not in response to the question asked. For example, if the source is asked, "Are you a member of the insurgent organization?" and he replies, "I support the opposition party in the legislature," he has truthfully answered a question, but not the question that was asked. This is a subtle form of deceit since the source is seemingly cooperative but is in fact evading providing complete answers.

ACTIONS UPON IDENTIFYING INDICATORS OF DECEIT

9-26. The exact actions by the HUMINT collector when identifying possible deceit are dependent on the type of collection, the circumstances of the collection, the specific sign of deceit observed, the type of approach used, and cultural factors. The HUMINT collector may—

- Question the topic in more detail looking for additional indicators.
- Reinforce the approach.
- Move to another topic and revisit the original topic later with repeat questions. Ask control questions (confirmed by known data) and questions to which the source should know the answer to see if he answers honestly.
- Point out the inconsistency to the source and ask for an explanation.
- Seek assistance from a more experienced HUMINT collector, analyst, or a technical expert on the culture or the topic being questioned.
- Conduct continuous assessments of source (see FM 34-5 (S//NF)).
- Research established databases.
- Ask yourself if the information makes sense; if not, conduct more research.
- Consider how the information was obtained.
- Compare the information provided to the source's placement and access.
- Compare answers with other sources with similar placement and access. Be aware that this method is merely a rough tool to check veracity and should not be used by the collector to confirm intelligence.
- Use the polygraph.
- Consider that a source motivated primarily by money will likely be tempted to fabricate information in order to get paid.
- Be aware that a source may read the local newspaper to report information that is already known or may also be providing information to another agency.

9-27. The one thing that the HUMINT collector cannot do is to ignore signs of deceit.

HUMINT COLLECTION AIDS

9-28. There are numerous procedural and recording aids that can assist the HUMINT collector in conducting rapid, accurate, yet systematic questioning. They include—

- HUMINT Collector's Guide. This guide is a pamphlet or notebook designed to guide the HUMINT collector through the questioning. The HUMINT team leader should ensure that team members prepare a HUMINT collector's guide, which could be included in the unit's SOP. The guide is made based on the AO and supported command intelligence requirements. The HUMINT collector and available intelligence analysts should jointly prepare the guide. Appendix G provides the basic topics and example questions that can be adapted to construct a HUMINT collector's guide. The guide must be updated for each interrogation as part of planning and preparation. The guide should contain information such as—
  - Intelligence requirements and ISR tasks.
  - Topical questioning sequence format.
  - Actual prepared questions to be used during questioning.
  - Guidelines for employing the various approach techniques.
  - Formats or samples of completed reports used by HUMINT collectors.
- Time Event Chart. A timeline, or event chart, is a graphic display upon which the HUMINT collector enters chronological information as it is collected. This facilitates the HUMINT collector in understanding and organizing the collected information. It also enables the HUMINT collector to identify gaps in information, to sequence events properly to facilitate follow-up questions, and to identify deception. The HUMINT collector can develop a basic timeline prior to questioning. The source should not be able to observe the timeline since doing so will help a deceptive source "keep his story straight." See Chapter 12 for how to create and use a time event chart.
- Organizational Chart. An organizational chart is a graphic representation of an organization. It is the equivalent of a military line-and-block chart. This is used to facilitate the questioning of organizations and in establishing their hierarchical and lateral linkages. A basic chart can be developed prior to the questioning based on the expected organizational questioning.

RECORDING TECHNIQUES

9-29. Accuracy and completeness are vital principles to reporting. However, it is usually not possible to completely record all information in a questioning session. Recording techniques may involve memory, handwritten or typed notes, tape recordings, and video recordings. Each has its advantage and corresponding disadvantage.

- Memory: Relying on one's memory has certain advantages. It does not require any equipment or extra time, and is the least intrusive method of recording information. It allows maximum interaction with the source and projects sincerity. An individual can train himself to remember highly detailed information. Often in elicitation, memory is the only viable recording method. However, in general, using the memory exclusively to record information is the most inaccurate methodology. Particularly in a long questioning session, details are forgotten and information tends to be generalized.

- Handwritten notes: Handwritten notes require minimal equipment (a pad and pencil), are not intimidating to most sources, and can be as detailed as the HUMINT collector desires. If an analyst or second interrogator is present, he should also take notes. This second set of notes can aid in report writing. The interrogator should not rely solely on an analyst's notes unless absolutely necessary. However, writing notes while questioning an individual often interferes with the rapport between the collector and the source. The collector loses eye contact and can easily miss subtle body language that might indicate lying. Detailed note taking can be extremely time consuming and many sources will, over time, begin to limit their responses so they do not have to repeat information or wait for the collector to write it down. It is somewhat intrusive and inhibiting to the source and is totally inappropriate in certain situations such as liaison and most casual source contacts. Handwritten notes can also be inaccurate, have limited details, and can be hard to read after the fact.

- Computer notes: With the proliferation of computer equipment, particularly laptops and handheld devices, note taking on computers is increasingly commonplace. A computer can provide access to data-based information that may support questioning such as foreign language dictionaries or technical support manuals, either through the Internet (if connected) or on its harddrive. If the computer is linked to a communications system, it also allows the HUMINT collector to transmit data, including SALUTE reports, during the course of the questioning. Notes taken on a computer, however, have many of the same disadvantages as handwritten notes. In addition, computer notetaking requires more equipment and technological support and access to either electricity or a plentiful supply of batteries. Computers may be intimidating to some sources and the fact that what the source says is being entered into a computer may cause the source to alter the information he is providing. Computers tend to isolate the collector from the source by dividing the collector's attention between the computer and the source, and again may cause the collector to miss critical body language clues. Finally, the computer is even more inappropriate to casual and controlled source operations than are handwritten notes.

- Audiotapes: If recording equipment is discrete and functioning properly, audiotapes can be extremely accurate. Use of tapes also allows the HUMINT collector to place his entire attention on the source. This not only enhances rapport but also allows the HUMINT collector to observe the source's body language. Taping a questioning session, if done overtly though, tends to be extremely inhibiting to the source and may seriously curtail the information obtained. Surreptitious taping can be illegal in some situations and dangerous in some situations as well. Consult your legal advisor to determine if taping is legal. Taped information can also be seriously affected by ambient noise and the relative positioning of the source and collector to the microphone. Writing a report based on a taped session can be extremely time consuming, since it takes as long to listen to a tape as it took to record it. This drawback can be reduced somewhat through the use of voice activated recording devices. Exclusive dependence on audiotapes tends to make the collector less attentive and more likely to miss follow-up questions. Also, if the tape is lost or damaged or does not function properly, the collector has no backup.

• Video recording: Video recording is possibly the most accurate method of recording a questioning session since it records not only the voices but also can be examined for details of body language and source and collector interaction. It is also the most resource intensive requiring proper lighting, cameras, viewing equipment, and possibly trained operators. If done overtly, video recording can be by far the most inhibiting to the source. Even if the source is willing to be videotaped, there is a tendency for both the source and the collector to "play to the camera," creating an artificiality to the questioning. Consult your legal advisor to determine the legality of overt or covert videotaping.

QUESTIONING WITH AN ANALYST OR A TECHNICAL EXPERT

9-30. The HUMINT collector may often find himself in the position where he needs to use an analyst or a technical expert, or both, in order to conduct questioning. Many of the techniques involved in using an analyst or technical expert are the same as those with using an interpreter (see Chapter 11). The HUMINT collector must pre-brief these supporting personnel. The degree to which the analyst or technical expert is involved in the actual questioning is dependent on the established relationship between the analyst or technical expert and the HUMINT collector. The HUMINT collector will always remain in charge of the questioning, be present throughout the questioning, and ensure that the questioning follows his questioning plan. He must ensure that the supporting analyst or technical expert has the proper security clearance.

9-31. An analyst or technical expert can participate in the questioning to various degrees listed below from least intrusive to most intrusive. As the degree of participation by the analyst or technical expert increases, the technical fidelity of the information collected usually increases but the rapport between the HUMINT collector and the source decreases as does the HUMINT collector's ability to control the content and judge the truthfulness of the information. The analyst or technical expert may provide—

• Advice Only: The HUMINT collector does the questioning. The expert provides information prior to the meeting and may review the collected information after the meeting. The technical expert is not present at the actual questioning.

• Remote Support: The HUMINT collector does the questioning. In addition to the above, the expert monitors the questioning and provides input to the HUMINT collector after the questioning as required. Based on the technological support, this can involve the expert sitting in on, but not participating in the questioning (which may make the source uncomfortable), or the expert viewing and listening to the questioning through a remote video and sound hook-up.

- Local Support: The HUMINT collector does the questioning. The expert sits in on the questioning and provides input to the HUMINT collector during the course of the questioning. This can break both the source's and the HUMINT collector's trains of thought and confuse the lines of control in the questioning.

- Expert Participation: The HUMINT collector initiates the questioning, but the expert participates throughout the questioning, asking for clarification and additional information as required. Unless properly trained, the expert can seriously taint the quality of the information through the use of poor questioning techniques. The HUMINT collector can lose rapport and control.

- Trained Expert Questioning: In rare instances, with particularly difficult technical topics or those areas of questioning that require a high degree of technical expertise, it may be easier to train the expert in basic questioning techniques than it is to train the HUMINT collector on the technical topic. In this instance, the HUMINT collector sits in on the questioning to ensure proper procedures and techniques are used and to advise the technical expert. The technical expert does most of the questioning.

9-32. In any case, if the source is to receive compensation for his time, it must come from the HUMINT collector, not the analyst or technical expert. This continues to reinforce that the HUMINT collector is in charge, and does not transfer the source's trust to the expert.

THIRD-PARTY OFFICIAL AND HEARSAY INFORMATION

9-33. The source may have information that he did not observe firsthand. While this information is not as reliable as firsthand knowledge, it is often significant enough to report. The HUMINT collector must be careful to identify this information as to its origin, type, and the time and manner that the information was obtained by the source. This information will be entered into the report as a source comment or a collector comment. This will include—

- The origin of the information. This may be the name, rank, and duty position or job of an individual or may be an official or unofficial document such as an OPORD, official memorandum, or party newspaper.

- The complete organization to which the person who provided the information belongs or the identity of the organization that produced the official or unofficial document from which the source obtained the information.

- Date-time group (DTG) when the source obtained the information.

- The circumstances under which the source obtained the information.

9-34. Comparing the details of the hearsay information, such as DTG, where the information was obtained and the circumstances under which the source claimed to have received it, to the source's known activities, may provide indications of truthfulness or deception on the part of the source.

CONDUCTING MAP TRACKING

9-35. Map tracking is a specific questioning skill that the HUMINT collector uses in all operations. It is a vital skill in supporting targeting and operational planning. Map tracking identifies and verifies key information by tracking the source's movement and activities within a specific area over a fixed period using a map or similar graphic aid. The area and the time involved are dependent on the collection requirements and the source's knowledge level. Map tracking can occur at any point in the questioning process. Normally, the HUMINT collector begins map tracking as soon as his questioning identifies a priority disposition or activity that the source's information can locate on the map.

9-36. Map-tracking techniques, if properly applied, can extract information from friendly, neutral, or threat sources and can be used with individuals ranging from those with detailed map skills to illiterates, and those who have never seen a map. Through map tracking, the HUMINT collector pinpoints locations of any threat activity, threat dispositions, or any other priority terrain-related information, such as trafficability, known to the source.

9-37. The HUMINT collector will determine these locations with the degree of fidelity needed to support operational requirements. The degree of detail needed may range from an 8-digit grid coordinate for unit locations to locations of specific buildings, rooms, or even items within a room. The HUMINT collector uses a variety of map-tracking aids including standard military maps, aerial photographs, commercial imagery, building blueprints and diagrams, and commercial road maps. Some advantages to map-tracking techniques include—

- The source is led through his memory in a logical manner.
- Discrepancies in the source's statements are easier to detect.
- Locations are identified to support targeting and battlefield visualization.
- Map tracking is a four-step process:
  - Step 1: Determine the source's map-reading skills.
  - Step 2: Establish and exploit common points of reference (CPRs).
  - Step 3: Establish routes of travel.
  - Step 4: Identify and exploit key dispositions.

DETERMINE THE SOURCE'S MAP-READING SKILLS

9-38. The first step in the map-tracking process is to determine the specific map-reading skills of the source. This step only occurs the first time that the HUMINT collector map tracks a particular source. This information will determine what methodology will be used for the rest of the process. In this step the HUMINT collector is determining existing skills; he should not attempt to teach the source additional map skills at this time. The HUMINT collector can use prior knowledge, such as the fact that the source is illiterate or cannot read a map, to skip some of the specific parts of the process. Below is a detailed description of the process to establish the map-reading skills of the source.

- The HUMINT collector asks the source if he can read the map being used. If the source answers in the affirmative, the HUMINT collector asks some key questions to verify this.

- If the source cannot read the map being used, the HUMINT collector determines if the source can read another type of available map or graphic representation. For example, a source may not be able to read a military map but might be able to use a commercial map or an imagery product.

- The HUMINT collector then establishes the method that will be used to describe movement (direction and distance) on the map. If the source knows how to use compass directions, that may be the most expedient method for determining direction. Again, the HUMINT collector must verify that the source knows how to use compass directions. This can be done best by having the source tell the compass directions between known points. Distance is normally determined by using the standard units of measurement with which the source is familiar, such as kilometers or miles. This can cause some problems, for example, if the map is measured in kilometers and the source normally expresses distance in miles. The HUMINT collector must make the adjustment rather than trying to teach the source the unfamiliar system.

- Compass directions and standard units of measure are not the only method or necessarily even the best method of indicating direction and distance in all circumstances. When using an urban map, direction and distance can often be described by indicating blocks traveled and turns made (right or left) at intersections. Direction of travel can be indicated in reference to key features such as going toward the downtown area or moving toward the river. When describing the interior of a building, references may be to upstairs, downstairs, floor number, or other descriptive terms. When map tracking in rural areas, especially when questioning someone who does not know how to use compass directions, terrain association is normally the best method of establishing direction of travel and distance. Questions such as "Were you traveling uphill at that time?" "What prominent terrain features could you see from that location?" "What was the nearest town?" or "Was the sun behind you?" help to identify locations on the map. The HUMINT collector should allow the source to use his own frames of reference. However, the HUMINT collector must ensure he understands the source.

ESTABLISH AND EXPLOIT COMMON POINTS OF REFERENCE

9-39. The second step of map tracking is to establish CPRs. It is important in accurate map tracking to talk the source through his past activities in the sequence in which they occurred and his movements in the direction in which they were traveled. Attempting to track a source backward in time and space is time consuming, inaccurate, and is often confusing to both the source and the HUMINT collector. Future activities should be tracked in the direction in which they are planned to occur. The HUMINT collector will normally establish various CPRs throughout the questioning of the source.

9-40. For certain sources such as friendly forces, tasked sources, or other instances where the starting and ending points of the mission being questioned are easily established, the first point of reference is normally where that source began the mission. For other sources such as detainees, line crossers, informers, and refugees, it is often more difficult to establish a "starting point." In these instances the HUMINT collector uses a sequential approach to the map tracking. He establishes a point of reference that is a logical end point for the subject being discussed. This may be, for example, the point of capture for a detainee, the point where a line crosser entered the friendly force area, or where a refugee left the area of intelligence interest. Second and subsequent points of reference are established during questioning when the source mentions a disposition, activity, or location of interest that can be located on the map. The HUMINT collector locates the reference point on the map through direct questioning and terrain association. He uses leading questions as necessary to establish an exact location. He then establishes the route of travel.

ESTABLISH ROUTES OF TRAVEL

9-41. Once the CPR is established, the HUMINT collector questions the source until he has extracted all pertinent information on the CPR and its immediate surroundings. For past missions and activities, the HUMINT collector then establishes the route the source traveled between the newly established CPR and a previously established CPR and exploits the route. For future missions or activities, the route is established from the previously established CPR toward the future mission CPR.

9-42. The HUMINT collector should establish the route traveled by determining the source's direction and physical description of the route of travel. The description should include details such as surface on which the source traveled and prominent terrain features along the route of travel and the distance the source traveled or, in the case of future locations, would travel. The HUMINT collector should also identify any pertinent dispositions or any activities of military significance, belonging to the opposition forces, along or in the vicinity of the route of travel. For longer routes, the HUMINT collector may divide the route into segments for ease of questioning.

IDENTIFY AND EXPLOIT KEY DISPOSITIONS

9-43. The HUMINT collector must obtain the exact location and description of every pertinent disposition known to the source. This includes the locations established as CPRs and any other pertinent disposition established during map tracking. At a minimum, the collector should—

- Establish a physical description of the disposition. The degree of fidelity will depend on the collection requirements. This may be as detailed as the physical layout of a room to the general description of a training area. This will include security measures and modus operandi at the location as appropriate.

- Identify and describe the significance of the disposition in terms of ongoing and future threat operations.
- Identify and describe key activities, equipment, or organizations at the location, as well as people and leaders.
- Identify and describe all pertinent collocated activities, locations, or organizations, as well as people and leaders.
- Identify the basis (hearsay or personal experience) and DTG of the source's knowledge of each disposition.

SPECIAL SOURCE CATEGORIES

9-44. Questioning of every source is unique and requires specific preparation. Special consideration and preparation must be made for some specific categories of sources. Some examples of special source categories include but are not limited to wounded or injured sources or illiterates.

WOUNDED OR INJURED SOURCES

9-45. HUMINT collectors may question (interrogate, debrief, or elicit information from) a wounded or injured source provided that they obtain certification from a competent medical authority that the questioning will not delay or hinder medical treatment or cause a worsening of the condition of the source. The HUMINT collector can question the source before, after, or during medical treatment. The HUMINT collector cannot at any time represent himself as being a doctor or any other type of medical personnel or member of the ICRC. Nor can he state, imply, or otherwise give the impression that any type of medical treatment is conditional on the source's cooperation in answering questions.

ILLITERATES

9-46. HUMINT collectors should never make the mistake of equating illiteracy with a lack of intelligence or an inability to provide meaningful information. In fact, many illiterates have developed extremely good memories to compensate for their inability to rely on the written word. An illiterate's frame of reference does not include street signs, mile markers, and calendars. It also will probably not include conventional time and distance measurements. The HUMINT collector must compensate for these differences. Map tracking, for example, must normally be accomplished by terrain association. If the source cannot tell time, time of day can be determined by the position of the sun.

Chapter 10

Reporting

10-1. Reporting is the final and in many cases the most vital phase in HUMINT collection. If the collected information is not reported accurately, in a timely manner, in the proper format, and to the correct recipient, it cannot become part of the all-source intelligence product or tip in time to affect operational decisions. Information that would support targeting must be reported by the fastest means possible.

REPORTING PRINCIPLES

10-2. The HUMINT collector must be able, in a written report, to convey to the user the information obtained from a source. Therefore, the following principles of good report writing are to be followed:

- Accuracy. Accurately reflect the information obtained from the source. Reporter comments and conclusions must be clearly identified as such.

- Brevity. Report all relevant information; however, the report should be brief, to the point, and avoid unnecessary words.

- Clarity. Use simple sentences and understandable language. Proper grammar and punctuation are a must. Another team member, if possible, should read the reports to ensure clarity.

- Coherence. Present the information in a logical pattern based on standard reporting formats.

- Completeness. Report all information collected. The collector should not filter information since all information is of interest to an analyst. Report negative responses to pertinent topics to prevent a misunderstanding or duplication of effort in subsequent questioning based on SDRs.

- Timeliness. Report information as soon as operationally feasible. Most collection requirements contain a LTIOV as part of the requirement. While written reports are preferable, critical or time-sensitive information is passed by the most expedient means available.

- Releasability. Include only releasable information in reports that are to be shared with multinational units. When possible, reports to be shared with multinational units should be kept to the appropriate classification to ensure the widest dissemination of the reported information.

REPORT TYPES

10-3. There are two major categories for reporting information: operational reports and source administrative reports. Figure 10-1 shows the HUMINT reporting channels. Refer to DIAM 58-11 (S//NF) and DIAM 58-12 (S//NF) for specific guidance in using these reports.

OPERATIONAL REPORTS

10-4. Operational reports is a broad category that encompasses all reports that do not contain information collected in response to intelligence requirements or the reporting of the technical, and usually sensitive, aspects of HUMINT collection. It includes but is not limited to all administrative and logistical reports. Unit SOPs and directives from higher headquarters establish operational reporting requirements, formats, and procedures. Operational reporting—

* Tells the commander where and when assets are conducting missions.
* Describes unit mission capability.
* Responds to administrative and logistical requirements.
* Describes support requirements.
* Includes but is not limited to unit status reports, mission planning reports, mission status reports, and equipment status.
* Reports ICF usage at any echelon where the use of ICFs is authorized.

SOURCE ADMINISTRATIVE REPORTS

10-5. Source administrative reports include intelligence reports that are used to pass or request information in order to answer intelligence requirements, and reports that address the HUMINT collector's contacts with the source. Intelligence reports include but are not limited to IIRs and SALUTE reports.

Intelligence Information Reports

10-6. The IIR is used to report all HUMINT information in response to collection requirements. It is used to expand on information previously reported by a SALUTE report or to report information that is either too extensive or not critical enough for SALUTE reporting. IIRs are written at any echelon and "released" by the appropriate authority before they enter the general Intelligence Community. Normally the G2X will be the release authority for IIRs.

10-7. At the tactical level, the HUMINT collectors will fill out the complete IIR; however, the requirements section may link the information collected against a unit requirement rather than against national requirements. In any case, the report will be forwarded to the OMT.

10-8. The team leader will review the IIR, place a copy of the IIR in the detainee's or source's local file and forward the IIR to the OMT. (When a detainee is transferred to another facility or evacuated to a higher echelon, a copy of each IIR written from interrogations of that detainee is forwarded with him.) The OMT reviews the report, requests additional information as necessary from the originator, adds additional administrative detail, and forwards the report to the HOC of the supporting C/J/G/S2X. The HOC and the 2X review the report, request additional information as required, add any final required information including linking it to national requirements, and then the 2X releases the report.

10-9. In addition to the above, the text information from the IIR can be forwarded to the unit's analytical elements and when it contains critical time-sensitive information, such as an impending attack, it is sent to units which may be affected by the information; however, it must be clearly marked "unevaluated information, not finally evaluated intelligence." The use of IIRs and the formats are covered in DIAM 58-12 (S//NF).

SALUTE Reports

10-10. The SALUTE report is a standard Army format used to report information of immediate interest by individuals at any echelon. (See Appendix H for a SALUTE report format.) The SALUTE report is the primary means used to report combat information to units that could be affected by that information. After review by the team leader, SALUTE reports are sent simultaneously to the supported unit S2, to the OMT in control of the HCT, and to the intelligence staff officer of any other tactical unit that may be affected by the information contained in the SALUTE report.

10-11. The OMT reviews the report and forwards it to the supporting HAT and supporting J/G/S2X for inclusion in the analysis picture. The supported S2 will—

- Review the information.
- Incorporate it into his unit intelligence products, as applicable.
- Forward the information to his higher echelon intelligence staff officer.
- Ensure that all affected units are notified.

10-12. Units must develop SOPs for the passing of information and intelligence to multinational units. Report writers and editors must ensure that reports that are to be shared with multinational units contain only releasable information. This will enable reports to have the widest dissemination. Arrangements are made through the C/J2X/LNO for distribution. When possible, reports to be shared with multinational units should be kept to the appropriate classification to ensure the widest dissemination of the reported information.

Basic Source Data Reports

10-13. The basic source data (BSD) reports provide the HUMINT chain with biographic and operational information related to a source. BSDs are used at all echelons to collect biographic information on all contacts. The use of BSDs and BSD formats are covered in DIAM 58-11 (S//NF).

Contact Reports

10-14. Collectors use contact reports to inform their technical chain (from OMT through J/G/S2X) of all relevant information concerning specific meetings with HUMINT sources. Information typically includes the circumstances of the contact (purpose, locations, time), the operational matters relative to the contact (topics discussed, taskings given), reports produced as a result of the contact, and logistics expended.

Other Reports

10-15. HUMINT collectors also use a number of other reports to administer source contacts and to report information. Copies of the following reports should be maintained in the detainee's permanent file for future reference. HUMINT collectors will review these reports when planning additional collection activities; release committees or tribunals can use the reports to help evaluate if a detainee can be released or not. These reports include—

- Screening Reports. Screening reports are used to report BSDs, knowledge areas and levels, cooperation, vulnerabilities to approaches, and other relevant source information between HUMINT collectors. It is normally filled out either electronically or manually by the initial HUMINT collector to speak to a source. The screening report is normally forwarded electronically to higher echelon HUMINT collection organizations and other MI organizations that might have interest in the source. Higher echelon organizations may add information to the screening sheet extracted through subsequent screenings. Available digital screening reports contained in the HUMINT collector's mission support software (for example, BAT or CI/HUMINT Automated Management System [CHAMS]) should be used whenever possible to ensure rapid transfer of data. If screening reports have to be handwritten, the information collected should conform to theater requirements and local SOPs.

- Knowledgeability Briefs. The KB is used to inform the Intelligence Community of a source's full identity, past history, and areas of knowledge, as well as to set a suspense date for the submission of intelligence requirements. It is normally only used at the strategic and operational echelons. When completed, a KB will be classified at least Confidential in accordance with the DIA Classification Guide to protect the identity of the source. The use of KBs and the formats are covered in DIAM 58-11 (S//NF). See Figure 10-2 for an example of a short form KB that can be used for screening at all echelons, and can also be prepared and published like the full KB. This allows the entire intelligence community to see who is either in custody or to whom US intelligence has access so that SDRs can be issued to help focus the intelligence collection effort.

- Notice of Intelligence Potential (NIP). A NIP is used to inform the US Intelligence Community of the availability of a source of potential interest and to notify them of what agency has responsibility for questioning that source and where to forward questions and requests for information from that agency. The use of NIPs and the formats are covered in DIAM 58-11 (S//NF).

- Lead Development Report (LDR). The LDR is used to inform the HUMINT chain of ongoing operations directed toward a specific source. It notifies them as to what element spotted the potential source, the current steps in assessing of the source, and the general information on the potential source.

- Interrogation Summary. An interrogation summary may be written to record relevant facts concerning the interrogation. The summary may include the attitude of the source; approach techniques that were tried and which ones were effective; incentives promised and whether or not they were delivered yet; recommended topics for further exploitation; and any other topics the HUMINT collector considers relevant. Local SOPs will dictate the use of the interrogation summary.

- Interrogation Plan. The interrogation plan is a report prepared by the HUMINT collector to organize his plan to approach and question a source. It lists collection objectives, approach techniques, preparation and liaison tasks, and interpreter usage plan. The interrogation plan also has approval blocks for interrogation supervisor approval of selected approaches and medical release for questioning. The last part of the form has termination, approach effectiveness, recommendations for further exploitation, and a summary of information obtained and reports expected to be published. Figure 10-3 is an example of an interrogation plan format.

- Termination Report. The termination report is used at all echelons to inform the technical chain of the termination of a contact relationship between a HUMINT collector and a source.

- Biographic Report. The biographic report is a formatted IIR used at all echelons to report information collected from one human source about another individual of actual or potential intelligence interest. The biographic report format is found in DIAM 58-11 (S//NF).

REPORTING ARCHITECTURE

10-16. There are three basic reporting channels (see Figure 10-1):

- The operational reporting chain consists of primarily the C2 elements for the HUMINT collection element. It includes the OMTs, unit commanders, and unit S3 and operation sections.

- The technical chain includes the OMTs, HOC, and the C/J/G/S2X, and in certain circumstances, the unit G2/S2s.

- The intelligence reporting chain includes the OMTs, HATS, C/J/G/S2Xs, and unit G2/S2s.

10-17. Many elements serve multiple and overlapping functions within the reporting architecture. Each element must be aware of its function within the architecture to ensure that information is disseminated expeditiously to the right place in the right format. This architecture should be established and published prior to implementation in order to avoid confusion.

OPERATIONAL REPORTING

10-18. Operational reporting is sent via the organic communications architecture (see Chapter 13). Operational reports are normally sent per unit SOP or based on direction from higher headquarters. HCTs normally send all operational reports through their OMT to the command element of the unit to which they are assigned. If an HCT is attached, it will normally send its operational reports to the unit to which it is attached with courtesy copies to their assigned unit as required. If there is an administrative or logistics relationship established with the supported unit, HCTs that are in DS send the principal copy of all related administrative and logistic reports to the supported unit with a courtesy copy to their parent unit. If the HCT is operating in GS, a courtesy copy of operational reports should be forwarded to all affected unit commanders in the supported AO.

TECHNICAL REPORTING

10-19. Technical reporting includes the forwarding of source information and technical parameters of collection operations from lower to higher and the passing of tasking specifics, source information, technical control measures, and other information from higher to lower. Technical reporting is conducted through the technical chain that extends from the HCT through the OMT and Operations Section (if one exists) to the C/J/G/S2X.

INTELLIGENCE REPORTING

10-20. The key to intelligence reporting is to balance the need for accurate reporting with the need to inform affected units as quickly as possible. The J/G/S2 and MI commander are key to ensuring the right balance.

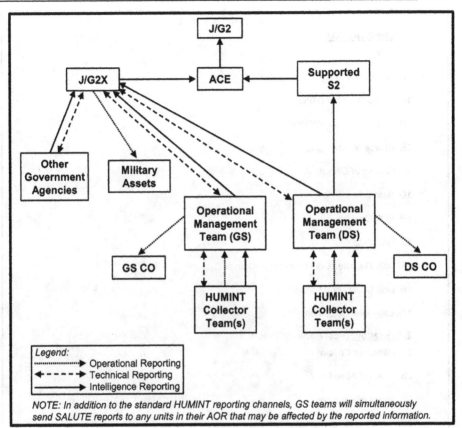

Figure 10-1. HUMINT Reporting Channels.

KB-EZ WORKSHEET

1. PERSONAL DATA:

1A. Name:

1B. Source Number (Capturing Unit):

1C. Source Number (MPs):

1D. Source Number (Other):

1E. Source Number (MI):

1F. Country of Citizenship:

1G. Birth City:

1H. Birth Country:

1I. Birth Date:

1K. Date Departed Country of Origin/Date of Capture:

1N. Last County of Residence:

1O. Language Competency:

2. Education: (Most Recent to Oldest)

2A. Military or Civilian:

2B. Dates of Attendance:

2C. Name of Institution:

2D. City Location of Institution:

2E. Country Location of Institution:

2F: Completion Status/Degree Type:

3. EMPLOYMENT: (Most Recent to Oldest)

3A. Dates of Employment:

3B. Name of Place of Employment:

3C. City Location of Place of Employment:

3D. Country of Place of Employment:

3E. Employment Duty Position:

3F. Security Clearance:

Figure 10-2. KB-EZ Worksheet.

4. MILITARY SERVICE: (Most Recent to Oldest)

4A. Dates of Service:

4B. Name of Post/Base:

4C. Armed Service Component:

4D. Rank of Equivalent:

4E. Name of Unit/Group:

4F. City Location of Unit/Group:

4G. Country Location of Unit/Group:

4H. Military/Group Duty Position/Title:

4I. Security Clearance:

5. Comments: (Character, intelligence, motivation, personality, cooperativeness)

5A. CIRCUMSTANCES OF CAPTURE: Capture date, capturing unit, circumstances, documents, weapons, and equipment.

5B. ASSESSMENT: Physical condition, mental condition, intelligence, cooperation (1, 2, 3), knowledgeability (A, B, C), personality.

5C. ADDITIONAL PERSONAL INFORMATION: (Skills, experience, marital status, other).

6. NAME OF SCREENER:

Theater-specific collection requirements may require modification of the KB-EZ format. Consider adding entries for:

• Race
• Ethnicity
• Tribal Affiliation
• Religion and Sect
• Language and Dialect Spoken

Entries for "Location" may need to include a village or even neighborhood.

Figure 10-2. KB-EZ Worksheet (continued).

```
INTERROGATION PLAN

PLANNING:                                          DTG:_____

Collector Name:_____FRN:_____

Detainee Name:_____

Detainee MP Number:_____ Other Identifying Numbers (specify): _____

Number of Times Interrogated:_____

Interrogation Objectives:_____
                               (Include PIR, SDR, RFI, IPSP, etc.)

Fit to undergo interrogation: YES_____ NO _____

Health concerns to be reported to the commander:_____

_____

Approach Strategies (Attach separate sheet if needed for additional approaches):

Initial Approach:                    _____

   Rationale:                        _____

Additional Approach:                 _____

   Rationale:                        _____

Additional Approach:                 _____

   Rationale:                        _____

REVIEW: Interrogation Supervisor_____ MI Unit Commander_____
         SJA_____ BSC (as appropriate) _____

APPROVAL AUTHORITY:    PRINTED NAME     DTG OF APPROVAL        SIGNATURE

First O-6 (as required):     _____   _____   _____

Interrogation Unit OIC       _____   _____   _____

Interrogation Supervisor     _____   _____   _____

Interpreter Name and Interpretation Method: _____

Other Participants:_____

Lead Agency: _____

Recording Method:_____Monitoring Method:_____
```

Figure 10-3. Interrogation Plan Format.

PREPARATION:

Coordinate with MP for access to the detainee.

Prepare for exploitation topics:
    Obtain appropriate map sheet(s)
    Obtain references
    Review previous reports, detainee correspondence
    Research collection topics
    Prepare questions

Prepare interrogation site (furnishings, lighting, climate, security, monitoring).

Ask Guard Questions.

Review Detainee Documentation:
    ID Card
    Capture Tag
    Documents captured with the detainee

- -

Post-Interrogation Report

Effectiveness of Approaches:

Attitude and Behavior of Detainee:

Summary of Topics Exploited:

Expected Reports Production in Response to Requirements:

Termination:
    Reason:
    Approach Reinforced:
    Incentive Promised:
    Delivered:

Recommendation for Further Interrogation and Rationale:

    Recommended Approach(es):

    Topics for Further Exploitation (Leads):

Disposition of Source:_____

Additional Comments:_____
_____
_____
_____
_____

Figure 10-3. Interrogation Plan Format (continued).

Chapter 11

HUMINT Collection With An Interpreter

11-1. The use of interpreters is an integral part of the HUMINT collection effort. It is vital that the HUMINT collection skills be paired up with a qualified interpreter. Use of an interpreter is time consuming and potentially confusing. Proper use and control of an interpreter is a skill that must be learned and practiced to maximize the potential of HUMINT collection. It is also vital for the HUMINT collector to confirm that the interpreter he intends to use holds the required clearance for the level of information that will be discussed or potentially collected, and is authorized access to the detainee. This chapter deals strictly with the use of interpreters to support HUMINT collection; it is not intended to be applied to more routine uses of interpreters in support of administrative, logistical, or other operational requirements.

ADVANTAGES AND DISADVANTAGES OF INTERPRETER USE

ADVANTAGES

11-2. Interpreters are frequently a necessary aid to HUMINT collection. There are certain advantages to using an interpreter. The most obvious is that without an interpreter, a HUMINT collector without the proper language or necessary proficiency in that language is severely limited. Furthermore, if properly trained, briefed, and assessed, the interpreter can be a valuable assistant to the HUMINT collector. The interpreter—

- Probably has a greater knowledge of the local culture and language usage than could be developed by the HUMINT collector.
- Can identify language and culturally based clues that can help the HUMINT collector confirm or refute the veracity of the source's statements.
- Can interpret not only the literal meaning of a statement but also the intent and emotion of a sentence.

DISADVANTAGES

11-3. There are, however, several significant disadvantages to using interpreters. Disadvantages may include—

- A significant increase in time to conduct the collection. Since the interpreter must repeat each phrase, the time for a given questioning session or meeting is normally at least doubled.
- Since there is now a third person in the communications loop, the potential for confusion or misunderstanding increases significantly. This is especially true when the interpreter is deficient in his command of either language.

- The establishment of rapport and the use of some approach techniques (see Chapter 8) are made difficult or even impossible when working through an interpreter.
- The ability of the HUMINT collector to interpret the source's veracity through the analysis of word usage, nuances of speech, and body language is curtailed.
- The interpreter will have his own set of biases that could influence the manner in which the dialogue is interpreted.
- The source may be culturally biased against the interpreter. This is especially possible if the interpreter was locally hired and is of a different ethnic, social, or religious group than the source.
- The interpreter may be culturally biased against the source and intentionally misinterpret the meaning to obtain a desired effect.
- There may be mission or subject matter classification problems involved.

CAUTIONS

11-4. Be careful of comments made in the presence of your interpreter. Although you plan comprehensively with your interpreter, you should only share information with your interpreter on a need-to-know basis. Obviously the exchange of information concerning the "what, where, when, with whom, and how" of each meeting must be discussed with your interpreter, but sometimes the "real why" is none of his business! You may be meeting with a source or contact because the commander believes this individual has lied. The real purpose (the why) of the meeting is to pose control questions and to determine whether the source or contact lied in the past or whether there was simply a miscommunication.

11-5. Be careful of sensitive or personal conversations when the interpreter is present. This applies to conversations en route to or from meetings, conversations over lunch or dinner in the operational area, and conversations in the team area. It is easy to get used to the presence of the interpreter and to overlook his presence. An interpreter is a necessary tool but we must remember that most are only very lightly screened for the sensitive access they have. If your interpreter turned out to be working for the other side, what information beyond "the necessary" could he provide?

METHODS OF INTERPRETER USE

11-6. There is a basic method and advanced method of interaction between the HUMINT collector and the interpreter. As the collector and the interpreter become experienced at working together and gain confidence in each other's abilities, they may use more advanced interactive techniques. It is the HUMINT collector's decision whether or not to use more advanced techniques.

BASIC METHOD

11-7. The basic method of interaction is used when—

- The interpreter and HUMINT collectors have not worked together extensively.
- The interpreter has language skills but no interpreter training or experience.
- The interpreter's skill in English or the target language is suspect.
- The HUMINT collector has limited experience using an interpreter.
- The interpreter's capabilities, loyalty, or cultural knowledge are not known or suspect.

11-8. Using the basic method, the interpreter is used solely as an interpretation device. When initial contact is made, the interpreter instructs the source to maintain eye contact with the HUMINT collector. The interpreter is briefed on the general course of the collection but usually is not advised of the specific purpose or collection goals. While the interpreter will be instructed to reflect the attitude, behavior, and tone of voice of both the collector and the source, he is told to *not* interpose comments or personal opinions at all in the conversation.

11-9. The questioning phase is conducted in the same way it would be if no interpreter were used with the obvious increase in time due to the interpretation. The interpreter uses the same person and tense as the HUMINT collector or source and neither adds nor subtracts anything from the dialogue. He does his best to fade into the background. When reports are written, the interpreter will only be asked questions based on the actual translation of the dialogue.

ADVANCED METHOD

11-10. The advanced method of interaction requires additional training on the part of the HUMINT collector and the interpreter, extensive experience working together, and a rapport between the HUMINT collector and the interpreter. The HUMINT collector must trust both the capabilities and the judgment of the interpreter. At this level of interaction, the interpreter becomes a more active participant in the HUMINT activities. The HUMINT collector remains in charge and makes it clear to the interpreter that he (the HUMINT collector) is responsible for the substance and direction of the questioning. The interpreter is normally briefed as to the specific goals of the collection.

11-11. The interpreter becomes a more active participant in the approach and termination phases to the point of even making planned comments to the source supportive of the HUMINT collector's approach. For example, if the HUMINT collector is using an incentive approach, the interpreter in an aside to the source can tell him that the HUMINT collector always keeps his promises. This type of technique should only be used if both planned and rehearsed.

11-12. During the questioning phase, the interpreter supports the collector by not only translating the word of the source but also cueing the collector when there are language or culturally based nuances to what the source is saying that might add credence or doubt as to the veracity of the statements. For example, the interpreter could point out that although the source claims to be a factory worker, his language use indicates that the source has a university education. In another example, the interpreter could indicate that the dialect or pronunciation that the source is using does not match the area that he claims to be from. During report writing, the interpreter supports the HUMINT collector by not only answering questions on the literal interpretation but also adds, when appropriate, comments on the significance of both what was said and how it was said.

SOURCES OF INTERPRETERS

11-13. There are almost never sufficient interpreters to meet all unit mission requirements. Interpreters in support of HUMINT collection require a security clearance and knowledge of the operational situation. While any qualified interpreter can be used to support HUMINT collection, the HUMINT collectors maximize the collection potential if the interpreter has received specific training. The number of interpreters needed to support a HUMINT collection mission is METT-TC driven based primarily on the number of HUMINT collectors, the dispersion of the HUMINT collectors in the AO, and the number of sources. Normally one interpreter for every two non-language qualified HUMINT collectors is sufficient; however, in situations where a large number of high-value sources must be questioned in a limited time, a ratio of 1 to 1 may be required. Interpreters are obtained from within the military and from the US and local civilian populations or other English-speaking countries.

MILITARY

11-14. There are many soldiers, including non-US citizens, who have native language abilities due to their upbringing. Their parent unit may identify these language abilities, or these soldiers may volunteer their abilities when a contingency arises. The ARNG, USAR, other US military services, and even coalition militaries, have language-trained and certified personnel in Military Intelligence MOSs, such as 98G or 09L, who may be called upon to serve as interpreters for the HUMINT collection effort.

CIVILIAN

11-15. Civilian corporations may be contracted by the military to provide interpreters for an operation. These interpreters are divided into three categories:

- CAT I Linguists - Locally hired personnel with an understanding of the English language. These personnel undergo a limited screening and are hired in-theater. They do not possess a security clearance and are used for unclassified work. During most operations, CAT I linguists are required to be re-screened on a scheduled basis. CAT I linguists should not be used for HUMINT collection operations.
- CAT II Linguists - US citizens who have native command of the target language and near-native command of the English language. These personnel undergo a screening process, which includes a national

agency check (NAC). Upon favorable findings, these personnel are granted an equivalent of a Secret collateral clearance. This is the category of linguist most used by HUMINT collectors.

- CAT III Linguists - US citizens who have native command of the target language and native command of the English language. These personnel undergo a screening process, which includes a special background investigation (SBI). Upon favorable findings, these personnel are granted an equivalent of a Top Secret (TS) clearance. CAT III linguists are used mostly for high-ranking official meetings and by strategic collectors.

INTERPRETATION TECHNIQUES

11-16. During the planning and preparation phase, the HUMINT collector, in collaboration with the interpreter, selects a method of interpretation. There are two methods:

- Alternate Interpretation. The interpreter listens to the entire phrase, sentence, or paragraph. The interpreter then translates it during natural pauses in speech.
- Simultaneous Interpretation. The interpreter listens to the source and translates what he says, just a phrase or a few words behind. The HUMINT collector should select the simultaneous method only if all the following criteria are met:

  - The sentence structure of the target language is parallel to English.
  - The interpreter can understand and speak English as well as the target language with ease.
  - The interpreter has special vocabulary skills for the topics to be covered.
  - The interpreter can easily imitate the HUMINT collector's tone of voice and attitude for the approaches and questioning technique selected.
  - Neither the collector nor the interpreter tends to get confused when using the simultaneous method of interpretation.

11-17. If any of the above-mentioned criteria in the simultaneous method cannot be met, the HUMINT collector should use the alternate method. The alternate method should also be used when a high degree of precision is required.

TRAINING AND BRIEFING THE INTERPRETER

11-18. The HUMINT collector will need to train an individual who has no interpreter experience as well as remind a trained and certified interpreter of the basic interpreter requirements. The requirements include—

- Statements made by the interpreter and the source should be interpreted in the first person, using the same content, tone of voice, inflection, and intent. The interpreter must not interject his or her own personality, ideas, or questions into the interview.

- The interpreter should inform the HUMINT collector if there are any inconsistencies in the language used by the source. The HUMINT collector will use this information in his assessment of the source.
- The interpreter needs to assist with the preparation of reports and administrative documents relevant to the source and meeting.

11-19. Once the HUMINT collector has chosen a method of interpretation, he must brief the interpreter. This briefing must cover—

- The current situation.
- Background information on the source (if available).
- The administrative particulars of the meeting such as where it will be held, the room setup, how long it will last.
- The specific positioning of the interpreter, collector, and source.
- The general or (if advanced method of interaction is being used) the specific collection objectives.
- The selected approach and possible alternate approaches that the HUMINT collector plans on using. If time allows, the collector and interpreter should rehearse the approaches.
- Any special topic or technical language that is anticipated. If time allows, the interpreter should research any anticipated technical vocabulary with which he is unfamiliar.

11-20. Throughout the briefing, the HUMINT collector fully and clearly answers questions the interpreter may have. This helps ensure the interpreter completely understands his role in the HUMINT collection process. With a more advanced interaction plan, the HUMINT collector and the interpreter should "wargame" their plan and rehearse their actions as necessary.

PLACEMENT OF THE INTERPRETER

11-21. The interpreter should be placed in a position that enhances the mood or general impression that the HUMINT collector wants to establish. When dealing with detainees or EPWs, the HUMINT collector generally wants to establish a dominant position, maintain a direct relationship with the source, and increase or at least maintain the anxiety level of the source. Having the HUMINT collector and the source facing each other with the interpreter located behind the source normally facilitates this. It allows the HUMINT collector to maximize control of both the source and interpreter. If desired, having the interpreter enter the room after the source, so the source never sees the interpreter, can further heighten the anxiety of the source.

11-22. Having the interpreter sit to the side of the HUMINT collector creates a more relaxed atmosphere. This is the norm for debriefings and official meetings. Having the interpreter at his side also facilities "off line" exchanges between the HUMINT collector and the interpreter. The collector should avoid having the interpreter sit beside the source since this has a tendency of establishing a stronger bond between the source and the interpreter and makes "off line" comments between the collector and the interpreter more difficult.

11-23. When conducting source meetings in a public setting, a more natural appearance is desirable. The seating needs to conform to the norm at the location where the meeting is taking place. For example, if meeting at a restaurant, the HUMINT collector, interpreter, and source will sit naturally around the table.

INTERACTIONS WITH AND CORRECTION OF THE INTERPRETER

11-24. The HUMINT collector must control the interpreter. He must be professional but firm and establish that he is in charge. During a questioning session, the HUMINT collector corrects the interpreter if he violates any standards that the pre-mission briefing covered. For example, if the interpreter interjects his own ideas into the meeting, he must be corrected. Corrections should be made in a low-key manner as to not alienate the interpreter, interrupt the flow of the questioning, or give the source the impression that there is an exploitable difference of opinion between the HUMINT collector and the interpreter. At no time should the HUMINT collector rebuke the interpreter sternly or loudly while they are with the source. The HUMINT collector should never argue with the interpreter in the presence of the source. If a major correction must be made, the HUMINT collector should temporarily terminate the meeting and leave the site temporarily to make the correction. The HUMINT collector needs to document any difficulties as part of his interpreter evaluation. The HUMINT collector must always ensure that the conduct and actions of the interpreter are within the bounds of applicable law and policy. Applicable law and policy include US law; the law of war; relevant international law; relevant directives including DOD Directive 3115.09, "DOD Intelligence Interrogations, Detainee Debriefings, and Tactical Questioning"; DOD Directive 2310.1E, "The Department of Defense Detainee Program"; DOD instructions; and military execute orders including FRAGOs.

11-25. The HUMINT collector must be alert for any signs that the interpreter is not performing as required. The following are some indicators of possible problems.

- Long-to-short. If you take 20 seconds to express yourself and the interpreter reduces it to a 3-second translation, it may indicate that something has been omitted, and you should not proceed until you have resolved the issue. There is nothing wrong with stating that you would "prefer" the interpreter translate everything that was just said. If you have trained your interpreter properly, this should not be an issue. If it arises even with the training you have given the interpreter, then it has significance and you must not let it pass.

- Short-to-long. If you take 5 seconds to express yourself and the interpreter expands it to a 30-second translation, it may indicate that something has been added, and you should not proceed until you have resolved the issue.

- Body-language shift. If the interpreter's body language suddenly has a significant shift from his normal behavior, you should look for the reason. (It is advisable for you to determine a base line of behavior for your interpreter to facilitate recognition of the changes.) Perhaps he is reluctant to translate what you just said. Be aware that the body shift means that something is happening—your task is to find out <u>what</u> it means.

- Unusual pauses. Look for a longer delay than usual before the translation begins. Unless it is a vocabulary or concept issue, the long delay means that the interpreter is "thinking" before he translates. Any thinking beyond what is needed to translate, as closely as possible, what was just said represents a potential problem. Again, you should establish a base line of behavior for your interpreter so you can recognize these unusual pauses.

- "Wrong" reactions. If you say something humorous that should provoke a positive response from the source, and you do not get that response, then you should wonder if the message got through. If the source becomes upset in response to something (positive) you said, then you should begin to wonder what message was passed by the interpreter. Did you fail to express yourself clearly, or was it an accidental or deliberate mistranslation?

11-26. A trusted linguist should periodically review the accuracy of the interpreter's translations by monitoring an interrogation or debriefing and critiquing the monitored interpreter's performance.

INTERPRETER SUPPORT IN REPORT WRITING

11-27. The interpreter assists the HUMINT collector in preparing all required reports. He may be able to fill gaps and unclear details in the HUMINT collector's notes. He may also assist in transliterating, translating, and explaining foreign terms.

EVALUATING THE INTERPRETER

11-28. After submitting all reports, the HUMINT collector evaluates the performance of his interpreter. This should be done in writing, and copies should be given to the interpreter and placed on file with the individual managing the HUMINT collection portion of the interpreter program. The interpreter program manager needs to develop a standard evaluation format for inclusion in the unit SOP. The evaluation forms should note at a minimum:

- Administrative data (for example, date, time, interpreter's name).
- Strengths and weaknesses of the interpreter with any problems and corrective actions taken.
- Type of interpretation used (simultaneous or alternate).
- Type of HUMINT operation the interpretation was supporting (that is, an interrogation, a debriefing, a liaison meeting).
- Ability or lack of ability of the interpreter to use specific technical language that may have been required.
- Name or collector number of the HUMINT collector.

11-29. The interpreter program manager uses these forms to decide on future use of the interpreters, to develop training programs for the interpreters, and to assign interpreters to make maximum use of their specific capabilities.

The HUMINT collector should also review these files before using an unfamiliar interpreter.

MANAGING AN INTERPRETER PROGRAM

11-30. Units requiring interpretation support need to identify an individual or individuals to manage the interpreter program. In most units, this will be someone in the G3/S3 section. Division and Corps-level units will have a language manager. In MI units whose specific function is HUMINT collection, it will normally be a senior Warrant Officer within that unit. The functions of the interpreter program manager include but are not limited to—

- Consolidating and prioritizing interpreter requirements.
- Coordinating with G2 or INSCOM to contract for qualified interpreters.
- Coordinating with the G1/S1 to identify personnel in the unit with language skills who can be used as interpreters.
- Coordinating with the G1/S1 and G5 to obtain qualified local-hire interpreters.
- Coordinating with G2/S2 for clearances.
- Coordinating with the G3/S3 to establish training for both the interpreters and those that will be using interpreters.
- Coordinating with the G3/S3 for language testing of the interpreters in both English and the target language as required.
- Coordinating with the G1/S1 and G4/S4 to ensure that all administrative and logistical requirements for the interpreters are met.
- Establishing and maintaining the administrative, operational, and evaluation files on the interpreters.
- Assigning or recommending the assignment of interpreters to operational missions based on their specific capabilities.

PART FOUR

Analysis and Tools

Part Four discusses HUMINT analysis and the automation and communication tools needed to support the HUMINT collection effort.

HUMINT analysis supports operational planning and provides direction to HUMINT collection operations. Analysts determine if information from a single human source is internally consistent based on factors such as placement and access of source, prior information from the source, and existing holdings. Source analysis is used to determine if the information from a source is complete, truthful, and responsive to collection requirements. Operational analysis consists of those actions taken to determine how to best meet requirements.

Modern automation and communications systems are vital to HUMINT collection. Real-time collaboration, detailed operational planning and ISR integration, as well as enhanced collection and source exploitation tools, must support team efforts. Emerging technology continues to allow the entire HUMINT collection system to operate more effectively. Commanders must be prepared to supply their HUMINT collection assets with the best possible technology.

Chapter 12

HUMINT Analysis and Production

12-1. Analytical processes provide information to support the commander, his staff, and his unit. Analysis is an integral part of HUMINT collection. Analysis occurs throughout the HUMINT collection process but can be divided into four primary categories: analytical support to operational planning and targeting, operational analysis and assessment, source analysis, and single-discipline HUMINT analysis and production.

ANALYTICAL SUPPORT TO OPERATIONAL PLANNING

12-2. Several elements provide analytical support at various echelons, including the following:

- The HAT is subordinate to the G2 ACE. The HAT supports the G2 in developing IPB products and in developing and tailoring SIRs to match HUMINT collection capabilities.

- The HAC is subordinate to the C/J/G/S2X and functions in the same capacity for the C/J/G/S2X as the HAT does for the ACE.
- The HOC of the C/J/G/S2X supports the C/J/G/S2 and C/J/G/S2X in the identification of HUMINT collection opportunities, the development of taskings and RFIs for HUMINT collection assets, as well as the development of a HUMINT database.

INTELLIGENCE PREPARATION OF THE BATTLEFIELD

12-3. The HAT assists the G2 in the identification and characterization of the human component of operations and its effects on friendly and enemy operations. As part of its assistance to the G2, the HAT compiles and analyzes data about the local civilian population including its political, ethnic, religious, cultural, tribal, economic, and other social components. It carefully examines the various component groups and their predicted reaction to friendly force operations.

12-4. The HAT also provides input to all-source analysis by identifying specific actions and motivational factors that should strengthen the local population's support of the US or at least weaken its support of the enemy and by providing information on transient (refugees, DPs, third-country nationals) population and its effects on friendly and enemy operations. In addition to the above, the HAT—

- Closely examines the current and potential threat to identify all factors, such as morale, motivation, training, and beliefs that would affect both positively and negatively on enemy and opposing force capabilities.
- Identifies formal and informal leaders of hostile, neutral, and friendly groups and how their influence is likely to affect operations.
- Develops overlays, databases, and matrices, as required, to support IPB. These overlays may represent a wide variety of intelligence issues, including battlefield infrastructure (for example, electrical power grid), population density, ethnic, religious, or tribal affiliation, and no-strike or collateral damage.
- Provides its products to the C/J/G/S2, the all-source analysts and CI analysts of the ACE, the HOC, the C/J/G/S2X, and HUMINT collection units as required.

ISR TASK DEVELOPMENT

12-5. The HAT and the C/J/G/S2X support the C/J/G/S2 by expanding the PIRs that can be answered through HUMINT collection into ISR tasks that can be answered by a human source and that can be tasked to a specific collection entity. The HAT and the C/J/G/S2X provide this information to support the development of the HUMINT collection plan and its integration into the overarching ISR plan. The HAT normally establishes a list of prioritized standing indicators, and supplements this with ISR tasks developed to answer specific PIRs. The standing indicators are incorporated into the ACE's all-source analysis team's list of indicators that point to a pattern or COA. Each standing indicator is integrated with other indicators and factors so that analysts can detect patterns and establish threat intentions.

SUPPORT TO HUMINT TASKING

12-6. The C/J/G/S2X supports the C/J/G/S2 and the C/J/G/S2 requirements manager by developing tasking for specific organic or attached HUMINT collection assets and by developing requests and coordinating for support from higher and lateral echelon HUMINT collection elements. The C/J/G/S2X matches specific collection requirements to collection units and sources. If current sources cannot answer the requirement, he develops profiles for new sources that guide the collection teams in the development of new sources. The C/J/G/S2X also determines the best method to achieve collection requirements. The C/J/G/S2X supplies any required technical support to the HUMINT operations section, OMTs, and HCTs.

HUMINT DATABASE DEVELOPMENT

12-7. The C/J/G/S2X maintains the source database, which receives input from HUMINT collection and CI operations. The C/J/G2X is responsible for maintaining the source database.

OPERATIONAL ANALYSIS AND ASSESSMENT

12-8. Operational analysis consists of those actions taken to determine how to best meet requirements. Assessment evaluates the effectiveness of the requirement effort. Operational analysis begins with the C/J/G/S2X when he determines the best collection assets and sources needed to answer collection requirements. For **analysis**, the C/J/G/S2X section and specifically the HOC—

- Provides higher echelon coordination and deconfliction for collection operations.
- Provides required technical support to the HUMINT operations sections, OMTs, and HCTs.
- Facilitates feedback and evaluations.

12-9. For **assessment**, the C/J/G/S2X section and specifically the HOC—

- Monitors all HUMINT reporting to ensure that requirements are being met.
- Refocuses efforts of assigned assets as needed.

12-10. The HUMINT operations section (if one exists) and the OMTs of tasked collection units determine which HCTs are best suited to meet specific collection requirements. They also conduct operational coordination for the HCT, provide technical support, and monitor all reporting to ensure that reports are properly formatted and support collection requirements. The team leader of the tasked HCT selects the specific collectors and sources to meet collection requirements, reviews the collection plan, provides technical support to the collectors, coordinates with the supported unit, and monitors all team reporting for format and content. He identifies additional lines of questioning and approaches required to fulfill collection requirements.

SOURCE ANALYSIS

12-11. Source analysis involves the actions taken to determine if information from a single human source is internally consistent based on factors such as placement and access of source, prior information from the source, and existing holdings. Source analysis is used to determine if the information from a source is complete, truthful, and responsive to collection requirements. Preliminarily source analysis is the responsibility of the individual HUMINT collector.

12-12. The HUMINT collector evaluates all source statements within the context of the information known about the source and the current situation to determine both the veracity and the validity of source statements. That is not to say that the HUMINT collector ignores any information that does not fit into expected pattern; rather that he treats that information with skepticism and uses appropriate questioning methodology to validate the statements (see Chapter 9). Source analysis is supported by the HCT leader, the OMTs, and the HOC. Each echelon reviews the intelligence and operational reports, identifies inconsistencies, evaluates veracity, and recommends additional lines of questioning as appropriate.

12-13. Analysts can assign an alphanumeric designator to collected information based on an analyst's determination of the reliability of the source and the probable accuracy of the information reported. A letter from A to F is assigned reflecting the level of reliability, with A representing the highest degree of reliability. The letter designation is then coupled with a number from 1 to 6. The number 1 represents confirmed intelligence and the numbers 2 through 6 represent increasing degrees of uncertainty as to the veracity of the information. A complete explanation if this rating system is contained in Appendix B. [NOTE: This system of labeling the reliability of sources and their reported information should not be confused with the alphanumeric rating scheme for source-screening operations as described in Chapter 6.]

SINGLE-DISCIPLINE HUMINT ANALYSIS AND PRODUCTION

12-14. Single-discipline HUMINT analysis involves the actions taken to evaluate the information provided by all HUMINT sources at a given echelon to determine interrelationships, trends, and contextual meaning. While called "single discipline," the analyst reviews and incorporates, as necessary, information from other disciplines and all-source analysis to provide a contextual basis for the HUMINT analysis. Single-discipline HUMINT analysis is conducted primarily by the HAT of the ACE. HUMINT operations sections and OMTs also conduct analysis to a lesser degree, based on the information from HUMINT sources at their echelon.

12-15. Analysis does more than simply restate facts. The analyst formulates a hypothesis based on available data, assesses the situation, and explains what the data means in logical terms that the user can understand. There are two basic thought processes used by analysts to study problems and reach conclusions: induction and deduction.

- Induction is the process of formulating hypotheses on the basis of observation or other evidence. It can best be characterized as a process of discovery when the analyst is able to establish a relationship between events under observation or study. Induction, or plausible reasoning, normally precedes deduction and is the type of reasoning analysts are required to perform most frequently.

- Deduction is the process of reasoning from general rules to particular cases. The analyst must draw out, or analyze, the premises to form a conclusion. Deductive reasoning is sometimes referred to as demonstrative reasoning because it is used to demonstrate the truth or validity of a conclusion based on certain premises.

ANALYTICAL TECHNIQUES AND TOOLS

12-16. There are three basic analytical techniques and automated tools that are particularly useful to single-discipline HUMINT analysis. Each of these tools takes fragmented bits of information and organizes them to create a chart or graph that can easily be read. They are the time event chart, matrices, and the link analysis diagram. HUMINT collectors and analysts can use automated computer programs such as Analyst Notebook or Crime Link to produce these tools or they can create them on paper. Computer programs are faster to use than previous methods and have the added advantage of producing a product that can be shared easily and rapidly over networks and portals. The diagrams in this chapter represent the tools that can be produced using automated programs.

Time Event Chart

12-17. A time event chart is a method for placing and representing individual or group actions in chronological form. It uses symbols to represent events, dates, and the flow of time. Normally, triangles are used to depict the beginning and end of the chart and may be used within the chart to indicate particularly critical events such as an ideological shift or change. Rectangles, used as event nodes, store administrative data and indicate significant events or activities. Drawing an "X" through the event node may highlight noteworthy or important events. Each of these symbols contains a sequence number, date (day, month, and year of the event), and may, if desired, contain a file reference number. The incident description written below the event node is a brief explanation of the incident and may include team size and type of incident. Arrows indicate time flow. By using these symbols and brief descriptions, it is possible to analyze the group's activities, transitions, trends, and particularly operational patterns in both time and activity. If desired, the event nodes may be color coded to indicate a particular event or type of event to aid in pattern recognition. The time event chart is the best analytical tool for pattern analysis. The example at Figure 12-1 depicts the history of the group, including most major players, which carried out the World Trade Center bombing in February 1993.

Matrices

12-18. Construction of a matrix is the easiest and simplest way to show the relationships between a number of similar or dissimilar associated items. The items can be anything that is important to a collection effort such as people, places, organizations, automobile license plates, weapons, telephone numbers, or locations. In analysis, matrices are often used to identify "who knows whom," or "who has been where or done what" in a clear concise manner. There are two types of matrices used in human analysis: the association matrix, used to determine existence of relationships between individual human beings, and the activities matrix, used to determine connectivity between individuals and any organization, event, address, activity, or any other non-personal entity. The graphics involved in constructing the two types of matrices differ slightly, but the principles are identical.

12-19. The association matrix (Figure 12-2) shows connections between key individuals involved in any event or activity. It shows associations within a group or associated activity. Normally, this type of matrix is constructed in the form of an equilateral triangle having the same number of rows and columns. Personalities must be listed in exactly the same order along both the rows and columns to ensure that all possible associations are correctly depicted. An alternate method is to list the names along the diagonal side of the matrix. This type of matrix does not show the nature, degree, or duration of a relationship, only that a relationship exists. The purpose of the matrix is to show the analyst who knows whom and who are suspected to know whom. In the event that a person of interest dies, a diamond is drawn next to his or her name on the matrix.

12-20. The analyst uses a dot or closed (filled-in) circle to depict a strong or known association. A known association is determined by direct contact between one or more persons. Direct contact is determined by several factors. *Direct associations* include—

- Face-to-face meetings.
- Telephonic conversations in which the analyst is sure who was conversing with whom.
- Members of a cell or other group who are involved in the same operations.

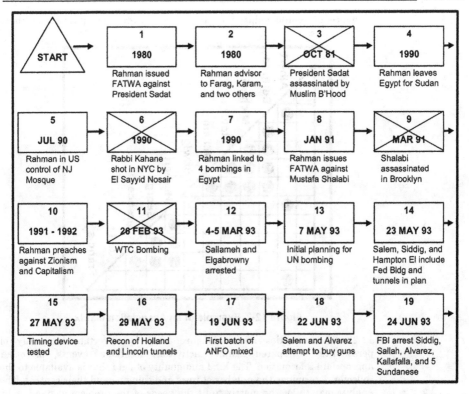

Figure 12-1. Example of a Time Event Chart.

12-21. Suspected or weak associations are those associations in which there are indicators that individuals may have had associations but there is no way to confirm that association; this is depicted with an open circle. Examples of *suspected associations* are—

- A known party calling a known telephone number (the analyst knows to whom the telephone number is listed) but it cannot be determined with certainty who answered the call.
- A face-to-face meeting where one party can be identified, but the other party can only be tentatively identified.

12-22. The rationale for depicting suspected associations is to get as close as possible to an objective analytic solution while staying as close as possible to known or confirmed facts. If a suspected association is later confirmed, the appropriate adjustment may be made on the association matrix. A secondary reason for depicting suspected associations is that it may give the analyst a focus for tasking limited intelligence collections assets in order to confirm the suspected association. An important point to remember about using the association matrix is that it will, without modification, show only the existence of relationships; not the nature, degree, or duration of those relationships.

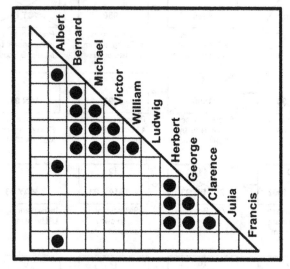

Figure 12-2. Example of an Association Matrix.

12-23. The activities matrix (Figure 12-3) is a rectangular array of personalities compared against activities, locations, events, or other appropriate information. The kind and quality of data that is available to the collector determines the number of rows and columns and their content. The analyst may tailor the matrix to fit the needs of the problem at hand or he may add to it as the problem expands in scope. This matrix normally is constructed with personalities arranged in a vertical listing on the left side of the matrix, and events, activities, organizations, addresses, or any other common denominator arranged along the bottom of the matrix. This matrix is critical for the study of a group's internal and external activities, external ties and linkages, and even modus operandi. As with the association matrix, confirmed or "strong" associations between individuals and non-personal entities are shown with a solid circle or dot, while suspected or "weak" associations are illustrated by an open circle.

12-24. Using matrices, the analyst can pinpoint the optimal targets for further intelligence collection, identify key personalities within an organization, and considerably increase the analyst's understanding of an organization and its structure. Matrices can be used to present briefings or to store information in a concise and understandable manner within a database. Matrices augment but cannot replace SOPs or standard database files. It is possible, and sometimes productive, to use one matrix for all associations. This is done routinely using the automated systems mentioned in paragraph 13-6.

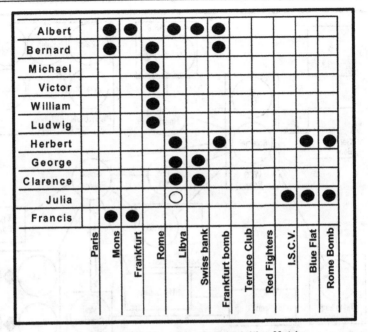

Figure 12-3. Example of an Activities Matrix.

12-25. The link analysis diagram (Figure 12-4) shows the connections between people, groups, or activities. The difference between matrices and link analysis is roughly the same as the difference between a mileage chart and a road map. The mileage chart (matrix) shows the connections between cities using numbers to represent travel distances. The map (link analysis diagram) uses symbols that represent cities, locations, and roads to show how two or more locations are linked to each other.

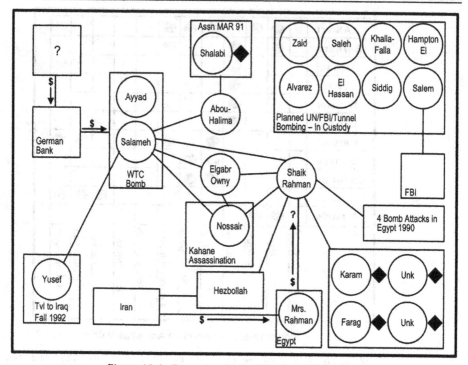

Figure 12-4. Example of a Link Analysis Diagram.

12-26. As with construction of association matrices, there are certain rules of graphics, symbology, and construction that must be followed. Standardization is critical to ensuring that everyone constructing, using, or reading a link analysis diagram understands exactly what the diagram depicts. Circles and lines are arranged so that no lines cross whenever possible. Often, especially when dealing with large groups, it is very difficult to construct a line diagram in which no lines cross. In these cases, every effort should be made to keep the number of crossings at an absolute minimum. The standard rules are as follows:

- Persons are shown as open circles with the name written inside the circle. Deceased persons are depicted in either open circles, with a diamond next to the circle representing that person (as in Figure 12-4) or as open diamonds with the name written inside the diamond.

- Persons known by more than one name (alias or AKA) are shown as overlapping circles with names in each circle (as shown below) or both names are simply listed in the same circle. If the alias is suspected, a dotted line is used to depict the intersection. If the alias is confirmed, the intersection is shown with a solid line.

- Non-personal entities (organizations, governments, events, locations) are shown as appropriately labeled rectangles.

- Solid lines denote confirmed linkages or associations and dotted lines show suspected linkages and associations.

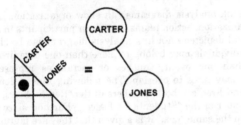 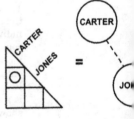

- Footnotes on the matrices can be shown as a brief legend on the connectivity line.

 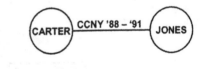

3. Attended CCNY '88-'91

- Each person or non-personal entity is depicted only once in a link analysis diagram.

12-27. The following diagram shows only connectivity between persons:

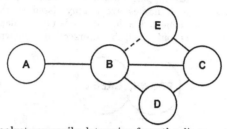

12-28. The analyst can easily determine from the diagram that Alpha knows Bravo, Bravo knows Charlie and Delta. Bravo is suspected of knowing Echo and Charlie knows Delta, Bravo, and Echo. Although the same information could be shown on a matrix, it is easier to understand when depicted on a link analysis diagram. As situations or investigations become more complex, the ease in understanding a link analysis diagram becomes more apparent. In almost all cases, the available information is first depicted and analyzed on both types of matrices, which are then used to construct a link analysis diagram for further analysis.

12-29. Link analysis diagrams can show organizations, membership within the organization, action teams or cells, or participants in an event. Since each individual depicted on a link analysis diagram can be shown only once, and some individuals may belong to more than one organization or take part in more than one event, squares or rectangles representing non-personal entities may have to overlap. The following illustration demonstrates that Ralph and Fred are both members of the "Red Fighters," and that Fred also is a member of the "Students for Peace." Further, since Ralph and Fred are shown in the same "box," it is a given that they are mutually associated.

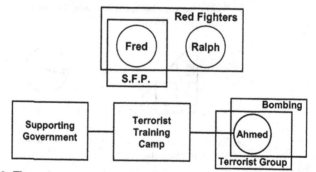

12-30. There is more to overlapping organizations than is immediately obvious. At first glance, the overlap indicates only that an individual may belong to more than one organization or has taken part in multiple activities. Further study and analysis would reveal connections between organizations, connections between events, or connections between organizations and events, either directly or through persons. The above diagram reveals a more complex connection between organizations, personal connections, and linkages.

12-31. The analysis diagram in paragraph 12-29 shows a connection between organizations and events to which an individual belongs or is associated. In this case, a national government runs a training camp for terrorists. Ahmed, a member of the terrorist group, is associated with the training camp, and participated in the bombing attack. From this diagram, one can link the supporting government to the bombing through the camp and the participant.

12-32. When, as is often the case, an organization or incident depicted in a link analysis diagram contains the names of more than one individual, it is not necessary to draw a solid line between those individuals to indicate connectivity. It is assumed that individual members of the same cell or participants in the same activity know each other, and the connection between them is therefore implied. If the persons are **not** mutually associated, they cannot be placed in the same "box." Another solution must be found to depict the situation; that is, show the persons as associated with a subordinate or different organization or activity.

12-33. A final set of rules for link analysis diagrams concerns connectivity between individuals who are not members of an organization or participants in an activity, but who are somehow connected to that entity. Two possibilities exist: First, the individual knows a member or members of the organization but is not associated with the organization itself; or second, the person is somehow connected with the organization or activity but cannot be directly linked with any particular member of that entity.

12-34. In the first case, the connectivity line is drawn only between the persons concerned as depicted here:

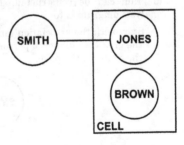

12-35. In the second case, where Smith is associated with the entity, but not the persons who are members of entity, the situation is shown as depicted here:

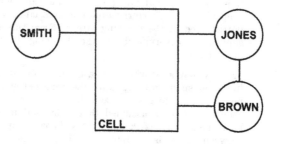

12-36. The steps in constructing a link analysis diagram are as follows:

- **Step 1.** Raw data or fragments of information are organized into logical order. Names of individuals, organizations, events, and locations are compiled on appropriate lists. At this point, a time event chart may be completed to assist in understanding the information and to arrange events into chronological order.
- **Step 2.** Information is entered onto the appropriate matrices, graphically displaying "who is associated with whom" and "who is associated with what."
- **Step 3.** Drawing information from the database and intelligence reports, and relationships from the matrices, the link analysis diagram can be constructed. The best method to start the link analysis diagram is to—
  - Start with the association matrix and determine which person has the greatest number of personal associations. Depict that person in the center of the page.

  - Determine which person has the next highest number of personal associations. Depict that person near the first person.

12-37. Use the association matrix and show all confirmed and suspected personal associations.

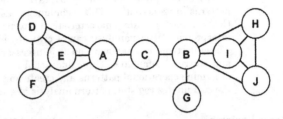

- After all personal associations have been shown on the link analysis diagram, the analyst uses the activities matrix to determine which activities, organizations, or other non-personal entities need to be depicted by appropriate rectangles. Having done so, the lines of connectivity between persons within the rectangles may be removed to prevent clutter. (It is assumed that participants in the same activity or members of the same cell are acquainted.)

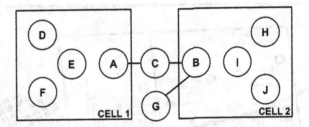

12-38. As shown in Figure 12-4, the link analysis diagram depicts the membership, organization, activities, and connections of the group that, under the leadership of Sheik Omar Abdul Rahman, carried out the bombing of the World Trade Center in New York City and planned other bombing attacks. Together with the time event chart (Figure 12-1), one can gain a basic understanding of the group and its activities, and develop working hypotheses for additional collection and analysis efforts.

12-39. After completion of the matrices and the link analysis diagram, the analyst makes recommendations about the group's structure, and areas can be identified for further collection. Collection assets are employed to verify suspected connections, ID key personalities, and substantiate or refute the conclusions and assessments drawn from the link analysis that has been done. The link analysis diagram and thorough analysis of the information it contains can reveal a great deal about an organization. It can identify the group's leadership, its strong and weak points, and operational patterns. The analyst can use these to predict future activities.

OTHER ANALYSIS TOOLS

12-40. Pattern analysis is the process of deducing the doctrine and TTP that threat forces prefer to employ by careful observation and evaluation of patterns in its activities. This technique is based on the premise that threat COAs reflect certain characteristic patterns that can be identified and interpreted. Pattern analysis can be critically important when facing a threat whose doctrine is unknown and it is necessary to create new threat model and doctrinal templates. Three additional tools that can help the analyst to determine operational patterns and create and update their threat model are the coordinates register, pattern analysis plot sheet, and OB factors.

Coordinates Register

12-41. The coordinates register, or incident map, is one type of pattern analysis tool (Figure 12-5). It illustrates cumulative events that have occurred within the AO and focuses on the "where" of an event. The analyst may use multiple coordinates registers that focus on a different subject or blend subjects. Normally, the coordinates register includes additional information such as notes or graphics. The analyst should use the coordinates register in conjunction with the pattern analysis plot sheet.

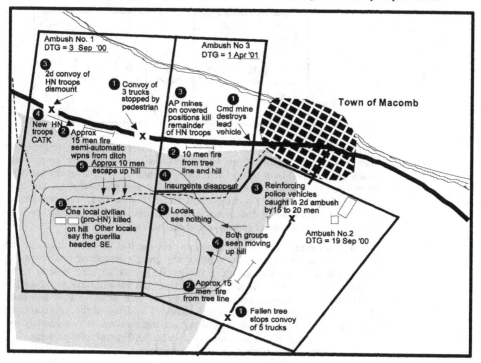

Figure 12-5. Coordinates Register.

Pattern Analysis Plot Sheet

12-42. The analyst uses a pattern analysis plot sheet to focus on the time and date of each serious incident that takes place within the AO (Figure 12-6) The rings depict days of the month; the segments depict the hours of the day As shown in the sheet's legend, the event itself is identified by using an alphanumeric designation and directly corresponds to the legend used on the coordinates register. Another type of the pattern analysis plot sheet helps distinguish patterns in activities that are tied to particular days, dates, or times. When used in conjunction with the coordinates register and any doctrinal templates, a pattern analysis plot sheet supplies the bulk of the data needed to complete an event template.

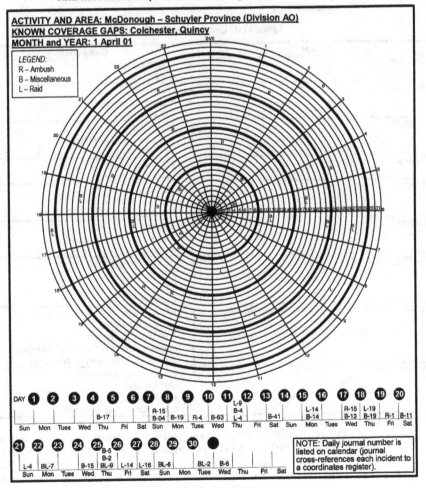

Figure 12-6. Pattern Analysis Plot Sheet.

OB Factors

12-43. A final analytical tool is OB. The OB does not predict enemy intentions or probable COAs but is a means of cataloging intelligence data that qualifies and quantifies certain aspects of threat units. The analyst uses nine factors shown in Table 12-1 to focus the analysis. These help the analyst determine threat capabilities, vulnerabilities, and COAs. Although the situation may dictate that one or more of the factors are given a higher priority, generally speaking, they are all of equal importance. The OB framework, while initially developed to support force-on-force offensive and defensive operations, is easily adapted to support stability and reconstruction operations and to depict "unconventional" forces or even civilian organizations.

Table 12-1. Order of Battle Factors.

Factor	Component
Composition	• Unit identification • Organization
Disposition	• Geographical Location • Tactical Deployment • Movements
Strength	• Personnel • Weapons and Equipment • Types of Units
Tactics	• Tactical Doctrine • Special Operations
Training	• Individual • Unit • Specialized
Logistics	• Systems • Status • Funding
Combat Effectiveness	• Combat Experience • Morale • Tactics • Logistics
Electronic Technical Data/Emitter Nomenclature	• Emitter Type • Mode of Emission • Frequency Range • Location Accuracy for Direction Finding • Associated Use (Units or Weapons)
Miscellaneous	• Personalities/Leadership • Unit History • Uniforms and Insignia • Code Names and Numbers

12-44. The OB is based on the premise that there are certain constants to any group activity. All groups whether they are conventional military forces, insurgent groups, or civilian organizations must have an organizational structure (composition). This structure may not be easily discernable but it will exist. Likewise, any organization has a location or locations in which it operates, personnel and equipment numbers, a system for training, getting supplies, judging efficiency and effectiveness of its operations, communicating, and taking care of other intangibles such as morale. The OB gives the analyst a framework to organize information. The analyst adapts the topic headings to match the particular environment.

HUMINT SOURCE SELECTION

12-45. HUMINT source selection involves identifying, researching, and actively locating a specific group, organization, or individual for the purpose of collecting information in response to intelligence requirements. The HUMINT source selection process includes the C/J/G/S2X, the HOC, the HAT, the OMT, and the HCT. The source selection process allows the HUMINT team leader to identify the most likely source of information, eliminating the uncertainty of the access and placement of sources. Source selection also helps optimize the HUMINT collection effort. Any individual, group, organization, or agency that can be approached for information regarding intelligence requirements is a potential source. Sources are chosen according to their reliability, level of cooperation, and placement and access. Selection is particularly important in stability and reconstruction operations where the HUMINT collectors have access to a large potential source pool. Source selection establishes which current sources can best answer requirements and establishes source profiles to support the screening and selection of new sources.

ESTABLISH DATABASES AND TARGET FOLDERS

12-46. The establishment of local databases, target folders, and personality files is normally the responsibility of the OMT. This is done in coordination with the supporting ACE or analysis control team (ACT). Databases are required to manage the information. By using databases one can identify gaps in the information. The HCT and OMT access higher databases through intelligence reach to share and deconflict locally maintained data with higher level databases. Local databases can be created and used to help track source production, knowledge, reliability, and accuracy, and they simplify cross-reference data that is of primarily local interest. It is ideal to review and update databases at least weekly.

12-47. A target folder provides the collector with up-to-date intelligence information about details of the target. It includes anything of HUMINT value including biographies, descriptions, photographs, and previous information reports. The information can be gained from the ACE or ACT, past reports, INTSUMs, and databases; it can then be organized into easily accessible automated folders. Information on people is categorized and recorded in a personality file. The file serves as reference material for collectors. Information on key military and civilian figures can be of significant value when establishing unit or group identification, tactics, and combat effectiveness. The file should not only provide information on cultural, religious, tribal, political, military, criminal, and governmental background but also contain specific personalities for collectors to focus their collection effort on. This allows the collectors to concentrate on mission planning and to conduct their mission rather than to research information.

COMPARE SOURCE LIST WITH REQUIREMENTS

12-48. As target folders are compiled, a list of high-value sources will emerge. It is a simple matter to compare the source list with the PIRs and/or SIRs. This will lead to efficient and time-saving missions for the collectors. Rather than spend time meeting with sources who may have information concerning certain subjects, the HCT is able to tackle the collection process with foresight. Upon receipt of the mission, the OMT conducts mission analysis to determine the optimal way to meet mission requirements. Proper mission analysis enables the collector to properly focus his assets (sources) to gain the maximum amount of intelligence from those sources most likely to possess the highest quality information.

12-49. The following products, which will focus the HCT's collection efforts, can be prepared in conjunction with the ACE and joint intelligence centers:

- Time event charts.
- Source coverage overlays or matrices (see Figure 12-7).
- Link analysis diagrams.
- HUMINT portions of OPORDs and situation reports.

12-50. The source coverage overlay or matrix helps tie in the source coverage to the requirements. It also helps identify gaps in collection. A collection matrix serves the same purpose. It supplies a quick reference when answering intelligence requirements. The matrices must cover both the geographical area and the placement and access of the source. A demographic overlay helps to identify ethnic groups in an area and to track events and patterns based on religious or ethnic differences. The overlay and matrix are examples of how source coverage can be tied to intelligence requirements.

12-51. Source profiles are vital to screening sources for HUMINT collection operations and to identifying personnel that might be of interest to other agencies such as CI and TECHINT. As the situation changes, the HCT might be tasked with new collection requirements that cannot be answered by the current sources. The HCT is constantly looking for new and better sources. When presented with new requirements, the OMT develops a source profile of the type of individual that would most likely be able to provide the information required. This profiling can include placement, access, age, ethnic type, gender, location, occupation, and military specialty. The OMT first searches through existing local databases to try to get a source match. If not, it passes the profile to the HCTs along with the requirements to facilitate their screening of potential sources.

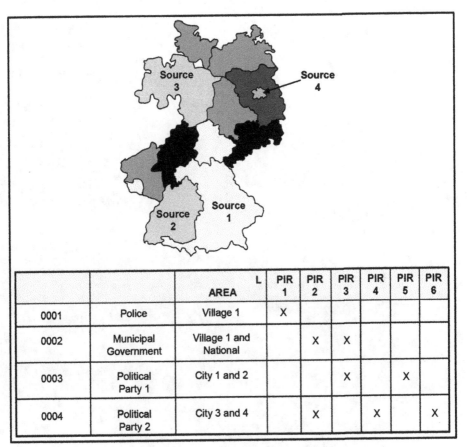

		AREA	L PIR 1	PIR 2	PIR 3	PIR 4	PIR 5	PIR 6
0001	Police	Village 1	X					
0002	Municipal Government	Village 1 and National		X	X			
0003	Political Party 1	City 1 and 2			X		X	
0004	Political Party 2	City 3 and 4		X		X		X

Figure 12-7. Example of a Source Coverage Overlay or Matrix.

Chapter 13

Automation and Communication

13-1. Modern automation and communications systems are vital to HUMINT collection. Real-time collaboration, detailed operational planning and ISR integration, as well as enhanced collection and source exploitation tools, must support team efforts. Emerging technology continues to allow the entire HUMINT collection system to operate more effectively. Commanders must be prepared to supply their HUMINT collection assets with the best possible technology not only to enhance collection but also to optimize the survivability of the collectors. (See Appendix L.) Commanders may not be able to rely solely on standard military equipment but must be prepared to bridge the inevitable technological development gap through the identification and adaptation of commercially available products and technologies. For specific system components and capabilities, see ST 2-50.

AUTOMATION

13-2. HUMINT automation uses common hardware and software solutions with a flexible interactive user interface to provide standardization of equipment and processes across all operational environments and conditions. HUMINT automation must be deployable and scalable to fit the mission or force package. System components must be capable of intelligence reach to support forward-deployed elements. HUMINT automation allows integration and interaction with existing intelligence operations, HUMINT operational systems, and databases. This integration allows operations personnel and analysts to develop plans and levy collection and operational requirements, as well as to manage, control, analyze, and report the information collected. HUMINT automation—

- Provides connectivity and reach capability between all echelons of HUMINT activity.
- Receives higher echelon requirements and transmits requests for information.
- Converts HUMINT reporting into formats for JTF or coalition task force (CTF), theater, or national consumption.
- Pushes requirements, requests, and plans for HUMINT operations in theater as required.
- Maintains the central HUMINT database for the theater or AO.
- Leverages JTF or CTF, theater, and national level requirements and products for strategic, operational, and tactical HUMINT assets in theater.

• Enables HUMINT to provide accurate and timely correlated information to supported commanders through established reporting channels.

• Provides automated analysis tools.

13-3. Systems such as Trusted Workstation (TWS) can convert HUMINT reporting into formats for JTF or CTF, theater, or national consumption. TWS can also connect between the SIPRNET and any lower level networks, such as coalition, multinational, or NATO, or unclassified networks such as NIPRNET or the Internet.

COLLECTION SUPPORT AUTOMATION REQUIREMENTS

BIOMETRICS

13-4. Biometrics is the study of measurable biological characteristics. In HUMINT collection, biometric devices, usually computer based, enable the HUMINT collector to use biological data to support the collection and analysis effort. Biometrics can also be used in non-HUMINT efforts to collect and maintain evidence for criminal prosecution. The two major types of biometric data that are useful to the HUMINT collector are identification data and data that indicate source truthfulness. Identification devices use biological information such as fingerprints, voiceprints, facial scans, and retinal scans to match an individual to a source database. They can verify the identity of a specific individual from the target population during screening.

13-5. HCTs may be equipped with portable equipment for collecting, storing, analyzing, forwarding, and retrieving biometric information. The BAT is able to identify personnel by using identifying characteristics of their irises, fingerprints, or facial photograph. The structured query language (SQL) server-based database links identifying characteristics with all previous reports related to the person. Once a person's identifying characteristics are entered into the database, if that person is again detained and scanned, the system has a probability of identifying them that approaches 100 percent. This ability is especially useful for determining if a source is providing the same information to multiple collectors; thereby avoiding false confirmation of information. HUMINT collectors primarily use BAT during screening operations at all echelons; from checkpoint screening, to screening at a DCP, to screening at a JIDC. MPs use the Detainee Reporting System (DRS) during in-processing at internment facilities. The DRS records data for detainee processing and tracking and is intended to interact with the BAT system to avoid duplication of effort.

13-6. The hardware that makes up the system, which is given to MI teams, consists of a commercial off-the-shelf (COTS) semi-hardened laptop computer running an operating system with a graphical user interface. It has a camera and an iris scanner, each of which is portable and can be used independent of the computer to collect and temporarily store information. The system also includes a fingerprint scanner that conforms to FBI requirements for admissible evidence. The fingerprint scanner must be attached to the computer during use.

13-7. Biometric devices such as voice stress analyzer and polygraph support the determination of the truthfulness of a source. The polygraph is of limited usefulness in general HUMINT collection due to the level of expertise needed to operate it and the lack of general availability of the device to the field. As devices are developed that can remotely collect and report information to the HUMINT collector on subtle changes in the source's respiration, heartbeat, perspiration, and eye movement that can be indicators of deceit, they can be used to support HUMINT collection.

MACHINE TRANSLATION AND INTERPRETATION

13-8. Understanding oral and written communication in a foreign language is often the center of effective HUMINT collection operations. The optimal solution is to have an individual who is a trained collector of native proficiency, totally versed in the local situation and US requirements, with the requisite security clearance, who is capable of reporting accurately in English. Commanders' access to such individuals is usually problematic. This requirement is met through a combination of MI linguists, contractors, native speakers within the DOD system, and locally hired civilian translators. Difficulties arise if the proficiency levels of MI linguists are not up to mission requirements, or if the linguists do not possess the proper language for the theater of operation. Using locally hired translators raises security problems. In light of these conditions, an increasingly viable solution for the commander is the use of machine translation devices to meet some of these requirements.

13-9. Voice and text translation machines or software are critical in augmenting available linguists. This includes natural language processing, artificial intelligence, and optical character recognition (OCR) capabilities. The basic application of machine translation, such as speech recognition and OCRs, dramatically increases the speed of processing information. Software programs are becoming widely available that allow a non-linguist to determine the intelligence significance of a foreign document, aid linguists with laborious tasks, and add consistency to human translation.

13-10. Machine interpretation is the use of a machine to interpret the spoken word between the HUMINT collectors and another individual speaking a foreign language. Linguists are in high demand during operations and usually limited in number. As machine interpretation devices that address this problem become available to the field, they will improve the communication ability of non-linguists.

ANALYTICAL AUTOMATION REQUIREMENTS

13-11. The requirement for a robust HUMINT single-discipline analytical capability extends through all echelons from national level to the OMTs. Communication between HUMINT analysts at the operational level and analysts at the staff level may best be accomplished through a web-based communication capability. Web-based visual analytical tools allow maximum analyst participation in the development of products geared to mission planning, targeting, and information analysis at all echelons.

Analytical products must be responsive to the special needs of a specific collection operation, project, or element.

13-12. HUMINT collectors run operations in terrain made up of persons, organizations, and installations of interest. Intelligence analysis support determines the specific terrain in each team area and how it differs from one team's named area of interest (NAI) to another. Specific products include studies on nominated targets (persons, organizations, and installations) and trends based on HUMINT reporting, as appropriate, and visual analysis products (time event charts, matrices, link analysis diagrams, and organizational diagrams).

AUTOMATED ANALYSIS TOOLS

13-13. Automation of HUMINT analytical tools such as time event charts, association matrices, activity matrices, and link analysis diagrams dramatically increase predictive analysis capability. Automation saves time and permits access to more complete information thus producing a more accurate, timely product. Automated analysis techniques, aided by computerized virtual-viewing programs, allow the analyst better battlefield visualization. Automated analysis, linked to data and databases, includes artificial intelligence programs. These programs assist the analyst in developing predictions and identifying information gaps to support targeting and collection. Automation and web-based tools allow the analyst to—

- Track and cross-cue HUMINT reports.
- Incorporate data extraction technology, retrieval, automated data organization, content analysis, and visualization.
- Share analytical conclusions with HUMINT teams and other analysts in real time.
- Apply multidimensional technologies, content analysis techniques, and web-based collaborations.
- Display analytical results and view HUMINT operations in real time.
- Share resources such as models, queries, visualizations, map overlays, geospatial images, and tool outputs through a common interface.
- Apply clustering (a nonlinear search that compiles the results based on search parameters) and rapid spatial graphical and geographic visualization tools to determine the meaning of large informational streams.
- Rapidly discover links, patterns, relationships, and trends in text to use in predictive analysis.
- Capture analytical conclusions and automatically transfer to intelligence databases and systems.

SEARCH ENGINES

13-14. Search engines provide access to previously collected or known information facilitating the development of comprehensive analytical and intelligence products and avoiding unnecessary collection tasking redundancy. A tool set for data visualization, search, and discovery is required, which is embedded with several software programs for manipulating data from multiple databases. The types of modules in visualization packages should include search engines and knowledge discovery (semantic clustering) for unformatted data, applications for extracting and organizing formatted data, and data labeling. The package should also include a model building tool to enable users to make their archives more efficient with respect to search, retrieval, and compatibility to other applications as well as archiving and maintenance tools to support what will eventually become a large data warehouse. Search engines should be—

- Multilingual and able to query multiple classified and unclassified databases.

- Capable of developing, querying, and manipulating stored information.

WEB-BASED REPORTING AND PORTALS

13-15. Web-based reporting employs current Internet portal technology. The web-based portal is an efficient and effective means of developing a repository of HUMINT information. It employs an interactive graphic interface using client browser technology, search engines, hyperlinks, and intelligent software agents for searching, finding, viewing, and maintaining databases and supporting HUMINT work, data, and information flows. It supports collaborative analysis at multiple echelons through connectivity on the SIPRNET. The following pertains to web-based reporting:

- Web-based databases work with any computer hardware, operating system, or software and can be made accessible through web portals.

- Firewalls and information access are controlled at each level with an approving systems administrator at each level conducting quality control through release authority procedures.

- Graphic user interface uses standard Army and DOD report formats.

- Graphic user interface walks the user through a critical task and is able to identify Army and DOD reports as required. Reports must be Army and DOD platform compatible and transferable through and to their respective systems.

- Multimedia supports applications for attaching, associating, and hyperlinking video, still photographs, voice, scanned objects, graphics, and maps to records and files.

13-16. Web-based reporting and web pages developed for specific products allow the user to—

- Leverage their effort and expertise against all requirements, not just the ones that must be met immediately.

- Identify timely intelligence gaps and the leads to fill those gaps.
- Ensure immediate analytical feedback on collector reports to—
  - Post questions directly to a web page to enable all HCTs to answer or be cued to the specific request.
  - Identify or request clarification on questionable data for quality control.
- Fuse HUMINT information and all-source information as required.
- Focus collection teams supporting maneuver commanders' requirements more effectively.
- Immediately extract information for crisis reaction.

13-17. If HCTs use web portals to submit reports directly to theater level, they must also send the reports through their OMT for submission to the 2X. Failure to do so may deny the 2X the ability to deconflict and cross-reference reports. HUMINT collectors must ensure that they follow the theater-specific methodology for access to the theater web portal.

DATABASES

13-18. Without databases, information is difficult or impossible to retrieve quickly, especially under adverse conditions. Databases allow access to data in a distributed environment and support many complex HUMINT functions and requirements, including—

- Mission deconfliction.
- RM.
- RFIs.
- HUMINT analysis.
- Summary, report, and assessment preparation.
- Threat and friendly situation tracking.
- Targeting.

13-19. Databases interact with other tools to support predictive analysis, prepare graphic analytical products, and provide situational understanding down to the HCT. These databases—

- Support time event charts, association matrices, link analysis, and other analysis tools.
- Require a designated systems administrator at each. To ensure a high degree of integrity, discrepancies must be verified for accuracy.
- Allow operators, managers, and analysts to—
  - Compartment (protect) source-sensitive operational database segments, files, records, and fields.
  - Create, update, and maintain databases from locally generated information.
  - Import complete or partial databases from larger or peer databases.
  - Export complete or partial databases to peer or larger databases.

- Share data and databases between peers, subordinates, or higher with appropriate access authorization.
- Provide systematic processing and automated parsing using standardized forms in intelligence operations, which are automatically parsed into appropriate databases for information storing, sharing, retrieval, and analysis.
- Allow query functions for decisionmaking as well as operational and analytical support.
- Provide analytical programs able to correlate data that facilitate information retrieval from any data repository.
- Incorporate information retrieval functions such as browsing (that is, point and click), key word searching, concepts, and similar functions.
- Support a suite of specialized decision support software (DSS)—a set of tools which supports HUMINT source administration, analysis, and risk management decisions. DSS tools should produce a set of HUMINT reports specifically tailored to the HUMINT decisionmaking, analysis, and assessment process.

13-20. HARMONY is the national intelligence database for foreign DOCEX and translations management. HARMONY is the single, comprehensive bibliographic reference for all available primary source foreign technical and military documents and their translations. This single database concept eliminates duplicate collection, translation, and reporting of primary source foreign technical and military documents and greatly streamlines the process of providing exploited documents to consumers. The HARMONY database application uses the DOD Information Infrastructure (DII) Common Operating Environment (COE) guidance. The HARMONY database is web-enabled and can be readily accessed, easily used, and responsive to the needs of analysts and other consumers within the US Government community.

AUTOMATION SYSTEMS

13-21. The HUMINT collection automation systems are normally shared systems used by both the HUMINT and CI communities. They must have connectivity with lateral units as well as higher and subordinate elements.

C/J/G/S2X, HOC, AND HAT AUTOMATION REQUIREMENTS

13-22. The HUMINT staff elements need to receive input from the OMTs and HCTs as well as input from higher and lateral echelons. They must be able to conduct HUMINT planning, RM, and report dissemination. They must transmit technical support information, interface with ACE and automated analysis systems, manipulate HUMINT databases, conduct reach, and have access to HUMINT analytical tools.

OMT AND COLLECTION TEAM LEADER REQUIREMENTS

13-23. The OMT must be able to track teams and team members; receive and transmit data including graphic data to and from higher, lateral, and lower HUMINT elements; create, receive, edit, and transmit reports; conduct single-discipline HUMINT analysis; receive and transmit technical support information and tasking information; conduct reach; and conduct mission planning.

INDIVIDUAL COLLECTOR AUTOMATION REQUIREMENTS

13-24. The key to effective HUMINT collection is unimpeded communication between the collector and the source of information. Any technological support to HUMINT collection must be as unobtrusive as possible to minimize the intimidation factor when dealing with human sources. The individual collector must be able to—

- Record (both video and voice) conversations with sources.
- Scan, translate, and transmit documents and photographs.
- Instantaneously locate themselves in both rural and urban environments.
- Immediately access local, theater, and even national level databases.
- Communicate instantaneously with other team elements.

HUMINT AND CI WORKSTATION REQUIREMENTS

13-25. The CI and HUMINT teams have organic computer and data processing equipment. These workstations provide HUMINT and CI teams with both productivity and management and analysis tools. They also provide SIPRNET connectivity and processing capability to identify requirements and facilitate reporting into other DOD systems as required. The HUMINT and CI workstation is able to use standard Army, DOD, and HUMINT and CI reporting programs, standard symbols, programs to produce map overlays, and map plotting software; all of which are included.

13-26. Teams use workstations to—

- Provide quality control and dissemination of reports from the subordinate HCTs.
- Direct activities of subordinate HCTs and provide management to them.
- Perform single-discipline HUMINT analysis for the supported commander.
- Transmit intelligence and administrative reports in NRT to higher headquarters.
- Receive tasking and administrative reports from higher headquarters and distribute to HCTs as required.
- Consolidate local databases and provide database input to higher headquarters.
- Receive database and digital information from higher headquarters and pass to lower and vice versa.

COMMUNICATIONS

13-27. Successful HUMINT operations must be supported by multi-echelon technical control and a communications system that provides internal team communications, links HCTs to OMTs, and links OMTs to higher headquarters, analytical elements, and theater and national agencies.

COMMUNICATION ARCHITECTURE

13-28. The HUMINT collection architecture requires operation on several communications and processing nets. These nets provide the framework needed to coordinate the tasking, reporting, C2, and service support of HUMINT collection units spread across the width and depth of the battlefield. Under most operational scenarios, HCTs are not stationary. They are constantly moving throughout their supported command's AO and are able to communicate on the move. They cannot rely on fixed communications nodes for support. Communications redundancy ensures the loss of any one system does not severely disrupt HUMINT operations. HCTs and OMTs normally operate at the collateral security level to ensure the timely dissemination of combat information and targeting data to organizations operating outside MI channels. The C/J/G/S2X normally requires access to Top Secret (SCI) communications capabilities to maintain coordination with national level agencies.

13-29. The HUMINT collection assets use three basic communications nets: the operations and intelligence (O/I) net, a command net, and a HUMINT-specific technical net. Dependent on their mission and battlefield location, the HCTs may also need to monitor the fire support element (FSE), aviation, or air defense artillery (ADA) communications nets.

- The O/I net links the collectors and producers of intelligence to the consumers of the intelligence information. It is used to pass information of immediate value to the affected unit and to analytical elements at the supported unit.

- The command nets exist at every echelon of command. They link the superior headquarters with its subordinate elements. Normally a unit will operate on two command nets; the one that links that unit to its higher headquarters and the one that links that unit to its subordinate elements. HUMINT elements will also use their unit's command net to coordinate logistic and administrative support.

- The technical nets link the control team to all of their subordinate collection teams and to the centers or organizations that provide the databases and technical guidance necessary for single-discipline collection and analysis. For example, the technical net would connect HCTs through their control teams to the S2X and higher echelon HUMINT analysis organizations.

MOBILE COMMUNICATIONS REQUIREMENTS

13-30. HUMINT mobile communications requirements augment the network connectivity that elements should have when at a base camp or facility; they vary with each element's mission and location as follows:

- Individual HUMINT collectors must maintain communications capability with the other team members and the team leader while dismounted. HUMINT collectors, especially when supporting offensive and defensive operations, may be deployed as individuals. They need to maintain contact with their team leader for technical and operational support.

- The HCT may operate anywhere within the supported unit's AO. They may operate mounted or dismounted. If supporting airmobile, airborne, amphibious, or other mobile operations, they may not have access to their vehicle-mounted communications systems for the critical early stages of these operations. They receive and report operational and technical information, as well as report intelligence information to the OMT using their unit's command net. They monitor their superior unit's O/I net. If in DS to a maneuver element, they also monitor the command net of the unit they are supporting.

- OMTs normally operate on the superior unit O/I net, their unit C2 net, and the HUMINT technical net. If the OMT is in DS, it must also operate on the C2 net of the supported unit.

- The C/J/G/S2X operates on the C2 net, monitors the O/I net, and controls its echelon HUMINT technical net. The 2X needs secure (SCI) communications capability to coordinate operations and pass data between themselves and higher HUMINT organizations.

Appendix A

Geneva Conventions

The articles in this section are extracted from the Geneva Convention Relative to the Treatment of Prisoners of War, 12 August 1949.

PART I GENERAL PROVISIONS

ARTICLE 1

The High Contracting Parties undertake to respect and to ensure respect for the present Convention in all circumstances.

ARTICLE 2

In addition to the provisions which shall be implemented in peace time, the present Convention shall apply to all cases of declared war or of any other armed conflict which may arise between two or more of the High Contracting Parties, even if the state of war is not recognized by one of them.

The Convention shall also apply to all cases of partial or total occupation of the territory of a High Contracting Party, even if the said occupation meets with no armed resistance.

Although one of the Powers in conflict may not be a party to the present Convention, the Powers who are parties thereto shall remain bound by it in their mutual relations. They shall furthermore be bound by the Convention in relation to the said Power, if the latter accepts and applies the provisions thereof.

ARTICLE 3

In the case of armed conflict not of an international character occurring in the territory of one of the High Contracting Parties, each party to the conflict shall be bound to apply, as a minimum, the following provisions:

1. Persons taking no active part in the hostilities, including members of armed forces who have laid down their arms and those placed hors de combat by sickness, wounds, detention, or any other cause, shall in all circumstances be treated humanely, without any adverse distinction founded on race, color, religion or faith, sex, birth or wealth, or any other similar criteria. To this end the following acts are and shall remain prohibited at any time and in any place whatsoever with respect to the above-mentioned persons:

(a) Violence to life and person, in particular murder of all kinds, mutilation, cruel treatment and torture;

(b) Taking of hostages;

(c) Outrages upon personal dignity, in particular, humiliating and degrading treatment;

(d) The passing of sentences and the carrying out of executions without previous judgment pronounced by a regularly constituted court affording all the judicial guarantees which are recognized as indispensable by civilized peoples.

2. The wounded and sick shall be collected and cared for.

An impartial humanitarian body, such as the International Committee of the Red Cross, may offer its services to the Parties to the conflict.

The Parties to the conflict should further endeavor to bring into force, by means of special agreements, all or part of the other provisions of the present Convention.

The application of the preceding provisions shall not affect the legal status of the Parties to the conflict.

ARTICLE 4

A. Prisoners of war, in the sense of the present Convention, are persons belonging to one of the following categories, who have fallen into the power of the enemy:

1. Members of the armed forces of a Party to the conflict as well as members of militias or volunteer corps forming part of such armed forces.

2. Members of other militias and members of other volunteer corps, including those of organized resistance movements, belonging to a Party to the conflict and operating in or outside their own territory, even if this territory is occupied, provided that such militias or volunteer corps, including such organized resistance movements, fulfill the following conditions:

(a) That of being commanded by a person responsible for his subordinates;

(b) That of having a fixed distinctive sign recognizable at a distance;

(c) That of carrying arms openly;

(d) That of conducting their operations in accordance with the laws and customs of war.

3. Members of regular armed forces who profess allegiance to a government or an authority not recognized by the Detaining Power.

4. Persons who accompany the armed forces without actually being members thereof, such as civilian members of military aircraft crews, war correspondents, supply contractors, members of labor units or of services responsible for the welfare of the armed forces, provided that they have received authorization from the armed forces which they accompany, who shall provide them for that purpose with an identity card similar to the annexed model.

5. Members of crews, including masters, pilots and apprentices, of the merchant marine and the crews of civil aircraft of the Parties to the conflict, who do not benefit by more favorable treatment under any other provisions of international law.

6. Inhabitants of a non-occupied territory, who on the approach of the enemy spontaneously take up arms to resist the invading forces, without having had time to form themselves into regular armed units, provided they carry arms openly and respect the laws and customs of war.

B. The following shall likewise be treated as prisoners of war under the present Convention:

1. Persons belonging, or having belonged, to the armed forces of the occupied country, if the occupying Power considers it necessary by reason of such allegiance to intern them, even though it has originally liberated them while hostilities were going on outside the territory it occupies, in particular where such persons have made an unsuccessful attempt to rejoin the armed forces to which they belong and which are engaged in combat, or where they fail to comply with a summons made to them with a view to internment.

2. The persons belonging to one of the categories enumerated in the present Article, who have been received by neutral or non-belligerent Powers on their territory and whom these Powers are required to intern under international law, without prejudice to any more favorable treatment which these Powers may choose to give and with the exception of Articles 8, 10, 15, 30, fifth paragraph, 58-67, 92, 126 and, where diplomatic relations exist between the Parties to the conflict and the neutral or non-belligerent Power concerned, those Articles concerning the Protecting Power. Where such diplomatic relations exist, the Parties to a conflict on whom these persons depend shall be allowed to perform towards them the functions of a Protecting Power as provided in the present Convention, without prejudice to the functions which these Parties normally exercise in conformity with diplomatic and consular usage and treaties.

C. This Article shall in no way affect the status of medical personnel and chaplains as provided for in Article 33 of the present Convention.

ARTICLE 5

The present Convention shall apply to the persons referred to in Article 4 from the time they fall into the power of the enemy and until their final release and repatriation.

Should any doubt arise as to whether persons, having committed a belligerent act and having fallen into the hands of the enemy, belong to any of the categories enumerated in Article 4, such persons shall enjoy the protection of the present Convention until such time as their status has been determined by a competent tribunal.

ARTICLE 6

In addition to the agreements expressly provided for in Articles 10, 23, 28, 33, 60, 65, 66, 67, 72, 73, 75, 109, 110, 118, 119, 122 and 132, the High Contracting Parties may conclude other special agreements for all matters concerning which they may deem it suitable to make separate provision. No special agreement shall adversely affect the situation of prisoners of war, as defined by the present Convention, nor restrict the rights which it confers upon them.

Prisoners of war shall continue to have the benefit of such agreements as long as the Convention is applicable to them, except where express provisions to the contrary are contained in the aforesaid or in subsequent agreements, or where more favorable measures have been taken with regard to them by one or other of the Parties to the conflict.

ARTICLE 7

Prisoners of war may in no circumstances renounce in part or in entirety the rights secured to them by the present Convention, and by the special agreements referred to in the foregoing Article, if such there be.

ARTICLE 8

The present Convention shall be applied with the cooperation and under the scrutiny of the Protecting Powers whose duty it is to safeguard the interests of the Parties to the conflict. For this purpose, the Protecting Powers may appoint, apart from their diplomatic or consular staff, delegates from amongst their own nationals or the nationals of other neutral Powers. The said delegates shall be subject to the approval of the Power with which they are to carry out their duties.

The Parties to the conflict shall facilitate to the greatest extent possible the task of the representatives or delegates of the Protecting Powers.

The representatives or delegates of the Protecting Powers shall not in any case exceed their mission under the present Convention. They shall, in particular, take account of the imperative necessities of security of the State wherein they carry out their duties.

ARTICLE 9

The provisions of the present Convention constitute no obstacle to the humanitarian activities which the International Committee of the Red Cross or any other impartial humanitarian organization may, subject to the consent of the Parties to the conflict concerned, undertake for the protection of prisoners of war and for their relief.

ARTICLE 10

The High Contracting Parties may at any time agree to entrust to an organization which offers all guarantees of impartiality and efficacy the duties incumbent on the Protecting Powers by virtue of the present Convention.

When prisoners of war do not benefit or cease to benefit, no matter for what reason, by the activities of a Protecting Power or of an organization provided for in the first paragraph above, the Detaining Power shall request a neutral State, or such an organization, to undertake the functions performed under the present Convention by a Protecting Power designated by the Parties to a conflict.

If protection cannot be arranged accordingly, the Detaining Power shall request or shall accept, subject to the provisions of this Article, the offer of the services of a humanitarian organization, such as the International Committee of the Red Cross, to assume the humanitarian functions performed by Protecting Powers under the present Convention.

Any neutral Power or any organization invited by the Power concerned or offering itself for these purposes, shall be required to act with a sense of responsibility towards the Party to the conflict on which persons protected by the present Convention depend, and shall be required to furnish sufficient assurances that it is in a position to undertake the appropriate functions and to discharge them impartially.

No derogation from the preceding provisions shall be made by special agreements between Powers one of which is restricted, even temporarily, in its freedom to negotiate with the other Power or its allies by reason of military events, more particularly where the whole, or a substantial part, of the territory of the said Power is occupied.

Whenever in the present Convention mention is made of a Protecting Power, such mention applies to substitute organizations in the sense of the present Article.

ARTICLE 11

In cases where they deem it advisable in the interest of protected persons, particularly in cases of disagreement between the Parties to the conflict as to the application or interpretation of the provisions of the present Convention, the Protecting Powers shall lend their good offices with a view to settling the disagreement.

For this purpose, each propose to the Parties of the Protecting Powers may, either at the invitation of one Party or on its own initiative, to the conflict a meeting of their representatives, and in particular of the authorities responsible for prisoners of war, possibly on neutral territory suitably chosen. The Parties to the conflict shall be bound to give effect to the proposals made to them for this purpose. The Protecting Powers may, if necessary, propose for approval by the Parties to the conflict a person belonging to a neutral Power, or delegated by the International Committee of the Red Cross, who shall be invited to take part in such a meeting.

PART II GENERAL PROTECTION OF PRISONERS OF WAR

ARTICLE 12

Prisoners of war are in the hands of the enemy Power, but not of the individuals or military units who have captured them. Irrespective of the individual responsibilities that may exist, the Detaining Power is responsible for the treatment given them.

Prisoners of war may only be transferred by the Detaining Power to a Power which is a party to the Convention and after the Detaining Power has satisfied itself of the willingness and ability of such transferee Power to apply the Convention. When prisoners of war are transferred under such circumstances, responsibility for the application of the Convention rests on the Power accepting them while they are in its custody.

Nevertheless if that Power fails to carry out the provisions of the Convention in any important respect, the Power by whom the prisoners of war were transferred shall, upon being notified by the PROTECTING Power, take effective measures to correct the situation or shall request the return of the prisoners of war. Such requests must be complied with.

ARTICLE 13

Prisoners of war must at all times be humanely treated. Any unlawful act or omission by the Detaining Power causing death or seriously endangering the health of a prisoner of war in its custody is prohibited, and will be regarded as a serious breach of the present Convention. In particular, no prisoner of war may be subjected to physical mutilation or to medical or scientific experiments of any kind which are not justified by the medical, dental or hospital treatment of the prisoner concerned and carried out in his interest.

Likewise, prisoners of war must at all times be protected, particularly against acts of violence or intimidation and against insults and public curiosity.

Measures of reprisal against prisoners of war are prohibited.

ARTICLE 14

Prisoners of war are entitled in all circumstances to respect for their persons and their honor. Women shall be treated with all the regard due to their sex and shall in all cases benefit by treatment as favorable as that granted to men. Prisoners of war shall retain the full civil capacity which they enjoyed at the time of their capture. The Detaining Power may not restrict the exercise, either within or without its own territory, of the rights such capacity confers except in so far as the captivity requires.

ARTICLE 15

The Power detaining prisoners of war shall be bound to provide free of charge for their maintenance and for the medical attention required by their state of health.

ARTICLE 16

Taking into consideration the provisions of the present Convention relating to rank and sex, and subject to any privileged treatment which may be accorded to them by reason of their state of health, age or professional qualifications, all prisoners of war shall be treated alike by the Detaining Power, without any adverse distinction based on race, nationality, religious belief or political opinions, or any other distinction founded on similar criteria.

PART III CAPTIVITY

SECTION I BEGINNING OF CAPTIVITY

ARTICLE 17

Every prisoner of war, when questioned on the subject, is bound to give only his surname, first names and rank, date of birth, and army, regimental, personal or serial number, or failing this, equivalent information. If he willfully infringes this rule, he may render himself liable to a restriction of the privileges accorded to his rank or status.

Each Party to a conflict is required to furnish the persons under its jurisdiction who are liable to become prisoners of war, with an identity card showing the owner's surname, first names, rank, army, regimental, personal or serial number or equivalent information, and date of birth. The identity card may, furthermore, bear the signature or the fingerprints, or both, of the owner, and may bear, as well, any other information the Party to the conflict may wish to add concerning persons belonging to its armed forces. As far as possible the card shall measure 6.5 x 10 cm. and shall be issued in duplicate. The identity card shall be shown by the prisoner of war upon demand, but may in no case be taken away from him.

No physical or mental torture, nor any other form of coercion, may be inflicted on prisoners of war to secure from them information of any kind whatever. Prisoners of war who refuse to answer may not be threatened, insulted, or exposed to any unpleasant or disadvantageous treatment of any kind.

Prisoners of war who, owing to their physical or mental condition, are unable to state their identity, shall be handed over to the medical service. The identity of such prisoners shall be established by all possible means, subject to the provisions of the preceding paragraph.

The questioning of prisoners of war shall be carried out in a language which they understand.

ARTICLE 18

All effects and articles of personal use, except arms, horses, military equipment and military documents shall remain in the possession of prisoners of war, likewise their metal helmets and gas masks and like articles issued for personal protection. Effects and articles used for their clothing or feeding shall likewise remain in their possession, even if such effects and articles belong to their regulation military equipment.

At no time should prisoners of war be without identity documents. The Detaining Power shall supply such documents to prisoners of war who possess none.

Badges of rank and nationality, decorations and articles having above all a personal or sentimental value may not be taken from prisoners of war.

Sums of money carried by prisoners of war may not be taken away from them except by order of an officer, and after the amount and particulars of the owner have been recorded in a special register and an itemized receipt has been given, legibly inscribed with the name, rank and unit of the person issuing the said receipt. Sums in the currency of the Detaining Power, or which are changed into such currency at the prisoner's request, shall be placed to the credit of the prisoner's account as provided in Article 64.

The Detaining Power may withdraw articles of value from prisoners of war only for reasons of security; when such articles are withdrawn, the procedure laid down for sums of money impounded shall apply.

Such objects, likewise the sums taken away in any currency other than that of the Detaining Power and the conversion of which has not been asked for by the owners, shall be kept in the custody of the Detaining Power and shall be returned in their initial shape to prisoners of war at the end of their captivity.

ARTICLE 19

Prisoners of war shall be evacuated, as soon as possible after their capture, to camps situated in an area far enough from the combat zone for them to be out of danger.

Only those prisoners of war who, owing to wounds or sickness, would run greater risks by being evacuated than by remaining where they are, may be temporarily kept back in a danger zone.

Prisoners of war shall not be unnecessarily exposed to danger while awaiting evacuation from a fighting zone.

ARTICLE 20

The evacuation of prisoners of war shall always be effected humanely and in conditions similar to those for the forces of the Detaining Power in their changes of station.

The Detaining Power shall supply prisoners of war who are being evacuated with sufficient food and potable water, and with the necessary clothing and medical attention. The Detaining Power shall take all suitable precautions to ensure their safety during evacuation, and shall establish as soon as possible a list of the prisoners of war who are evacuated.

If prisoners of war must, during evacuation, pass through transit camps, their stay in such camps shall be as brief as possible.

SECTION II INTERNMENT OF PRISONERS OF WAR

CHAPTER I GENERAL OBSERVATIONS

ARTICLE 21

The Detaining Power may subject prisoners of war to internment. It may impose on them the obligation of not leaving, beyond certain limits, the camp where they are interned, or if the said camp is fenced in, of not going outside its perimeter. Subject to the provisions of the present Convention relative to penal and disciplinary sanctions, prisoners of war may not be held in close confinement except where necessary to safeguard their health and then only during the continuation of the circumstances which make such confinement necessary.

Prisoners of war may be partially or wholly released on parole or promise, in so far as is allowed by the laws of the Power on which they depend. Such measures shall be taken particularly in cases where this may contribute to the improvement of their state of health. No prisoner of war shall be compelled to accept liberty on parole or promise.

Upon the outbreak of hostilities, each Party to the conflict shall notify the adverse Party of the laws and regulations allowing or forbidding its own nationals to accept liberty on parole or promise. Prisoners of war who are paroled or who have given their promise in conformity with the laws and regulations so notified, are bound on their personal honor scrupulously to fulfil, both towards the Power on which they depend and towards the Power which has captured them, the engagements of their paroles or promises. In such cases, the Power on which they depend is bound neither to require nor to accept from them any service incompatible with the parole or promise given.

ARTICLE 22

Prisoners of war may be interned only in premises located on land and affording every guarantee of hygiene and healthfulness. Except in particular cases which are justified by the interest of the prisoners themselves, they shall not be interned in penitentiaries.

Prisoners of war interned in unhealthy areas, or where the climate is injurious for them, shall be removed as soon as possible to a more favorable climate.

The Detaining Power shall assemble prisoners of war in camps or camp compounds according to their nationality, language and customs, provided that such prisoners shall not be separated from prisoners of war belonging to the armed forces with which they were serving at the time of their capture, except with their consent.

ARTICLE 23

No prisoner of war may at any time be sent to or detained in areas where he may be exposed to the fire of the combat zone, nor may his presence be used to render certain points or areas immune from military operations.

Prisoners of war shall have shelters against air bombardment and other hazards of war, to the same extent as the local civilian population. With the exception of those engaged in the protection of their quarters against the aforesaid hazards, they may enter such shelters as soon as possible after the giving of the alarm. Any other protective measure taken in favor of the population shall also apply to them.

Detaining Powers shall give the Powers concerned, through the intermediary of the Protecting Powers, all useful information regarding the geographical location of prisoner of war camps.

Whenever military considerations permit, prisoner of war camps shall be indicated in the daytime by the letters PW or PG, placed so as to be clearly visible from the air. The Powers concerned may, however, agree upon any other system of marking. Only prisoner of war camps shall be marked as such.

ARTICLE 24

Transit or screening camps of a permanent kind shall be fitted out under conditions similar to those described in the present Section, and the prisoners therein shall have the same treatment as in other camps.

CHAPTER II QUARTERS, FOOD AND CLOTHING OF PRISONERS OF WAR

ARTICLE 25

Prisoners of war shall be quartered under conditions as favorable as those for the forces of the Detaining Power who are billeted in the same area. The said conditions shall make allowance for the habits and customs of the prisoners and shall in no case be prejudicial to their health.

The foregoing provisions shall apply in particular to the dormitories of prisoners of war as regards both total surface and minimum cubic space, and the general installations, bedding and blankets.

The premises provided for the use of prisoners of war individually or collectively, shall be entirely protected from dampness and adequately heated and lighted, in particular between dusk and lights out. All precautions must be taken against the danger of fire.

In any camps in which women prisoners of war, as well as men, are accommodated, separate dormitories shall be provided for them.

ARTICLE 26

The basic daily food rations shall be sufficient in quantity, quality and variety to keep prisoners of war in good health and to prevent loss of weight or the development of nutritional deficiencies. Account shall also be taken of the habitual diet of the prisoners.

The Detaining Power shall supply prisoners of war who work with such additional rations as are necessary for the labor on which they are employed.

Sufficient drinking water shall be supplied to prisoners of war. The use of tobacco shall be permitted.

Prisoners of war shall, as far as possible, be associated with the preparation of their meals; they may be employed for that purpose in the kitchens. Furthermore, they shall be given the means of preparing, themselves, the additional food in their possession.

Adequate premises shall be provided for messing.

Collective disciplinary measures affecting food are prohibited.

ARTICLE 27

Clothing, underwear and footwear shall be supplied to prisoners of war in sufficient quantities by the Detaining Power, which shall make allowance for the climate of the region where the prisoners are detained. Uniforms of enemy armed forces captured by the Detaining Power should, if suitable for the climate, be made available to clothe prisoners of war.

The regular replacement and repair of the above articles shall be assured by the Detaining Power. In addition, prisoners of war who work shall receive appropriate clothing, wherever the nature of the work demands.

ARTICLE 28

Canteens shall be installed in all camps, where prisoners of war may procure foodstuffs, soap and tobacco and ordinary articles in daily use. The tariff shall never be in excess of local market prices. The profits made by camp canteens shall be used for the benefit of the prisoners; a special fund shall be created for this purpose. The prisoners' representative shall have the right to collaborate in the management of the canteen and of this fund.

When a camp is closed down, the credit balance of the special fund shall be handed to an international welfare organization, to be employed for the benefit of prisoners of war of the same nationality as those who have contributed to the fund. In case of a general repatriation, such profits shall be kept by the Detaining Power, subject to any agreement to the contrary between the Powers concerned

CHAPTER III HYGIENE AND MEDICAL ATTENTION

ARTICLE 29

The Detaining Power shall be bound to take all sanitary measures necessary to ensure the cleanliness and healthfulness of camps and to prevent epidemics.

Prisoners of war shall have for their use, day and night, conveniences which conform to the rules of hygiene and are maintained in a constant state of cleanliness. In any camps in which women prisoners of war are accommodated, separate conveniences shall be provided for them.

Also, apart from the baths and showers with which the camps shall be furnished, prisoners of war shall be provided with sufficient water and soap for their personal toilet and for washing their personal laundry; the necessary installations, facilities and time shall be granted them for that purpose.

ARTICLE 30

Every camp shall have an adequate infirmary where prisoners of war may have the attention they require, as well as appropriate diet. Isolation wards shall, if necessary, be set aside for cases of contagious or mental disease.

Prisoners of war suffering from serious disease, or whose condition necessitates special treatment, a surgical operation or hospital care, must be admitted to any military or civilian medical unit where such treatment can be given, even if their repatriation is contemplated in the near future. Special facilities shall be afforded for the care to be given to the disabled, in particular to the blind, and for their rehabilitation, pending repatriation.

Prisoners of war shall have the attention, preferably, of medical personnel of the Power on which they depend and, if possible, of their nationality.

Prisoners of war may not be prevented from presenting themselves to the medical authorities for examination. The detaining authorities shall, upon request, issue to every prisoner who has undergone treatment, an official certificate indicating the nature of his illness or injury, and the duration and kind of treatment received. A duplicate of this certificate shall be forwarded to the Central Prisoners of War Agency.

The costs of treatment, including those of any apparatus necessary for the maintenance of prisoners of war in good health, particularly dentures and other artificial appliances, and spectacles, shall be borne by the Detaining Power.

ARTICLE 31

Medical inspections of prisoners of war shall be held at least once a month. They shall include the checking and the recording of the weight of each prisoner of war. Their purpose shall be, in particular, to supervise the general state of health, nutrition and cleanliness of prisoners and to detect contagious diseases, especially tuberculosis, malaria and venereal disease. For this purpose the most efficient methods available shall be employed, e.g. periodic mass miniature radiography for the early detection of tuberculosis.

ARTICLE 32

Prisoners of war who, though not attached to the medical service of their armed forces, are physicians, surgeons, dentists, nurses or medical orderlies, may be required by the Detaining Power to exercise their medical functions in the interests of prisoners of war dependent on the same Power. In that case they shall continue to be prisoners of war, but shall receive the same treatment as corresponding medical personnel retained by the Detaining Power. They shall be exempted from any other work under Article 49.

Chapter IV Medical Personnel and Chaplains Retained To Assist Prisoners Of War

ARTICLE 33

Members of the medical personnel and chaplains while retained by the Detaining Power with a view to assisting prisoners of war, shall not be considered as prisoners of war. They shall, however, receive as a minimum the benefits and protection of the present Convention, and shall also be granted all facilities necessary to provide for the medical care of, and religious ministration to, prisoners of war.

They shall continue to exercise their medical and spiritual functions for the benefit of prisoners of war, preferably those belonging to the armed forces upon which they depend, within the scope of the military laws and regulations of the Detaining Power and under the control of its competent services, in accordance with their professional etiquette. They shall also benefit by the following facilities in the exercise of their medical or spiritual functions:

(a) They shall be authorized to visit periodically prisoners of war situated in working detachments or in hospitals outside the camp. For this purpose, the Detaining Power shall place at their disposal the necessary means of transport.

(b) The senior medical officer in each camp shall be responsible to the camp military authorities for everything connected with the activities of retained medical personnel. For this purpose, Parties to the conflict shall agree at the outbreak of hostilities on the subject of the corresponding ranks of the medical personnel, including that of societies mentioned in Article 26 of the Geneva Convention for the Amelioration of the Condition of the Wounded and Sick in Armed Forces in the Field of August 12, 1949. This senior medical officer, as well as chaplains, shall have the right to deal with the competent authorities of the camp on all questions relating to their duties. Such authorities shall afford them all necessary facilities for correspondence relating to these questions.

(c) Although they shall be subject to the internal discipline of the camp in which they are retained, such personnel may not be compelled to carry out any work other than that concerned with their medical or religious duties.

During hostilities, the Parties to the conflict shall agree concerning the possible relief of retained personnel and shall settle the procedure to be followed.

None of the preceding provisions shall relieve the Detaining Power of its obligations with regard to prisoners of war from the medical or spiritual point of view.

CHAPTER V RELIGIOUS, INTELLECTUAL AND PHYSICAL ACTIVITIES

ARTICLE 34

Prisoners of war shall enjoy complete latitude in the exercise of their religious duties, including attendance at the service of their faith, on condition that they comply with the disciplinary routine prescribed by the military authorities.

Adequate premises shall be provided where religious services may be held.

ARTICLE 35

Chaplains who fall into the hands of the enemy Power and who remain or are retained with a view to assisting prisoners of war, shall be allowed to minister to them and to exercise freely their ministry amongst prisoners of war of the same religion, in accordance with their religious conscience. They shall be allocated among the various camps and labor detachments containing prisoners of war belonging to the same forces, speaking the same language or practicing the same religion. They shall enjoy the necessary facilities, including the means of transport provided for in Article 33, for visiting the prisoners of war outside their camp. They shall be free to correspond, subject to censorship, on matters concerning their religious duties with the ecclesiastical authorities in the country of detention and with international religious organizations. Letters and cards which they may send for this purpose shall be in addition to the quota provided for in Article 71.

ARTICLE 36

Prisoners of war who are ministers of religion, without having officiated as chaplains to their own forces, shall be at liberty, whatever their denomination, to minister freely to the members of their community. For this purpose, they shall receive the same treatment as the chaplains retained by the Detaining Power. They shall not be obliged to do any other work.

ARTICLE 37

When prisoners of war have not the assistance of a retained chaplain or of a prisoner of war minister of their faith, a minister belonging to the prisoners' or a similar denomination, or in his absence a qualified layman, if such a course is feasible from a confessional point of view, shall be appointed, at the request of the prisoners concerned, to fill this office. This appointment, subject to the approval of the Detaining Power, shall take place with the agreement of the community of prisoners concerned and, wherever necessary, with the approval of the local religious authorities of the same faith. The person thus appointed shall comply with all regulations established by the Detaining Power in the interests of discipline and military security.

ARTICLE 38

While respecting the individual preferences of every prisoner, the Detaining Power shall encourage the practice of intellectual, educational, and recreational pursuits, sports and games amongst prisoners, and shall take the measures necessary to ensure the exercise thereof by providing them with adequate premises and necessary equipment.

Prisoners shall have opportunities for taking physical exercise, including sports and games, and for being out of doors. Sufficient open spaces shall be provided for this purpose in all camps.

CHAPTER VI DISCIPLINE

ARTICLE 39

Every prisoner of war camp shall be put under the immediate authority of a responsible commissioned officer belonging to the regular armed forces of the Detaining Power. Such officer shall have in his possession a copy of the present Convention; he shall ensure that its provisions are known to the camp staff and the guard and shall be responsible, under the direction of his government, for its application.

Prisoners of war, with the exception of officers, must salute and show to all officers of the Detaining Power the external marks of respect provided for by the regulations applying in their own forces.

Officer prisoners of war are bound to salute only officers of a higher rank of the Detaining Power; they must, however, salute the camp commander regardless of his rank.

ARTICLE 40

The wearing of badges of rank and nationality, as well as of decorations, shall be permitted.

ARTICLE 41

In every camp the text of the present Convention and its Annexes and the contents of any special agreement provided for in Article 6, shall be posted, in the prisoners' own language, at places where all may read them. Copies shall be supplied, on request, to the prisoners who cannot have access to the copy which has been posted.

Regulations, orders, notices and publications of every kind relating to the conduct of prisoners of war shall be issued to them in a language which they understand. Such regulations, orders and publications shall be posted in the manner described above and copies shall be handed to the prisoners' representative. Every order and command addressed to prisoners of war individually must likewise be given in a language which they understand.

ARTICLE 42

The use of weapons against prisoners of war, especially against those who are escaping or attempting to escape, shall constitute an extreme measure, which shall always be preceded by warnings appropriate to the circumstances.

CHAPTER VII RANK OF PRISONERS OF WAR

ARTICLE 43

Upon the outbreak of hostilities, the Parties to the conflict shall communicate to one another the titles and ranks of all the persons mentioned in Article 4 of the present Convention, in order to ensure equality of treatment between prisoners of equivalent rank. Titles and ranks which are subsequently created shall form the subject of similar communications.

The Detaining Power shall recognize promotions in rank which have been accorded to prisoners of war and which have been duly notified by the Power on which these prisoners depend.

ARTICLE 44

Officers and prisoners of equivalent status shall be treated with the regard due to their rank and age.

In order to ensure service in officers' camps, other ranks of the same armed forces who, as far as possible, speak the same language, shall be assigned in sufficient numbers, account being taken of the rank of officers and prisoners of equivalent status. Such orderlies shall not be required to perform any other work.

Supervision of the mess by the officers themselves shall be facilitated in every way.

ARTICLE 45

Prisoners of war other than officers and prisoners of equivalent status shall be treated with the regard due to their rank and age.

Supervision of the mess by the prisoners themselves shall be facilitated in every way.

CHAPTER VIII TRANSFER OF PRISONERS OF WAR AFTER THEIR ARRIVAL IN CAMP

ARTICLE 46

The Detaining Power, when deciding upon the transfer of prisoners of war, shall take into account the interests of the prisoners themselves, more especially so as not to increase the difficulty of their repatriation.

The transfer of prisoners of war shall always be effected humanely and in conditions not less favorable than those under which the forces of the Detaining Power are transferred. Account shall always be taken of the climatic conditions to which the prisoners of war are accustomed and the conditions of transfer shall in no case be prejudicial to their health.

The Detaining Power shall supply prisoners of war during transfer with sufficient food and drinking water to keep them in good health, likewise with the necessary clothing, shelter and medical attention. The Detaining Power shall take adequate precautions especially in case of transport by sea or by air, to ensure their safety during transfer, and shall draw up a complete list of all transferred prisoners before their departure.

ARTICLE 47

Sick or wounded prisoners of war shall not be transferred as long as their recovery may be endangered by the journey, unless their safety imperatively demands it.

If the combat zone draws closer to a camp, the prisoners of war in the said camp shall not be transferred unless their transfer can be carried out in adequate conditions of safety, or if they are exposed to greater risks by remaining on the spot than by being transferred.

ARTICLE 48

In the event of transfer, prisoners of war shall be officially advised of their departure and of their new postal address. Such notifications shall be given in time for them to pack their luggage and inform their next of kin.

They shall be allowed to take with them their personal effects, and the correspondence and parcels which have arrived for them. The weight of such baggage may be limited, if the conditions of transfer so require, to what each prisoner can reasonably carry, which shall in no case be more than twenty-five kilograms per head.

Mail and parcels addressed to their former camp shall be forwarded to them without delay. The camp commander shall take, in agreement with the prisoners' representative, any measures needed to ensure the transport of the prisoners' community property and of the luggage they are unable to take with them in consequence of restrictions imposed by virtue of the second paragraph of this Article.

The costs of transfers shall be borne by the Detaining Power.

SECTION III LABOR OF PRISONERS OF WAR

ARTICLE 49

The Detaining Power may utilize the labor of prisoners of war who are physically fit, taking into account their age, sex, rank and physical aptitude, and with a view particularly to maintaining them in a good state of physical and mental health.

Non-commissioned officers who are prisoners of war shall only be required to do supervisory work. Those not so required may ask for other suitable work which shall, so far as possible, be found for them.

If officers or persons of equivalent status ask for suitable work, it shall be found for them, so far as possible, but they may in no circumstances be compelled to work.

ARTICLE 50

Besides work connected with camp administration, installation or maintenance, prisoners of war may be compelled to do only such work as is included in the following classes:

(a) Agriculture;

(b) Industries connected with the production or the extraction of raw materials, and manufacturing industries, with the exception of metallurgical, machinery and chemical industries; public works and building operations which have no military character or purpose;

(c) Transport and handling of stores which are not military in character or purpose;

(d) Commercial business, and arts and crafts;

(e) Domestic service;

(f) Public utility services having no military character or purpose.

Should the above provisions be infringed, prisoners of war shall be allowed to exercise their right of complaint, in conformity with Article 78.

ARTICLE 51

Prisoners of war must be granted suitable working conditions, especially as regards accommodation, food, clothing and equipment; such conditions shall not be inferior to those enjoyed by nationals of the Detaining Power employed in similar work; account shall also be taken of climatic conditions.

The Detaining Power, in utilizing the labor of prisoners of war, shall ensure that in areas in which prisoners are employed, the national legislation concerning the protection of labor, and, more particularly, the regulations for the safety of workers, are duly applied.

Prisoners of war shall receive training and be provided with the means of protection suitable to the work they will have to do and similar to those accorded to the nationals of the Detaining Power. Subject to the provisions of Article 52, prisoners may be submitted to the normal risks run by these civilian workers.

Conditions of labor shall in no case be rendered more arduous by disciplinary measures.

ARTICLE 52

Unless he be a volunteer, no prisoner of war may be employed on labor which is of an unhealthy or dangerous nature.

No prisoner of war shall be assigned to labor which would be looked upon as humiliating for a member of the Detaining Power's own forces.

The removal of mines or similar devices shall be considered as dangerous labor.

ARTICLE 53

The duration of the daily labor of prisoners of war, including the time of the journey to and fro, shall not be excessive, and must in no case exceed that permitted for civilian workers in the district, who are nationals of the Detaining Power and employed on the same work.

Prisoners of war must be allowed, in the middle of the day's work, a rest of not less than one hour. This rest will be the same as that to which workers of the Detaining Power are entitled, if the latter is of longer duration. They shall be allowed in addition a rest of twenty-four consecutive hours every week, preferably on Sunday or the day of rest in their country of origin. Furthermore, every prisoner who has worked for one year shall be granted a rest of eight consecutive days, during which his working pay shall be paid him.

If methods of labor such as piecework are employed, the length of the working period shall not be rendered excessive thereby.

ARTICLE 54

The working pay due to prisoners of war shall be fixed in accordance with the provisions of Article 62 of the present Convention.

Prisoners of war who sustain accidents in connection with work, or who contract a disease in the course, or in consequence of their work, shall receive all the care their condition may require. The Detaining Power shall furthermore deliver to such prisoners of war a medical certificate enabling them to submit their claims to the Power on which they depend, and shall send a duplicate to the Central Prisoners of War Agency provided for in Article 123.

ARTICLE 55

The fitness of prisoners of war for work shall be periodically verified by medical examinations at least once a month. The examinations shall have particular regard to the nature of the work which prisoners of war are required to do.

If any prisoner of war considers himself incapable of working, he shall be permitted to appear before the medical authorities of his camp. Physicians or surgeons may recommend that the prisoners who are, in their opinion, unfit for work, be exempted therefrom.

ARTICLE 56

The organization and administration of labor detachments shall be similar to those of prisoner of war camps.

Every labor detachment shall remain under the control of and administratively part of a prisoner of war camp. The military authorities and the commander of the said camp shall be responsible, under the direction of their government, for the observance of the provisions of the present Convention in labor detachments.

The camp commander shall keep an up-to-date record of the labor detachments dependent on his camp, and shall communicate it to the delegates of the Protecting Power, of the International Committee of the Red Cross, or of other agencies giving relief to prisoners of war, who may visit the camp.

ARTICLE 57

The treatment of prisoners of war who work for private persons, even if the latter are responsible for guarding and protecting them, shall not be inferior to that which is provided for by the present Convention. The Detaining Power, the military authorities and the commander of the camp to which such prisoners belong shall be entirely responsible for the maintenance, care, treatment, and payment of the working pay of such prisoners of war.

Such prisoners of war shall have the right to remain in communication with the prisoners' representatives in the camps on which they depend.

SECTION IV FINANCIAL RESOURCES OF PRISONERS OF WAR

ARTICLE 58

Upon the outbreak of hostilities, and pending an arrangement on this matter with the Protecting Power, the Detaining Power may determine the maximum amount of money in cash or in any similar form that prisoners may have in their possession. Any amount in excess, which was properly in their possession and which has been taken or withheld from them, shall be placed to their account, together with any monies deposited by them, and shall not be converted into any other currency without their consent.

If prisoners of war are permitted to purchase services or commodities outside the camp against payment in cash, such payments shall be made by the prisoner himself or by the camp administration who will charge them to the accounts of the prisoners concerned. The Detaining Power will establish the necessary rules in this respect.

ARTICLE 59

Cash which was taken from prisoners of war, in accordance with Article 18, at the time of their capture, and which is in the currency of the Detaining Power, shall be placed to their separate accounts, in accordance with the provisions of Article 64 of the present Section.

The amounts, in the currency of the Detaining Power, due to the conversion of sums in other currencies that are taken from the prisoners of war at the same time, shall also be credited to their separate accounts.

ARTICLE 60

The Detaining Power shall grant all prisoners of war a monthly advance of pay, the amount of which shall be fixed by conversion, into the currency of the said Power, of the following amounts:

Category I: Prisoners ranking below sergeant: eight Swiss francs.

Category II: Sergeants and other non-commissioned officers, or prisoners of equivalent rank: twelve Swiss francs.

Category III: Warrant officers and commissioned officers below the rank of major or prisoners of equivalent rank: fifty Swiss francs.

Category IV: Majors, lieutenant colonels, colonels or prisoners of equivalent rank: sixty Swiss francs.

Category V: General officers or prisoners of equivalent rank: seventy-five Swiss francs.

However, the Parties to the conflict concerned may by special agreement modify the amount of advances of pay due to prisoners of the preceding categories.

Furthermore, if the amounts indicated in the first paragraph above would be unduly high compared with the pay of the Detaining Power's armed forces or would, for any reason, seriously embarrass the Detaining Power, then, pending the conclusion of a special agreement with the Power on which the prisoners depend to vary the amounts indicated above, the Detaining Power:

(a) Shall continue to credit the accounts of the prisoners with the amounts indicated in the first paragraph above;

(b) May temporarily limit the amount made available from these advances of pay to prisoners of war for their own use, to sums which are reasonable, but which, for Category I, shall never be inferior to the amount that the Detaining Power gives to the members of its own armed forces.

The reasons for any limitations will be given without delay to the Protecting Power.

ARTICLE 61

The Detaining Power shall accept for distribution as supplementary pay to prisoners of war sums which the Power on which the prisoners depend may forward to them, on condition that the sums to be paid shall be the same for each prisoner of the same category, shall be payable to all prisoners of that category depending on that Power, and shall be placed in their separate accounts, at the earliest opportunity, in accordance with the provisions of Article 64. Such supplementary pay shall not relieve the Detaining Power of any obligation under this Convention.

ARTICLE 62

Prisoners of war shall be paid a fair working rate of pay by the detaining authorities direct. The rate shall be fixed by the said authorities, but shall at no time be less than one-fourth of one Swiss franc for a full working day. The Detaining Power shall inform prisoners of war, as well as the Power on which they depend, through the intermediary of the Protecting Power, of the rate of daily working pay that it has fixed.

Working pay shall likewise be paid by the detaining authorities to prisoners of war permanently detailed to duties or to a skilled or semi-skilled occupation in connection with the administration, installation or maintenance of camps, and to the prisoners who are required to carry out spiritual or medical duties on behalf of their comrades.

The working pay of the prisoners' representative, of his advisers, if any, and of his assistants, shall be paid out of the fund maintained by canteen profits. The scale of this working pay shall be fixed by the prisoners' representative and approved by the camp commander. If there is no such fund, the detaining authorities shall pay these prisoners a fair working rate of pay.

ARTICLE 63

Prisoners of war shall be permitted to receive remittances of money addressed to them individually or collectively.

Every prisoner of war shall have at his disposal the credit balance of his account as provided for in the following Article, within the limits fixed by the Detaining Power, which shall make such payments as are requested. Subject to financial or monetary restrictions which the Detaining Power regards as essential, prisoners of war may also have payments made abroad. In this case payments addressed by prisoners of war to dependants shall be given priority.

In any event, and subject to the consent of the Power on which they depend, prisoners may have payments made in their own country, as follows: the Detaining Power shall send to the aforesaid

Power through the Protecting Power a notification giving all the necessary particulars concerning the prisoners of war, the beneficiaries of the payments, and the amount of the sums to be paid, expressed in the Detaining Power's currency. The said notification shall be signed by the prisoners and countersigned by the camp commander. The Detaining Power shall debit the prisoners' account by a corresponding amount; the sums thus debited shall be placed by it to the credit of the Power on which the prisoners depend.

To apply the foregoing provisions, the Detaining Power may usefully consult the Model Regulations in Annex V of the present Convention.

ARTICLE 64

The Detaining Power shall hold an account for each prisoner of war, showing at least the following:

1. The amounts due to the prisoner or received by him as advances of pay, as working pay or derived from any other source; the sums in the currency of the Detaining Power which were taken from him; the sums taken from him and converted at his request into the currency of the said Power.

2. The payments made to the prisoner in cash, or in any other similar form; the payments made on his behalf and at his request; the sums transferred under Article 63, third paragraph.

ARTICLE 65

Every item entered in the account of a prisoner of war shall be countersigned or initialed by him, or by the prisoners' representative acting on his behalf.

Prisoners of war shall at all times be afforded reasonable facilities for consulting and obtaining copies of their accounts, which may likewise be inspected by the representatives of the Protecting Powers at the time of visits to the camp.

When prisoners of war are transferred from one camp to another, their personal accounts will follow them. In case of transfer from one Detaining Power to another, the monies which are their property and are not in the currency of the Detaining Power will follow them. They shall be given certificates for any other monies standing to the credit of their accounts.

The Parties to the conflict concerned may agree to notify to each other at specific intervals through the Protecting Power, the amount of the accounts of the prisoners of war.

ARTICLE 66

On the termination of captivity, through the release of a prisoner of war or his repatriation, the Detaining Power shall give him a statement, signed by an authorized officer of that Power, showing the credit balance then due to him. The Detaining Power shall also send through the Protecting Power to the government upon which the prisoner of war depends, lists giving all appropriate particulars of all prisoners of war whose captivity has been terminated by repatriation, release, escape, death or any other means, and showing the amount of their credit balances. Such lists shall be certified on each sheet by an authorized representative of the Detaining Power.

Any of the above provisions of this Article may be varied by mutual agreement between any two Parties to the conflict.

The Power on which the prisoner of war depends shall be responsible for settling with him any credit balance due to him from the Detaining Power on the termination of his captivity.

ARTICLE 67

Advances of pay, issued to prisoners of war in conformity with Article 60, shall be considered as made on behalf of the Power on which they depend. Such advances of pay, as well as all payments made by the said Power under Article 63, third paragraph, and Article 68, shall form the subject of arrangements between the Powers concerned, at the close of hostilities.

ARTICLE 68

Any claim by a prisoner of war for compensation in respect of any injury or other disability arising out of work shall be referred to the Power on which he depends, through the Protecting Power. In accordance with Article 54, the Detaining Power will, in all cases, provide the prisoner of war concerned with a statement showing the nature of the injury or disability, the circumstances in which it arose and particulars of medical or hospital treatment given for it. This statement will be signed by a responsible officer of the Detaining Power and the medical particulars certified by a medical officer.

Any claim by a prisoner of war for compensation in respect of personal effects, monies or valuables impounded by the Detaining Power under Article 18 and not forthcoming on his repatriation, or in respect of loss alleged to be due to the fault of the Detaining Power or any of its servants, shall likewise be referred to the Power on which he depends. Nevertheless, any such personal effects required for use by the prisoners of war whilst in captivity shall be replaced at the expense of the Detaining Power. The Detaining Power will, in all cases, provide the prisoner of war with a statement, signed by a responsible officer, showing all available information regarding the reasons why such effects, monies or valuables have not been restored to him. A copy of this statement will be forwarded to the Power on which he depends through the Central Prisoners of War Agency provided for in Article 123.

SECTION V RELATIONS OF PRISONERS OF WAR WITH THE EXTERIOR

ARTICLE 69

Immediately upon prisoners of war falling into its power, the Detaining Power shall inform them and the Powers on which they depend, through the Protecting Power, of the measures taken to carry out the provisions of the present Section. They shall likewise inform the parties concerned of any subsequent modifications of such measures.

ARTICLE 70

Immediately upon capture, or not more than one week after arrival at a camp, even if it is a transit camp, likewise in case of sickness or transfer to hospital or another camp, every prisoner of war shall be enabled to write direct to his family, on the one hand, and to the Central

Prisoners of War Agency provided for in Article 123, on the other hand, a card similar, if possible, to the model annexed to the present Convention, informing his relatives of his capture, address and state of health. The said cards shall be forwarded as rapidly as possible and may not be delayed in any manner.

ARTICLE 71

Prisoners of war shall be allowed to send and receive letters and cards. If the Detaining Power deems it necessary to limit the number of letters and cards sent by each prisoner of war, the said number shall not be less than two letters and four cards monthly, exclusive of the capture cards provided for in Article 70, and conforming as closely as possible to the models annexed to the present Convention. Further limitations may be imposed only if the Protecting Power is satisfied that it would be in the interests of the prisoners of war concerned to do so owing to difficulties of translation caused by the Detaining Power's inability to find sufficient qualified linguists to carry out the necessary censorship. If limitations must be placed on the correspondence addressed to prisoners of war, they may be ordered only by the Power on which the prisoners depend, possibly at the request of the Detaining Power. Such letters and cards must be conveyed by the most rapid method at the disposal of the Detaining Power; they may not be delayed or retained for disciplinary reasons.

Prisoners of war who have been without news for a long period, or who are unable to receive news from their next of kin or to give them news by the ordinary postal route, as well as those who are at a great distance from their homes, shall be permitted to send telegrams, the fees being charged against the prisoners of war's accounts with the Detaining Power or paid in the currency at their disposal. They shall likewise benefit by this measure in cases of urgency.

As a general rule, the correspondence of prisoners of war shall be written in their native language. The Parties to the conflict may allow correspondence in other languages.

Sacks containing prisoner of war mail must be securely sealed and labeled so as clearly to indicate their contents, and must be addressed to offices of destination.

ARTICLE 72

Prisoners of war shall be allowed to receive by post or by any other means individual parcels or collective shipments containing, in particular, foodstuffs, clothing, medical supplies and articles of a religious, educational or recreational character which may meet their needs, including books, devotional articles, scientific equipment, examination papers, musical instruments, sports outfits and materials allowing prisoners of war to pursue their studies or their cultural activities.

Such shipments shall in no way free the Detaining Power from the obligations imposed upon it by virtue of the present Convention.

The only limits which may be placed on these shipments shall be those proposed by the Protecting Power in the interest of the prisoners themselves, or by the International Committee of the Red Cross or any other organization giving assistance to the prisoners, in respect of their own shipments only, on account of exceptional strain on transport or communications.

The conditions for the sending of individual parcels and collective relief shall, if necessary, be the subject of special agreements between the Powers concerned, which may in no case delay the receipt by the prisoners of relief supplies. Books may not be included in parcels of clothing and foodstuffs. Medical supplies shall, as a rule, be sent in collective parcels.

ARTICLE 73

In the absence of special agreements between the Powers concerned on the conditions for the receipt and distribution of collective relief shipments, the rules and regulations concerning collective shipments, which are annexed to the present Convention, shall be applied.

The special agreements referred to above shall in no case restrict the right of prisoners' representatives to take possession of collective relief shipments intended for prisoners of war, to proceed to their distribution or to dispose of them in the interest of the prisoners.

Nor shall such agreements restrict the right of representatives of the Protecting Power, the International Committee of the Red Cross or any other organization giving assistance to prisoners of war and responsible for the forwarding of collective shipments, to supervise their distribution to the recipients.

ARTICLE 74

All relief shipments for prisoners of war shall be exempt from import, customs and other dues.

Correspondence, relief shipments and authorized remittances of money addressed to prisoners of war or dispatched by them through the post office, either direct or through the Information Bureau provided for in Article 122 and the Central Prisoners of War Agency provided for in Article 123, shall be exempt from any postal dues, both in the countries of origin and destination, and in intermediate countries.

If relief shipments intended for prisoners of war cannot be sent through the post office by reason of weight or for any other cause, the cost of transportation shall be borne by the Detaining Power in all the territories under its control. The other Powers party to the Convention shall bear the cost of transport in their respective territories.

In the absence of special agreements between the Parties concerned, the costs connected with transport of such shipments, other than costs covered by the above exemption, shall be charged to the senders.

The High Contracting Parties shall endeavor to reduce, so far as possible, the rates charged for telegrams sent by prisoners of war, or addressed to them.

ARTICLE 75

Should military operations prevent the Powers concerned from fulfilling their obligation to assure the transport of the shipments referred to in Articles 70, 71, 72 and 77, the Protecting Powers concerned, the International Committee of the Red Cross or any other organization duly approved by the Parties to the conflict may undertake to ensure the conveyance of such shipments by suitable means (railway wagons, motor vehicles, vessels or aircraft, etc.). For this purpose, the High Contracting Parties shall endeavor to supply them with such transport and to allow its circulation, especially by granting the necessary safe-conducts.

Such transport may also be used to convey:

(a) Correspondence, lists and reports exchanged between the Central Information Agency referred to in Article 123 and the National Bureau referred to in Article 122;

(b) Correspondence and reports relating to prisoners of war which the Protecting Powers, the International Committee of the Red Cross or any other body assisting the prisoners, exchange either with their own delegates or with the Parties to the conflict.

These provisions in no way detract from the right of any Party to the conflict to arrange other means of transport, if it should so prefer, nor preclude the granting of safe-conducts, under mutually agreed conditions, to such means of transport.

In the absence of special agreements, the costs occasioned by the use of such means of transport shall be borne proportionally by the Parties to the conflict whose nationals are benefited thereby.

ARTICLE 76

The censoring of correspondence addressed to prisoners of war or dispatched by them shall be done as quickly as possible. Mail shall be censored only by the dispatching State and the receiving State, and once only by each.

The examination of consignments intended for prisoners of war shall not be carried out under conditions that will expose the goods contained in them to deterioration; except in the case of written or printed matter, it shall be done in the presence of the addressee, or of a fellow-prisoner duly delegated by him. The delivery to prisoners of individual or collective consignments shall not be delayed under the pretext of difficulties of censorship.

Any prohibition of correspondence ordered by Parties to the conflict, either for military or political reasons, shall be only temporary and its duration shall be as short as possible.

ARTICLE 77

The Detaining Powers shall provide all facilities for the transmission, through the Protecting Power or the Central Prisoners of War Agency provided for in Article 123, of instruments, papers or documents intended for prisoners of war or dispatched by them, especially powers of attorney and wills.

In all cases they shall facilitate the preparation and execution of such documents on behalf of prisoners of war; in particular, they shall allow them to consult a lawyer and shall take what measures are necessary for the authentication of their signatures.

SECTION VI RELATIONS BETWEEN PRISONERS OF WAR AND THE AUTHORITIES

CHAPTER I COMPLAINTS OF PRISONERS OF WAR RESPECTING THE CONDITIONS OF CAPTIVITY

ARTICLE 78

Prisoners of war shall have the right to make known to the military authorities in whose power they are, their requests regarding the conditions of captivity to which they are subjected.

They shall also have the unrestricted right to apply to the representatives of the Protecting Powers either through their prisoners' representative or, if they consider it necessary, direct, in order to draw their attention to any points on which they may have complaints to make regarding their conditions of captivity.

These requests and complaints shall not be limited nor considered to be a part of the correspondence quota referred to in Article 71. They must be transmitted immediately. Even if they are recognized to be unfounded, they may not give rise to any punishment.

Prisoners' representatives may send periodic reports on the situation in the camps and the needs of the prisoners of war to the representatives of the Protecting Powers.

CHAPTER II PRISONER OF WAR REPRESENTATIVES

ARTICLE 79

In all places where there are prisoners of war, except in those where there are officers, the prisoners shall freely elect by secret ballot, every six months, and also in case of vacancies, prisoners' representatives entrusted with representing them before the military authorities, the Protecting Powers, the International Committee of the Red Cross and any other organization which may assist them. These prisoners' representatives shall be eligible for re-election.

In camps for officers and persons of equivalent status or in mixed camps, the senior officer among the prisoners of war shall be recognized as the camp prisoners' representative. In camps for officers, he shall be assisted by one or more advisers chosen by the officers; in mixed camps, his assistants shall be chosen from among the prisoners of war who are not officers and shall be elected by them.

Officer prisoners of war of the same nationality shall be stationed in labor camps for prisoners of war, for the purpose of carrying out the camp administration duties for which the prisoners of war are responsible. These officers may be elected as prisoners' representatives under the first paragraph of this Article. In such a case the assistants to the prisoners' representatives shall be chosen from among those prisoners of war who are not officers.

Every representative elected must be approved by the Detaining Power before he has the right to commence his duties. Where the Detaining Power refuses to approve a prisoner of war elected by his fellow prisoners of war, it must inform the Protecting Power of the reason for such refusal.

In all cases the prisoners' representative must have the same nationality, language and customs as the prisoners of war whom he represents. Thus, prisoners of war distributed in different sections of a camp, according to their nationality, language or customs, shall have for each section their own prisoners' representative, in accordance with the foregoing paragraphs.

ARTICLE 80

Prisoners' representatives shall further the physical, spiritual and intellectual well being of prisoners of war.

In particular, where the prisoners decide to organize amongst themselves a system of mutual assistance, this organization will be within the province of the prisoners' representative, in addition to the special duties entrusted to him by other provisions of the present Convention.

Prisoners' representatives shall not be held responsible, simply by reason of their duties, for any offences committed by prisoners of war.

ARTICLE 81

Prisoners' representatives shall not be required to perform any other work, if the accomplishment of their duties is thereby made more difficult.

Prisoners' representatives may appoint from amongst the prisoners such assistants as they may require. All material facilities shall be granted them, particularly a certain freedom of movement necessary for the accomplishment of their duties (inspection of labor detachments, receipt of supplies, etc.).

Prisoners' representatives shall be permitted to visit premises where prisoners of war are detained, and every prisoner of war shall have the right to consult freely his prisoners' representative.

All facilities shall likewise be accorded to the prisoners' representatives for communication by post and telegraph with the detaining authorities, the Protecting Powers, the International Committee of the Red Cross and their delegates, the Mixed Medical Commissions and with the bodies which give assistance to prisoners of war. Prisoners' representatives of labor detachments shall enjoy the same facilities for communication with the prisoners' representatives of the principal camp. Such communications shall not be restricted, nor considered as forming a part of the quota mentioned in Article 71.

Prisoners' representatives who are transferred shall be allowed a reasonable time to acquaint their successors with current affairs.

In case of dismissal, the reasons therefor shall be communicated to the Protecting Power.

CHAPTER III PENAL AND DISCIPLINARY SANCTIONS

I. GENERAL PROVISIONS

ARTICLE 82

A prisoner of war shall be subject to the laws, regulations and orders in force in the armed forces of the Detaining Power; the Detaining Power shall be justified in taking judicial or disciplinary measures in respect of any offence committed by a prisoner of war against such laws, regulations or orders. However, no proceedings or punishments contrary to the provisions of this Chapter shall be allowed.

If any law, regulation or order of the Detaining Power shall declare acts committed by a prisoner of war to be punishable, whereas the same acts would not be punishable if committed by a member of the forces of the Detaining Power, such acts shall entail disciplinary punishments only.

ARTICLE 83

In deciding whether proceedings in respect of an offence alleged to have been committed by a prisoner of war shall be judicial or disciplinary, the Detaining Power shall ensure that the competent authorities exercise the greatest leniency and adopt, wherever possible, disciplinary rather than judicial measures.

ARTICLE 84

A prisoner of war shall be tried only by a military court, unless the existing laws of the Detaining Power expressly permit the civil courts to try a member of the armed forces of the Detaining Power in respect of the particular offence alleged to have been committed by the prisoner of war.

In no circumstances whatever shall a prisoner of war be tried by a court of any kind which does not offer the essential guarantees of independence and impartiality as generally recognized, and, in particular, the procedure of which does not afford the accused the rights and means of defense provided for in Article 105.

ARTICLE 85

Prisoners of war prosecuted under the laws of the Detaining Power for acts committed prior to capture shall retain, even if convicted, the benefits of the present Convention.

ARTICLE 86

No prisoner of war may be punished more than once for the same act, or on the same charge.

ARTICLE 87

Prisoners of war may not be sentenced by the military authorities and courts of the Detaining Power to any penalties except those provided for in respect of members of the armed forces of the said Power who have committed the same acts.

When fixing the penalty, the courts or authorities of the Detaining Power shall take into consideration, to the widest extent possible, the fact that the accused, not being a national of the Detaining Power, is not bound to it by any duty of allegiance, and that he is in its power as the result of circumstances independent of his own will. The said courts or authorities shall be at liberty to reduce the penalty provided for the violation of which the prisoner of war is accused, and shall therefore not be bound to apply the minimum penalty prescribed.

Collective punishment for individual acts, corporal punishments, imprisonment in premises without daylight and, in general, any form of torture or cruelty, are forbidden.

No prisoner of war may be deprived of his rank by the Detaining Power, or prevented from wearing his badges.

ARTICLE 88

Officers, non-commissioned officers and men who are prisoners of war undergoing a disciplinary or judicial punishment, shall not be subjected to more severe treatment than that applied in respect of the same punishment to members of the armed forces of the Detaining Power of equivalent rank.

A woman prisoner of war shall not be awarded or sentenced to a punishment more severe, or treated whilst undergoing punishment more severely, than a woman member of the armed forces of the Detaining Power dealt with for a similar offence.

In no case may a woman prisoner of war be awarded or sentenced to a punishment more severe, or treated whilst undergoing punishment more severely, than a male member of the armed forces of the Detaining Power dealt with for a similar offence.

Prisoners of war who have served disciplinary or judicial sentences may not be treated differently from other prisoners of war.

II. DISCIPLINARY SANCTIONS

ARTICLE 89

The disciplinary punishments applicable to prisoners of war are the following:

1. A fine which shall not exceed 50 per cent of the advances of pay and working pay which the prisoner of war would otherwise receive under the provisions of Articles 60 and 62 during a period of not more than thirty days.

2. Discontinuance of privileges granted over and above the treatment provided for by the present Convention.

3. Fatigue duties not exceeding two hours daily.

4. Confinement.

The punishment referred to under (3) shall not be applied to officers.

In no case shall disciplinary punishments be inhuman, brutal or dangerous to the health of prisoners of war.

ARTICLE 90

The duration of any single punishment shall in no case exceed thirty days. Any period of confinement awaiting the hearing of a disciplinary offence or the award of disciplinary punishment shall be deducted from an award pronounced against a prisoner of war.

The maximum of thirty days provided above may not be exceeded, even if the prisoner of war is answerable for several acts at the same time when he is awarded punishment, whether such acts are related or not.

The period between the pronouncing of an award of disciplinary punishment and its execution shall not exceed one month.

When a prisoner of war is awarded a further disciplinary punishment, a period of at least three days shall elapse between the execution of any two of the punishments, if the duration of one of these is ten days or more.

ARTICLE 91

The escape of a prisoner of war shall be deemed to have succeeded when:

1. He has joined the armed forces of the Power on which he depends, or those of an allied Power;

2. He has left the territory under the control of the Detaining Power, or of an ally of the said Power;

3. He has joined a ship flying the flag of the Power on which he depends, or of an allied Power, in the territorial waters of the Detaining Power, the said ship not being under the control of the last-named Power.

Prisoners of war who have made good their escape in the sense of this Article and who are recaptured, shall not be liable to any punishment in respect of their previous escape.

ARTICLE 92

A prisoner of war who attempts to escape and is recaptured before having made good his escape in the sense of Article 91 shall be liable only to a disciplinary punishment in respect of this act, even if it is a repeated offence.

A prisoner of war who is recaptured shall be handed over without delay to the competent military authority.

Article 88, fourth paragraph, notwithstanding, prisoners of war punished as a result of an unsuccessful escape may be subjected to special surveillance. Such surveillance must not affect the state of their health, must be undergone in a prisoner of war camp, and must not entail the suppression of any of the safeguards granted them by the present Convention.

ARTICLE 93

Escape or attempt to escape, even if it is a repeated offence, shall not be deemed an aggravating circumstance if the prisoner of war is subjected to trial by judicial proceedings in respect of an offence committed during his escape or attempt to escape.

In conformity with the principle stated in Article 83, offences committed by prisoners of war with the sole intention of facilitating their escape and which do not entail any violence against life or limb, such as offences against public property, theft without intention of self-enrichment, the drawing up or use of false papers, the wearing of civilian clothing, shall occasion disciplinary punishment only.

Prisoners of war who aid or abet an escape or an attempt to escape shall be liable on this count to disciplinary punishment only.

ARTICLE 94

If an escaped prisoner of war is recaptured, the Power on which he depends shall be notified thereof in the manner defined in Article 122, provided notification of his escape has been made.

ARTICLE 95

A prisoner of war accused of an offence against discipline shall not be kept in confinement pending the hearing unless a member of the armed forces of the Detaining Power would be so kept if he were accused of a similar offence, or if it is essential in the interests of camp order and discipline.

Any period spent by a prisoner of war in confinement awaiting the disposal of an offence against discipline shall be reduced to an absolute minimum and shall not exceed fourteen days.

The provisions of Articles 97 and 98 of this Chapter shall apply to prisoners of war who are in confinement awaiting the disposal of offences against discipline.

ARTICLE 96

Acts which constitute offences against discipline shall be investigated immediately.

Without prejudice to the competence of courts and superior military authorities, disciplinary punishment may be ordered only by an officer having disciplinary powers in his capacity as camp commander, or by a responsible officer who replaces him or to whom he has delegated his disciplinary powers.

In no case may such powers be delegated to a prisoner of war or be exercised by a prisoner of war.

Before any disciplinary award is pronounced, the accused shall be given precise information regarding the offences of which he is accused, and given an opportunity of explaining his conduct and of defending himself. He shall be permitted, in particular, to call witnesses and to have recourse, if necessary, to the services of a qualified interpreter. The decision shall be announced to the accused prisoner of war and to the prisoners' representative.

A record of disciplinary punishments shall be maintained by the camp commander and shall be open to inspection by representatives of the Protecting Power.

ARTICLE 97

Prisoners of war shall not in any case be transferred to penitentiary establishments (prisons, penitentiaries, convict prisons, etc.) to undergo disciplinary punishment therein.

All premises in which disciplinary punishments are undergone shall conform to the sanitary requirements set forth in Article 25. A prisoner of war undergoing punishment shall be enabled to keep himself in a state of cleanliness, in conformity with Article 29.

Officers and persons of equivalent status shall not be lodged in the same quarters as non-commissioned officers or men.

Women prisoners of war undergoing disciplinary punishment shall be confined in separate quarters from male prisoners of war and shall be under the immediate supervision of women.

ARTICLE 98

A prisoner of war undergoing confinement as a disciplinary punishment, shall continue to enjoy the benefits of the provisions of this Convention except in so far as these are necessarily rendered inapplicable by the mere fact that he is confined. In no case may he be deprived of the benefits of the provisions of Articles 78 and 126.

A prisoner of war awarded disciplinary punishment may not be deprived of the prerogatives attached to his rank.

Prisoners of war awarded disciplinary punishment shall be allowed to exercise and to stay in the open air at least two hours daily.

They shall be allowed, on their request, to be present at the daily medical inspections. They shall receive the attention which their state of health requires and, if necessary, shall be removed to the camp infirmary or to a hospital.

They shall have permission to read and write, likewise to send and receive letters. Parcels and remittances of money, however, may be withheld from them until the completion of the punishment; they shall meanwhile be entrusted to the prisoners' representative, who will hand over to the infirmary the perishable goods contained in such parcels.

III. JUDICIAL PROCEEDINGS

ARTICLE 99

No prisoner of war may be tried or sentenced for an act which is not forbidden by the law of the Detaining Power or by international law, in force at the time the said act was committed.

No moral or physical coercion may be exerted on a prisoner of war in order to induce him to admit himself guilty of the act of which he is accused.

No prisoner of war may be convicted without having had an opportunity to present his defense and the assistance of a qualified advocate or counsel.

ARTICLE 100

Prisoners of war and the Protecting Powers shall be informed as soon as possible of the offences which are punishable by the death sentence under the laws of the Detaining Power.

Other offences shall not thereafter be made punishable by the death penalty without the concurrence of the Power upon which the prisoners of war depend.

The death sentence cannot be pronounced on a prisoner of war unless the attention of the court has, in accordance with Article 87, second paragraph, been particularly called to the fact that since the accused is not a national of the Detaining Power, he is not bound to it by any duty of allegiance, and that he is in its power as the result of circumstances independent of his own will.

ARTICLE 101

If the death penalty is pronounced on a prisoner of war, the sentence shall not be executed before the expiration of a period of at least six months from the date when the Protecting Power receives, at an indicated address, the detailed communication provided for in Article 107.

ARTICLE 102

A prisoner of war can be validly sentenced only if the sentence has been pronounced by the same courts according to the same procedure as in the case of members of the armed forces of the Detaining Power, and if, furthermore, the provisions of the present Chapter have been observed.

ARTICLE 103

Judicial investigations relating to a prisoner of war shall be conducted as rapidly as circumstances permit and so that his trial shall take place as soon as possible. A prisoner of war shall not be confined while awaiting trial unless a member of the armed forces of the Detaining Power would be so confined if he were accused of a similar offence, or if it is essential to do so in the interests of national security. In no circumstances shall this confinement exceed three months.

Any period spent by a prisoner of war in confinement awaiting trial shall be deducted from any sentence of imprisonment passed upon him and taken into account in fixing any penalty.

The provisions of Articles 97 and 98 of this Chapter shall apply to a prisoner of war whilst in confinement awaiting trial.

ARTICLE 104

In any case in which the Detaining Power has decided to institute judicial proceedings against a prisoner of war, it shall notify the Protecting Power as soon as possible and at least three weeks before the opening of the trial. This period of three weeks shall run as from the day on which such notification reaches the Protecting Power at the address previously indicated by the latter to the Detaining Power.

The said notification shall contain the following information:

1. Surname and first names of the prisoner of war, his rank, his army, regimental, personal or serial number, his date of birth, and his profession or trade, if any;

2. Place of internment or confinement;

3. Specification of the charge or charges on which the prisoner of war is to be arraigned, giving the legal provisions applicable;

4 . Designation of the court which will try the case, likewise the date and place fixed for the opening of the trial.

The same communication shall be made by the Detaining Power to the prisoners' representative.

If no evidence is submitted, at the opening of a trial, that the notification referred to above was received by the Protecting Power, by the prisoner of war and by the prisoners' representative concerned, at least three weeks before the opening of the trial, then the latter cannot take place and must be adjourned.

ARTICLE 105

The prisoner of war shall be entitled to assistance by one of his prisoner comrades, to defense by a qualified advocate or counsel of his own choice, to the calling of witnesses and, if he deems necessary, to the services of a competent interpreter. He shall be advised of these rights by the Detaining Power in due time before the trial.

Failing a choice by the prisoner of war, the Protecting Power shall find him an advocate or counsel, and shall have at least one week at its disposal for the purpose. The Detaining Power shall deliver to the said Power, on request, a list of persons qualified to present the defense. Failing a choice of an advocate or counsel by the prisoner of war or the Protecting Power, the Detaining Power shall appoint a competent advocate or counsel to conduct the defense.

The advocate or counsel conducting the defense on behalf of the prisoner of war shall have at his disposal a period of two weeks at least before the opening of the trial, as well as the necessary facilities to prepare the defense of the accused. He may, in particular, freely visit the accused and interview him in private. He may also confer with any witnesses for the defense, including prisoners of war. He shall have the benefit of these facilities until the term of appeal or petition has expired.

Particulars of the charge or charges on which the prisoner of war is to be arraigned, as well as the documents which are generally communicated to the accused by virtue of the laws in force in the armed forces of the Detaining Power, shall be communicated to the accused prisoner of war in a language which he understands, and in good time before the opening of the trial. The same communication in the same circumstances shall be made to the advocate or counsel conducting the defense on behalf of the prisoner of war.

The representatives of the Protecting Power shall be entitled to attend the trial of the case, unless, exceptionally, this is held in camera in the interest of State security. In such a case the Detaining Power shall advise the Protecting Power accordingly.

ARTICLE 106

Every prisoner of war shall have, in the same manner as the members of the armed forces of the Detaining Power, the right of appeal or petition from any sentence pronounced upon him, with a view to the quashing or revising of the sentence or the reopening of the trial. He shall be fully informed of his right to appeal or petition and of the time limit within which he may do so.

ARTICLE 107

Any judgment and sentence pronounced upon a prisoner of war shall be immediately reported to the Protecting Power in the form of a summary communication, which shall also indicate whether he has the right of appeal with a view to the quashing of the sentence or the reopening of the trial. This communication shall likewise be sent to the prisoners' representative concerned. It shall also be sent to the accused prisoner of war in a language he understands, if the sentence was not pronounced in his presence. The Detaining Power shall also immediately communicate to the Protecting Power the decision of the prisoner of war to use or to waive his right of appeal.

Furthermore, if a prisoner of war is finally convicted or if a sentence pronounced on a prisoner of war in the first instance is a death sentence, the Detaining Power shall as soon as possible address to the Protecting Power a detailed communication containing:

1. The precise wording of the finding and sentence;

2. A summarized report of any preliminary investigation and of the trial, emphasizing in particular the elements of the prosecution and the defense;

3. Notification, where applicable, of the establishment where the sentence will be served.

The communications provided for in the foregoing subparagraphs shall be sent to the Protecting Power at the address previously made known to the Detaining Power.

ARTICLE 108

Sentences pronounced on prisoners of war after a conviction has become duly enforceable, shall be served in the same establishments and under the same conditions as in the case of members of the armed forces of the Detaining Power. These conditions shall in all cases conform to the requirements of health and humanity.

A woman prisoner of war on whom such a sentence has been pronounced shall be confined in separate quarters and shall be under the supervision of women.

In any case, prisoners of war sentenced to a penalty depriving them of their liberty shall retain the benefit of the provisions of Articles 78 and 126 of the present Convention. Furthermore, they shall be entitled to receive and dispatch correspondence, to receive at least one relief parcel monthly, to take regular exercise in the open air, to have the medical care required by their state of health, and the spiritual assistance they may desire. Penalties to which they may be subjected shall be in accordance with the provisions of Article 87, third paragraph.

PART IV TERMINATION OF CAPTIVITY

SECTION I DIRECT REPATRIATION AND ACCOMMODATION IN NEUTRAL COUNTRIES

ARTICLE 109

Subject to the provisions of the third paragraph of this Article, Parties to the conflict are bound to send back to their own country, regardless of number or rank, seriously wounded and seriously sick prisoners of war, after having cared for them until they are fit to travel, in accordance with the first paragraph of the following Article.

Throughout the duration of hostilities, Parties to the conflict shall endeavor, with the cooperation of the neutral Powers concerned, to make arrangements for the accommodation in neutral countries of the sick and wounded prisoners of war referred to in the second paragraph of the following Article. They may, in addition, conclude agreements with a view to the direct repatriation or internment in a neutral country of able-bodied prisoners of war who have undergone a long period of captivity.

No sick or injured prisoner of war who is eligible for repatriation under the first paragraph of this Article, may be repatriated against his will during hostilities.

ARTICLE 110

The following shall be repatriated direct:

1. Incurably wounded and sick whose mental or physical fitness seems to have been gravely diminished.

2. Wounded and sick who, according to medical opinion, are not likely to recover within one year, whose condition requires treatment and whose mental or physical fitness seems to have been gravely diminished.

3. Wounded and sick who have recovered, but whose mental or physical fitness seems to have been gravely and permanently diminished.

The following may be accommodated in a neutral country:

1. Wounded and sick whose recovery may be expected within one year of the date of the wound or the beginning of the illness, if treatment in a neutral country might increase the prospects of a more certain and speedy recovery.

2. Prisoners of war whose mental or physical health, according to medical opinion, is seriously threatened by continued captivity, but whose accommodation in a neutral country might remove such a threat.

The conditions which prisoners of war accommodated in a neutral country must fulfil in order to permit their repatriation shall be fixed, as shall likewise their status, by agreement between the Powers concerned. In general, prisoners of war who have been accommodated in a neutral country, and who belong to the following categories, should be repatriated:

1. Those whose state of health has deteriorated so as to fulfil the conditions laid down for direct repatriation;

2. Those whose mental or physical powers remain, even after treatment, considerably impaired.

If no special agreements are concluded between the Parties to the conflict concerned, to determine the cases of disablement or sickness entailing direct repatriation or accommodation in a neutral country, such cases shall be settled in accordance with the principles laid down in the Model Agreement concerning direct repatriation and accommodation in neutral countries of wounded and sick prisoners of war and in the Regulations concerning Mixed Medical Commissions annexed to the present Convention.

ARTICLE 111

The Detaining Power, the Power on which the prisoners of war depend, and a neutral Power agreed upon by these two Powers, shall endeavor to conclude agreements which will enable prisoners of war to be interned in the territory of the said neutral Power until the close of hostilities.

ARTICLE 112

Upon the outbreak of hostilities, Mixed Medical Commissions shall be appointed to examine sick and wounded prisoners of war, and to make all appropriate decisions regarding them. The appointment, duties and functioning of these Commissions shall be in conformity with the provisions of the Regulations annexed to the present Convention.

However, prisoners of war who, in the opinion of the medical authorities of the Detaining Power, are manifestly seriously injured or seriously sick, may be repatriated without having to be examined by a Mixed Medical Commission.

ARTICLE 113

Besides those who are designated by the medical authorities of the Detaining Power, wounded or sick prisoners of war belonging to the categories listed below shall be entitled to present themselves for examination by the Mixed Medical Commissions provided for in the foregoing Article:

1. Wounded and sick proposed by a physician or surgeon who is of the same nationality, or a national of a Party to the conflict allied with the Power on which the said prisoners depend, and who exercises his functions in the camp.

2. Wounded and sick proposed by their prisoners' representative.

3. Wounded and sick proposed by the Power on which they depend, or by an organization duly recognized by the said Power and giving assistance to the prisoners.

Prisoners of war who do not belong to one of the three foregoing categories may nevertheless present themselves for examination by Mixed Medical Commissions, but shall be examined only after those belonging to the said categories.

The physician or surgeon of the same nationality as the prisoners who present themselves for examination by the Mixed Medical Commission, likewise the prisoners' representative of the said prisoners, shall have permission to be present at the examination.

ARTICLE 114

Prisoners of war who meet with accidents shall, unless the injury is self-inflicted, have the benefit of the provisions of this Convention as regards repatriation or accommodation in a neutral country.

ARTICLE 115

No prisoner of war on whom a disciplinary punishment has been imposed and who is eligible for repatriation or for accommodation in a neutral country, may be kept back on the plea that he has not undergone his punishment.

Prisoners of war detained in connection with a judicial prosecution or conviction and who are designated for repatriation or accommodation in a neutral country, may benefit by such measures before the end of the proceedings or the completion of the punishment, if the Detaining Power consents.

Parties to the conflict shall communicate to each other the names of those who will be detained until the end of the proceedings or the completion of the punishment.

ARTICLE 116

The costs of repatriating prisoners of war or of transporting them to a neutral country shall be borne, from the frontiers of the Detaining Power, by the Power on which the said prisoners depend.

ARTICLE 117

No repatriated person may be employed on active military service.

SECTION II RELEASE AND REPATRIATION OF PRISONERS OF WAR AT THE CLOSE OF HOSTILITIES

ARTICLE 118

Prisoners of war shall be released and repatriated without delay after the cessation of active hostilities.

In the absence of stipulations to the above effect in any agreement concluded between the Parties to the conflict with a view to the cessation of hostilities, or failing any such agreement, each of the Detaining Powers shall itself establish and execute without delay a plan of repatriation in conformity with the principle laid down in the foregoing paragraph.

In either case, the measures adopted shall be brought to the knowledge of the prisoners of war.

The costs of repatriation of prisoners of war shall in all cases be equitably apportioned between the Detaining Power and the Power on which the prisoners depend. This apportionment shall be carried out on the following basis:

(a) If the two Powers are contiguous, the Power on which the prisoners of war depend shall bear the costs of repatriation from the frontiers of the Detaining Power.

(b) If the two Powers are not contiguous, the Detaining Power shall bear the costs of transport of prisoners of war over its own territory as far as its frontier or its port of embarkation nearest to the territory of the Power on which the prisoners of war depend. The Parties concerned shall agree between themselves as to the equitable apportionment of the remaining costs of the repatriation. The conclusion of this agreement shall in no circumstances justify any delay in the repatriation of the prisoners of war.

ARTICLE 119

Repatriation shall be effected in conditions similar to those laid down in Articles 46 to 48 inclusive of the present Convention for the transfer of prisoners of war, having regard to the provisions of Article 118 and to those of the following paragraphs.

On repatriation, any articles of value impounded from prisoners of war under Article 18, and any foreign currency which has not been converted into the currency of the Detaining Power, shall be restored to them. Articles of value and foreign currency which, for any reason whatever, are not restored to prisoners of war on repatriation, shall be dispatched to the Information Bureau set up under Article 122.

Prisoners of war shall be allowed to take with them their personal effects, and any correspondence and parcels which have arrived for them. The weight of such baggage may be limited, if the conditions of repatriation so require, to what each prisoner can reasonably carry. Each prisoner shall in all cases be authorized to carry at least twenty-five kilograms.

The other personal effects of the repatriated prisoner shall be left in the charge of the Detaining Power which shall have them forwarded to him as soon as it has concluded an agreement to this effect, regulating the conditions of transport and the payment of the costs involved, with the Power on which the prisoner depends.

Prisoners of war against whom criminal proceedings for an indictable offence are pending may be detained until the end of such proceedings, and, if necessary, until the completion of the punishment. The same shall apply to prisoners of war already convicted for an indictable offence.

Parties to the conflict shall communicate to each other the names of any prisoners of war who are detained until the end of the proceedings or until punishment has been completed.

By agreement between the Parties to the conflict, commissions shall be established for the purpose of searching for dispersed prisoners of war and of assuring their repatriation with the least possible delay.

SECTION III DEATH OF PRISONERS OF WAR

ARTICLE 120

Wills of prisoners of war shall be drawn up so as to satisfy the conditions of validity required by the legislation of their country of origin, which will take steps to inform the Detaining Power of its requirements in this respect. At the request of the prisoner of war and, in all cases, after death, the will shall be transmitted without delay to the Protecting Power; a certified copy shall be sent to the Central Agency.

Death certificates in the form annexed to the present Convention, or lists certified by a responsible officer, of all persons who die as prisoners of war shall be forwarded as rapidly as possible to the Prisoner of War Information Bureau established in accordance with Article 122. The death certificates or certified lists shall show particulars of identity as set out in the third paragraph of Article 17, and also the date and place of death, the cause of death, the date and place of burial and all particulars necessary to identify the graves.

The burial or cremation of a prisoner of war shall be preceded by a medical examination of the body with a view to confirming death and enabling a report to be made and, where necessary, establishing identity.

The detaining authorities shall ensure that prisoners of war who have died in captivity are honorably buried, if possible according to the rites of the religion to which they belonged, and that their graves are respected, suitably maintained and marked so as to be found at any time. Wherever possible, deceased prisoners of war who depended on the same Power shall be interred in the same place.

Deceased prisoners of war shall be buried in individual graves unless unavoidable circumstances require the use of collective graves. Bodies may be cremated only for imperative reasons of hygiene, on account of the religion of the deceased or in accordance with his express wish to this effect. In case of cremation, the fact shall be stated and the reasons given in the death certificate of the deceased.

In order that graves may always be found, all particulars of burials and graves shall be recorded with a Graves Registration Service established by the Detaining Power. Lists of graves and particulars of the prisoners of war interred in cemeteries and elsewhere shall be transmitted to the Power on which such prisoners of war depended. Responsibility for the care of these graves and for records of any subsequent moves of the bodies shall rest on the Power controlling the territory, if a Party to the present Convention. These provisions shall also apply to the ashes, which shall be kept by the Graves Registration Service until proper disposal thereof in accordance with the wishes of the home country.

ARTICLE 121

Every death or serious injury of a prisoner of war caused or suspected to have been caused by a sentry, another prisoner of war, or any other person, as well as any death the cause of which is unknown, shall be immediately followed by an official enquiry by the Detaining Power.

A communication on this subject shall be sent immediately to the Protecting Power. Statements shall be taken from witnesses, especially from those who are prisoners of war, and a report including such statements shall be forwarded to the Protecting Power.

If the enquiry indicates the guilt of one or more persons, the Detaining Power shall take all measures for the prosecution of the person or persons responsible.

PART V INFORMATION BUREAU AND RELIEF SOCIETIES FOR PRISONERS OF WAR

ARTICLE 122

Upon the outbreak of a conflict and in all cases of occupation, each of the Parties to the conflict shall institute an official Information Bureau for prisoners of war who are in its power. Neutral or non-belligerent Powers who may have received within their territory persons belonging to one of the categories referred to in Article 4, shall take the same action with respect to such persons. The Power concerned shall ensure that the Prisoners of War Information Bureau is provided with the necessary accommodation, equipment and staff to ensure its efficient working. It shall be at liberty to employ prisoners of war in such a Bureau under the conditions laid down in the Section of the present Convention dealing with work by prisoners of war.

Within the shortest possible period, each of the Parties to the conflict shall give its Bureau the information referred to in the fourth, fifth and sixth paragraphs of this Article regarding any enemy person belonging to one of the categories referred to in Article 4, who has fallen into its power. Neutral or non-belligerent Powers shall take the same action with regard to persons belonging to such categories whom they have received within their territory.

The Bureau shall immediately forward such information by the most rapid means to the Powers concerned, through the intermediary of the Protecting Powers and likewise of the Central Agency provided for in Article 123.

This information shall make it possible quickly to advise the next of kin concerned. Subject to the provisions of Article 17, the information shall include, in so far as available to the Information Bureau, in respect of each prisoner of war, his surname, first names, rank, army, regimental, personal or serial number, place and full date of birth, indication of the Power on which he depends, first name of the father and maiden name of the mother, name and address of the person to be informed and the address to which correspondence for the prisoner may be sent.

The Information Bureau shall receive from the various departments concerned information regarding transfers, releases, repatriations, escapes, admissions to hospital, and deaths, and shall transmit such information in the manner described in the third paragraph above.

Likewise, information regarding the state of health of prisoners of war who are seriously ill or seriously wounded shall be supplied regularly, every week if possible.

The Information Bureau shall also be responsible for replying to all inquiries sent to it concerning prisoners of war, including those who have died in captivity; it will make any inquiries necessary to obtain the information which is asked for if this is not in its possession.

All written communications made by the Bureau shall be authenticated by a signature or a seal.

The Information Bureau shall furthermore be charged with collecting all personal valuables, including sums in currencies other than that of the Detaining Power and documents of importance to the next of kin, left by prisoners of war who have been repatriated or released, or who have escaped or died, and shall forward the said valuables to the Powers concerned. Such articles shall be sent by the Bureau in sealed packets which shall be accompanied by statements giving clear and full particulars of the identity of the person to whom the articles belonged, and by a complete list of the contents of the parcel. Other personal effects of such prisoners of war shall be transmitted under arrangements agreed upon between the Parties to the conflict concerned.

ARTICLE 123

A Central Prisoners of War Information Agency shall be created in a neutral country. The International Committee of the Red Cross shall, if it deems necessary, propose to the Powers concerned the organization of such an Agency.

The function of the Agency shall be to collect all the information it may obtain through official or private channels respecting prisoners of war, and to transmit it as rapidly as possible to the country of origin of the prisoners of war or to the Power on which they depend. It shall receive from the Parties to the conflict all facilities for effecting such transmissions.

The High Contracting Parties, and in particular those whose nationals benefit by the services of the Central Agency, are requested to give the said Agency the financial aid it may require.

The foregoing provisions shall in no way be interpreted as restricting the humanitarian activities of the International Committee of the Red Cross, or of the relief Societies provided for in Article 125.

ARTICLE 124

The national Information Bureau and the Central Information Agency shall enjoy free postage for mail, likewise all the exemptions provided for in Article 74, and further, so far as possible, exemption from telegraphic charges or, at least, greatly reduced rates.

ARTICLE 125

Subject to the measures which the Detaining Powers may consider essential to ensure their security or to meet any other reasonable need, the representatives of religious organizations, relief societies, or any other organization assisting prisoners of war, shall receive from the said Powers, for themselves and their duly accredited agents, all necessary facilities for visiting the prisoners, distributing relief supplies and material, from any source, intended for religious, educational or recreative purposes, and for assisting them in organizing their leisure time within the camps. Such societies or organizations may be constituted in the territory of the Detaining Power or in any other country, or they may have an international character.

The Detaining Power may limit the number of societies and organizations whose delegates are allowed to carry out their activities in its territory and under its supervision, on condition, however, that such limitation shall not hinder the effective operation of adequate relief to all prisoners of war.

The special position of the International Committee of the Red Cross in this field shall be recognized and respected at all times.

As soon as relief supplies or material intended for the above-mentioned purposes are handed over to prisoners of war, or very shortly afterwards, receipts for each consignment, signed by the prisoners' representative, shall be forwarded to the relief society or organization making the shipment. At the same time, receipts for these consignments shall be supplied by the administrative authorities responsible for guarding the prisoners.

PART VI EXECUTION OF THE CONVENTION

SECTION I GENERAL PROVISIONS

ARTICLE 126

Representatives or delegates of the Protecting Powers shall have permission to go to all places where prisoners of war may be, particularly to places of internment, imprisonment and labor, and shall have access to all premises occupied by prisoners of war; they shall also be allowed to go to the places of departure, passage and arrival of prisoners who are being transferred. They shall be able to interview the prisoners, and in particular the prisoners' representatives, without witnesses, either personally or through an interpreter.

Representatives and delegates of the Protecting Powers shall have full liberty to select the places they wish to visit. The duration and frequency of these visits shall not be restricted. Visits may not be prohibited except for reasons of imperative military necessity, and then only as an exceptional and temporary measure.

The Detaining Power and the Power on which the said prisoners of war depend may agree, if necessary, that compatriots of these prisoners of war be permitted to participate in the visits.

The delegates of the International Committee of the Red Cross shall enjoy the same prerogatives. The appointment of such delegates shall be submitted to the approval of the Power detaining the prisoners of war to be visited.

ARTICLE 127

The High Contracting Parties undertake, in time of peace as in time of war, to disseminate the text of the present Convention as widely as possible in their respective countries, and, in particular, to include the study thereof in their programs of military and, if possible, civil instruction, so that the principles thereof may become known to all their armed forces and to the entire population.

Any military or other authorities, who in time of war assume responsibilities in respect of prisoners of war, must possess the text of the Convention and be specially instructed as to its provisions.

ARTICLE 128

The High Contracting Parties shall communicate to one another through the Swiss Federal Council and, during hostilities, through the Protecting Powers, the official translations of the present Convention, as well as the laws and regulations which they may adopt to ensure the application thereof.

ARTICLE 129

The High Contracting Parties undertake to enact any legislation necessary to provide effective penal sanctions for persons committing, or ordering to be committed, any of the grave breaches of the present Convention defined in the following Article.

Each High Contracting Party shall be under the obligation to search for persons alleged to have committed, or to have ordered to be committed, such grave breaches, and shall bring such persons, regardless of their nationality, before its own courts. It may also, if it prefers, and in accordance with the provisions of its own legislation, hand such persons over for trial to another High Contracting Party concerned, provided such High Contracting Party has made out a prima facie case.

Each High Contracting Party shall take measures necessary for the suppression of all acts contrary to the provisions of the present Convention other than the grave breaches defined in the following Article.

In all circumstances, the accused persons shall benefit by safeguards of proper trial and defense, which shall not be less favorable than those provided by Article 105 and those following of the present Convention.

ARTICLE 130

Grave breaches to which the preceding Article relates shall be those involving any of the following acts, if committed against persons or property protected by the Convention: willful killing, torture or inhuman treatment, including biological experiments, willfully causing great suffering or serious injury to body or health, compelling a prisoner of war to serve in the forces of the hostile Power, or willfully depriving a prisoner of war of the rights of fair and regular trial prescribed in this Convention.

ARTICLE 131

No High Contracting Party shall be allowed to absolve itself or any other High Contracting Party of any liability incurred by itself or by another High Contracting Party in respect of breaches referred to in the preceding Article.

ARTICLE 132

At the request of a Party to the conflict, an enquiry shall be instituted, in a manner to be decided between the interested Parties, concerning any alleged violation of the Convention.

If agreement has not been reached concerning the procedure for the enquiry, the Parties should agree on the choice of an umpire who will decide upon the procedure to be followed.

Once the violation has been established, the Parties to the conflict shall put an end to it and shall repress it with the least possible delay.

SECTION II FINAL PROVISIONS

ARTICLE 133

The present Convention is established in English and in French. Both texts are equally authentic. The Swiss Federal Council shall arrange for official translations of the Convention to be made in the Russian and Spanish languages.

ARTICLE 134

The present Convention replaces the Convention of 27 July 1929, in relations between the High Contracting Parties.

ARTICLE 135

In the relations between the Powers which are bound by The Hague Convention respecting the Laws and Customs of War on Land, whether that of July 29, 1899, or that of October 18, 1907, and which are parties to the present Convention, this last Convention shall be complementary to Chapter II of the Regulations annexed to the above-mentioned Conventions of The Hague.

ARTICLE 136

The present Convention, which bears the date of this day, is open to signature until February 12, 1950, in the name of the Powers represented at the Conference which opened at Geneva on April 21, 1949; furthermore, by Powers not represented at that Conference, but which are parties to the Convention of July 27, 1929.

ARTICLE 137

The present Convention shall be ratified as soon as possible and the ratifications shall be deposited at Berne.

A record shall be drawn up of the deposit of each instrument of ratification and certified copies of this record shall be transmitted by the Swiss Federal Council to all the Powers in whose name the Convention has been signed, or whose accession has been notified.

ARTICLE 138

The present Convention shall come into force six months after not less than two instruments of ratification have been deposited.

Thereafter, it shall come into force for each High Contracting Party six months after the deposit of the instrument of ratification.

ARTICLE 139

From the date of its coming into force, it shall be open to any Power in whose name the present Convention has not been signed, to accede to this Convention.

ARTICLE 140

Accessions shall be notified in writing to the Swiss Federal Council, and shall take effect six months after the date on which they are received.

The Swiss Federal Council shall communicate the accessions to all the Powers in whose name the Convention has been signed, or whose accession has been notified.

ARTICLE 141

The situations provided for in Articles 2 and 3 shall give immediate effect to ratifications deposited and accessions notified by the Parties to the conflict before or after the beginning of hostilities or occupation. The Swiss Federal Council shall communicate by the quickest method any ratifications or accessions received from Parties to the conflict.

ARTICLE 142

Each of the High Contracting Parties shall be at liberty to denounce the present Convention.

The denunciation shall be notified in writing to the Swiss Federal Council, which shall transmit it to the Governments of all the High Contracting Parties.

The denunciation shall take effect one year after the notification thereof has been made to the Swiss Federal Council. However, a denunciation of which notification has been made at a time when the denouncing Power is involved in a conflict shall not take effect until peace has been concluded, and until after operations connected with the release and repatriation of the persons protected by the present Convention have been terminated.

The denunciation shall have effect only in respect of the denouncing Power. It shall in no way impair the obligations which the Parties to the conflict shall remain bound to fulfill by virtue of the principles of the law of nations, as they result from the usages established among civilized peoples, from the laws of humanity and the dictates of the public conscience.

ARTICLE 143

The Swiss Federal Council shall register the present Convention with the Secretariat of the United Nations. The Swiss Federal Council shall also inform the Secretariat of the United Nations of all ratifications, accessions and denunciations received by it with respect to the present Convention.

IN WITNESS WHEREOF the undersigned, having deposited their respective full powers, have signed the present Convention.

DONE at Geneva this twelfth day of August 1949, in the English and French languages. The original shall be deposited in the Archives of the Swiss Confederation. The Swiss Federal Council shall transmit certified copies thereof to each of the signatory and acceding States.

SECTION II. GENEVA CONVENTIONS RELATIVE TO THE PROTECTION OF CIVILIAN PERSONS IN TIME OF WAR (FOURTH GENEVA CONVENTION)

The following excerpted articles from the Geneva Conventions contain requirements concerning the treatment of civilians during time of war. Violations of these articles can constitute war crimes and should be treated as such.

PART I

GENERAL PROVISIONS

Article 1

The High Contracting Parties undertake to respect and to ensure respect for the present Convention in all circumstances.

Article 2

In addition to the provisions which shall be implemented in peacetime, the present Convention shall apply to all cases of declared war or of any other armed conflict which may arise between two or more of the High Contracting Parties, even if the state of war is not recognized by one of them.

The Convention shall also apply to all cases of partial or total occupation of the territory of a High Contracting Party, even if the said occupation meets with no armed resistance.

Although one of the Powers in conflict may not be a party to the present Convention, the Powers who are parties thereto shall remain bound by it in their mutual relations. They shall furthermore be bound by the Convention in relation to the said Power, if the latter accepts and applies the provisions thereof.

Article 3

In the case of armed conflict not of an international character occurring in the territory of one of the High Contracting Parties, each Party to the conflict shall be bound to apply, as a minimum, the following provisions:

1. Persons taking no active part in the hostilities, including members of armed forces who have laid down their arms and those placed hors de combat by sickness, wounds, detention, or any other cause, shall in all circumstances be treated humanely, without any adverse distinction founded on race, colour, religion or faith, sex, birth or wealth, or any other similar criteria.

To this end, the following acts are and shall remain prohibited at any time and in any place whatsoever with respect to the above-mentioned persons:

(a) Violence to life and person, in particular murder of all kinds, mutilation, cruel treatment and torture;

(b) Taking of hostages;

(c) Outrages upon personal dignity, in particular humiliating and degrading treatment;

(d) The passing of sentences and the carrying out of executions without previous judgment pronounced by a regularly constituted court, affording all the judicial guarantees which are recognized as indispensable by civilized peoples.

2. The wounded and sick shall be collected and cared for.

An impartial humanitarian body, such as the International Committee of the Red Cross, may offer its services to the Parties to the conflict.

The Parties to the conflict should further endeavour to bring into force, by means of special agreements, all or part of the other provisions of the present Convention.

The application of the preceding provisions shall not affect the legal status of the Parties to the conflict.

Article 4

Persons protected by the Convention are those who, at a given moment and in any manner whatsoever, find themselves, in case of a conflict or occupation, in the hands of a Party to the conflict or Occupying Power of which they are not nationals.
Nationals of a State which is not bound by the Convention are not protected by it.

Nationals of a neutral State who find themselves in the territory of a belligerent State, and nationals of a co-belligerent State, shall not be regarded as protected persons while the State of which they are nationals has normal diplomatic representation in the State in whose hands they are.

The provisions of Part II are, however, wider in application, as defined in Article 13.

Persons protected by the Geneva Convention for the Amelioration of the Condition of the Wounded and Sick in Armed Forces in the Field of August 12, 1949, or by the Geneva Convention for the Amelioration of the Condition of Wounded, Sick and Shipwrecked Members of Armed Forces at Sea of August 12, 1949, or by the Geneva Convention relative to the Treatment of Prisoners of War of August 12, 1949, shall not be considered as protected persons within the meaning of the present Convention.

Article 5

Where, in the territory of a Party to the conflict, the latter is satisfied that an individual protected person is definitely suspected of or engaged in activities hostile to the security of the State, such individual person shall not be entitled to claim such rights and privileges under the present Convention as would, if exercised in the favour of such individual person, be prejudicial to the security of such State.

Where in occupied territory an individual protected person is detained as a spy or saboteur, or as a person under definite suspicion of activity hostile to the security of the Occupying Power, such person shall, in those cases where absolute military security so requires, be regarded as having forfeited rights of communication under the present Convention.

In each case, such persons shall nevertheless be treated with humanity, and in case of trial, shall not be deprived of the rights of fair and regular trial prescribed by the present Convention. They shall also be granted the full rights and privileges of a protected person under the present Convention at the earliest date consistent with the security of the State or Occupying Power, as the case may be.

Article 25

All persons in the territory of a Party to the conflict, or in a territory occupied by it, shall be enabled to give news of a strictly personal nature to members of their families, wherever they may be, and to receive news from them. This correspondence shall be forwarded speedily and without undue delay.

If, as a result of circumstances, it becomes difficult or impossible to exchange family correspondence by the ordinary post, the Parties to the conflict concerned shall apply to a neutral intermediary, such as the Central Agency provided for in Article 140, and shall decide in consultation with it how to ensure the fulfillment of their obligations under the best possible conditions, in particular with the cooperation of the National Red Cross (Red Crescent, Red Lion and Sun) Societies.

If the Parties to the conflict deem it necessary to restrict family correspondence, such restrictions shall be confined to the compulsory use of standard forms containing twenty-five freely chosen words, and to the limitation of the number of these forms dispatched to one each month.

PART III

STATUS AND TREATMENT OF PROTECTED PERSONS

SECTION I

PROVISIONS COMMON TO THE TERRITORIES OF THE PARTIES TO THE CONFLICT AND TO OCCUPIED TERRITORIES

Article 27

Protected persons are entitled, in all circumstances, to respect for their persons, their honour, their family rights, their religious convictions and practices, and their manners and customs. They shall at all times be humanely treated, and shall be protected especially against all acts of violence or threats thereof and against insults and public curiosity.

Women shall be especially protected against any attack on their honour, in particular against rape, enforced prostitution, or any form of indecent assault.

Without prejudice to the provisions relating to their state of health, age and sex, all protected persons shall be treated with the same consideration by the Party to the conflict in whose power they are, without any adverse distinction based, in particular, on race, religion or political opinion.

However, the Parties to the conflict may take such measures of control and security in regard to protected persons as may be necessary as a result of the war.

Article 28

The presence of a protected person may not be used to render certain points or areas immune from military operations.

Article 29

The Party to the conflict in whose hands protected persons may be is responsible for the treatment accorded to them by its agents, irrespective of any individual responsibility which may be incurred.

Article 30

Protected persons shall have every facility for making application to the Protecting Powers, the International Committee of the Red Cross, the National Red Cross (Red Crescent, Red Lion and Sun) Society of the country where they may be, as well as to any organization that might assist them.

These several organizations shall be granted all facilities for that purpose by the authorities, within the bounds set by military or security considerations.

Apart from the visits of the delegates of the Protecting Powers and of the International Committee of the Red Cross, provided for by Article 143, the Detaining or Occupying Powers shall facilitate as much as possible visits to protected persons by the representatives of other organizations whose object is to give spiritual aid or material relief to such persons.

Article 31

No physical or moral coercion shall be exercised against protected persons, in particular to obtain information from them or from third parties.

Article 32

The High Contracting Parties specifically agree that each of them is prohibited from taking any measure of such a character as to cause the physical suffering or extermination of protected persons in their hands. This prohibition applies not only to murder, torture, corporal punishment, mutilation and medical or scientific experiments not necessitated by the medical treatment of a protected person but also to any other measures of brutality whether applied by civilian or military agents.

Article 33

No protected person may be punished for an offence he or she has not personally committed. Collective penalties and likewise all measures of intimidation or of terrorism are prohibited.

Pillage is prohibited.

Reprisals against protected persons and their property are prohibited.

Article 34

The taking of hostages is prohibited.

SECTION II

ALIENS IN THE TERRORITY OF A PARTY TO THE CONFLICT

Article 41

Should the Power in whose hands protected persons may be consider the measures of control mentioned in the present Convention to be inadequate, it may not have recourse to any other measure of control more severe than that of assigned residence or internment, in accordance with the provisions of Articles 42 and 43.

In applying the provisions of Article 39, second paragraph, to the cases of persons required to leave their usual places of residences by virtue of a decision placing them in assigned residence elsewhere, the Detaining Power shall be guided as closely as possible by the standards of welfare set forth in Part III, Section IV of this Convention.

Article 42

The internment or placing in assigned residence of protected persons may be ordered only if the security of the Detaining Power makes it absolutely necessary.

If any person, acting through the representatives of the Protecting Power, voluntarily demands internment, and if his situation renders this step necessary, he shall be interned by the Power in whose hands he may be.

Article 43

Any protected person who has been interned or placed in assigned residence shall be entitled to have such action reconsidered as soon as possible by an appropriate court or administrative board designated by the Detaining Power for that purpose. If the internment or placing in assigned residence is maintained, the court or administrative board shall periodically, and at least twice yearly, give consideration to his or her case, with a view to the favourable amendment of the initial decision, if circumstances permit.

Unless the protected persons concerned object, the Detaining Power shall, as rapidly as possible, give the Protecting Power the names of any protected persons who have been interned or subjected to assigned residence, or who have been released from internment or assigned residence. The decisions of the courts or boards mentioned in the first paragraph of the present Article shall also, subject to the same conditions, be notified as rapidly as possible to the Protecting Power.

Article 44

In applying the measures of control mentioned in the present Convention, the Detaining Power shall not treat as enemy aliens exclusively on the basis of their nationality de jure of an enemy State, refugees who do not, in fact, enjoy the protection of any government.

Appendix B

Source and Information Reliability Matrix

SOURCE RELIABILITY

B-1. Reliability ratings range from "Reliable" (A) to "Unreliable" (E) as shown in Table B-1. In every instance the rating is based on previous reporting from that source. If there has been no previous reporting, the source must be rated as "F". [NOTE: An "F" rating does not necessarily mean that the source cannot be trusted, but that there is no reporting history and therefore no basis for making a determination.]

Table B-1. Evaluation of Source Reliability.

A	Reliable	**No doubt** of authenticity, trustworthiness, or competency; has a history of complete reliability
B	Usually Reliable	**Minor doubt** about authenticity, trustworthiness, or competency; has a history of valid information most of the time
C	Fairly Reliable	**Doubt** of authenticity, trustworthiness, or competency but has provided valid information in the past
D	Not Usually Reliable	**Significant doubt** about authenticity, trustworthiness, or competency but has provided valid information in the past
E	Unreliable	**Lacking** in authenticity, trustworthiness, and competency; history of invalid information
F	Cannot Be Judged	**No basis** exists for evaluating the reliability of the source

INFORMATION CONTENT

B-2. The highest degree of confidence in reported information is given to that which has been confirmed by outside sources, "1". Table B-2 shows evaluation of information content. The degree of confidence decreases if the information is not confirmed, and/or does not seem to make sense. The lowest evaluated rating of "5" means that the information is considered to be false. [NOTE: A rating of "6" does not necessarily mean false information, but is generally used to indicate that no determination can be made since the information is completely new.]

Table B-2. Evaluation of Information Content.

1	Confirmed	**Confirmed** by other independent sources; **logical** in itself; **Consistent** with other information on the subject
2	Probably True	Not confirmed; **logical** in itself; **consistent** with other information on the subject
3	Possibly True	Not confirmed; **reasonably logical** in itself; **agrees with some** other information on the subject
4	Doubtfully True	Not confirmed; possible but **not logical**; **no other information** on the subject
5	Improbable	Not confirmed; **not logical** in itself; **contradicted** by other information on the subject
6	Cannot Be Judged	**No basis** exists for evaluating the validity of the information

Appendix C

Pre-Deployment Planning

HUMINT leaders must anticipate, identify, consider, and evaluate all potential threats. They must take advantage of enhanced information flow through hierarchical and nonhierarchical networks (computer, communications, and personnel). HCTs should—

- Review available databases on assigned contingency AOIs; review existing IPB products, conduct IPB on these AOIs; and develop appropriate IPB products. Information on databases created for specific contingencies can be gotten from the J2X.
- Continually monitor and update their OPLANs to reflect the evolving situation.
- Be aware of higher headquarters SOPs and DIA manuals for specific ISR management guidance.
- Prepare and practice an intelligence "surge" on likely contingency crises.
- Prepare and practice coordination from pre-deployment through redeployment with personnel from HUMINT, IMINT, MASINT, SIGINT, CA, PSYOP, SOF units, MP, and Engineers.
- Plan for requirements to support 24-hour operations: automation, communications capacity, and personnel necessary to provide continuous intelligence information collection and RM, processing, and reporting.
- Plan and coordinate for linguistic support.
- Forward all RFIs to higher headquarters in accordance with SOPs.
- Prepare and practice coordination with units they will support during pre-deployment exercises. Liaison must be conducted with commanders, S2s, administrative support personnel, logistical support personnel, communications personnel, and others. Obtain supported unit's briefing slide formats.
- Obtain copies of the supported unit's SOPs and ensure all team members are aware of the procedures governing HUMINT interface.
- Ensure that team data processing equipment is compatible with the supported unit's network structure and that appropriate interfaces are available.
- Exercise unit deployment SOPs, load plans, and packing lists.

Helpful Unclassified Links:

- https://portal.rccb.osis.gov/index.jsp Foreign Broadcast Information Service home page.
- http://wnc.fedworld.gov/ International news compiled by the US Department of Commerce.

- http://www.state.gov/s/inr/ Department of State's Bureau of Intelligence and Research home page. Contains country and region-specific information, policies, and warnings.
- http://ebird.afis.osd.mil/ Early Bird News Service of the Armed Forces Information Service.
- http://164.214.2.59/ National Geospatial-Intelligence Agency (NGA) (formerly National Imagery and Mapping Agency).
- http://memory.loc.gov/frd/cs/cshome.html#toc Country Studies from the Federal Research Division of the Library of Congress.

Appendix D

S2 Guide for Handling Detainees, Captured Enemy Documents, and Captured Enemy Equipment

D-1. Detainees, retained personnel, CEDs, and CEE are critical sources of combat intelligence. Often the Maneuver Battalion S2 is the first MI officer to encounter these sources. His actions are critical to the exploitation system. Information from these items is time sensitive, and these information sources need to be exploited at as low an echelon as possible. The S2 should anticipate requirements for support based on planned missions and request HUMINT collector support as necessary. If unable to receive HUMINT collector support, the S2 must be prepared to exploit these sources of information to the best of his ability and more importantly expedite their evacuation to locations and units where they can be exploited.

PURPOSE

D-2. This guide is for battalion and brigade S2s. It explains standard procedures on what the S2 should do when his unit—

- Captures an enemy soldier or other detainee.
- Encounters a civilian on the battlefield.
- Finds or captures an enemy document.
- Discovers an unusual enemy weapon or other unusual piece of equipment during tactical operations.

PERSONNEL HANDLING

D-3. The GPW defines persons entitled to treatment as prisoners of war upon capture, in Article 4 (see Appendix A, Section I).

D-4. The GC defines the civilian population (exclusive of those civilian persons listed in GPW, Article 4) who benefit to varying degrees from the provisions of the Geneva Conventions. (See Appendix A, Sections II and III.)

D-5. Persons in each of these categories have distinct rights, duties, and restrictions. Persons who are not members of the Armed Forces, as defined in Article 4, GPW, who bear arms or engage in other conduct hostile to the enemy thereby deprive themselves of many of the privileges attaching to the members of the civilian population. The capturing unit treats all combatants and noncombatants who are suspected of being part of the threat force as EPWs or retained personnel until their status can be determined. This determination normally occurs at the detainee collection point or at a higher echelon. Noncombatants are handled, questioned, detained, evacuated, and released in accordance with theater policy. In all cases, detainees are treated humanely.

D-6. Detainees are treated humanely but with firmness at all times. High standards of discipline are required not only of detainees but also of capturing and escort forces. Fraternization with detainees or mistreatment or abuse of them is not only a violation but also is not conducive to good discipline. In addition to not being conducive to good discipline, the mistreatment or abuse of detainees is a violation of the UCMJ for which violators may be punished. The control of detainees is exercised through the issuance and firm enforcement of necessary instructions in their own language. Instructions relating to their control during evacuation from the combat zone should be as brief as possible. Care must be taken to ensure that detainees have a clear understanding of all instructions to them.

D-7. At the capture point, the capturing element performs the following steps on detainees. The senior soldier will ensure that the steps are performed. The steps are referred to as the "Five S's and a T".

STEP 1. SEARCH

D-8. The capturing unit's first job is to disarm, search, and maintain positive control over all detainees. The detainees are disarmed and searched for concealed weapons and for equipment and documents of particular intelligence value immediately upon capture, unless the number of detainees captured, enemy action, or other circumstances make such a search impracticable. Until each detainee is searched, the responsible forces must be alert to prevent the use of concealed weapons or destruction of documents or equipment.

D-9. The capturing unit gathers all loose CEDs and CEE in the area. Identification documents and protective military equipment such as helmets or NBC gear stay with the detainee unless otherwise directed by the battalion S2.

- Equipment. Items of personal or individual equipment that are new or appear to be of a type not previously observed may be of intelligence value and should be processed and reported in accordance with the unit's SOP, specific evacuation instructions in Annex B (Intelligence) of the OPORD, and theater policy. Equipment for personal protection such as protective masks or protective clothing may not be taken unless replaced with equivalent equipment.

- Documents. A CED is any piece of recorded information that has been in the hands of the enemy. CEDs include but are not limited to maps, sketches, photographs, orders, tactical and technical manuals, and instructions, code books, log books, maintenance records, shipping and packing slips and lists, war and field diaries, personal diaries, pay books, newspapers, service records, postal savings books, payrolls, postcards and letters, and any written, printed, engraved, or photographic matter that may contain information relative to the enemy and to weather and terrain data. A capturing unit is normally not able to accurately determine the potential intelligence value of any documents found on the detainee. It is therefore normally expedient to remove all documents, with the exception of the detainee's primary identification document. These documents are sealed in a waterproof container and tagged with part C of the capture tag. If capture tags are not available, the document bag must be marked at a minimum to identify the detainee to whom the documents belong (see Appendix I).

• Personal effects. Except as provided in Step 1, detainees should be permitted to retain all of their personal effects including money; valuables; protective equipment, such as helmets, protective masks, and like items; effects and articles used for clothing or eating, except knives and forks; identification cards or tags; badges of grade and nationality; and articles having a personal or sentimental value. When items of equipment issued for the personal protection of detainees are taken from them, they must be replaced with equivalent items serving the same purpose. Although money and other valuables may be taken from detainees as a security measure, they must then be receipted for and a record thereof maintained in a special register. These administrative steps normally are not practical to accomplish prior to arrival of the detainee at an EPW camp.

STEP 2. SILENCE

D-10. Detainees are kept silent so that they cannot plan deception or encourage each other to resist. Keeping the detainees silent also prevents them from relieving the stress and shock of capture by talking with others. If the shock of capture is preserved, HUMINT collectors can take advantage of it in an approach. The capturing unit instructs or signals the detainees to be silent. If that does not work, the detainee is gagged. Guards give orders to detainees, but do not converse with them or give them any comfort items.

STEP 3. SAFEGUARD

D-11. All detainees are promptly evacuated out of the "danger" zone. Their presence may not be used to render points or areas immune to attack, nor should they be retained for participation in psychological warfare or other activities. The capturing forces must protect detainees from reprisals. Detainees will not be denied food, potable water, or appropriate clothing and shelter. Necessary medical attention will not be delayed. Those detainees held in an area should be provided protective facilities and equipment and should be oriented as to procedures to be followed in case of chemical, biological, and radiological agent attack.

STEP 4. SEGREGATE

D-12. The capturing unit separates officers from enlisted, senior from junior, male from female, and civilian from military within their capabilities to both guard and safeguard the detainees. (Physical segregation at this point is not always possible.) Deserters and people of different nationalities and ideologies should be further segregated. The capturing unit prepares a capture tag and puts one on each detainee (see Figure D-1). Tagging procedures are discussed in paragraph D-16.

STEP 5. SPEED TO THE REAR

D-13. The capturing unit moves detainees and CEDs to the unit supply point or other area where transportation to the rear is available for evacuation. Evacuation of detainees from the combat zone should be effected within the minimum time after capture. While in the combat zone, not only may detainees become casualties as the result of enemy fire but also the fluidity of operations, the wide dispersion of units, and the austerity of facilities may necessitate their rapid evacuation.

D-14. The normal evacuation channel is from the detainee collection point through intermediate detainee holding areas to an internment facility at a higher echelon. Available returning transportation, however, may bypass any intermediate detainee holding area and proceed directly to a corps or theater internment facility. Detainees will then be processed directly into the corps or theater internment facility. Evacuation may be by foot, vehicle, rail, aircraft, or ship. Evacuate detainees who are litter patients through medical channels.

D-15. The command (brigade and above) from which the detainees are being evacuated is responsible to provide transportation and rations and for coordinating all other matters related to the evacuation. Escort guards are furnished by the command (division and above) to which the detainees are being evacuated.

STEP 6. TAG

D-16. When the detainees have been searched and segregated, the capturing unit prepares a capture tag and puts one on each detainee. It is very important that the capturing unit fill out the Capture Tag as accurately and completely as possible. HUMINT collectors will use the information from the tag when preparing to interrogate detainees. The "capturing unit" and "location of capture" information will be used to provide direct feedback to the capturing unit when information of immediate tactical value is obtained. Each EPW tag has a different serial number used for the purpose of accountability and cannot be reproduced. The EPW tag is perforated into three parts: Part A is attached to the detainee, Part B is retained by the capturing unit, and Part C is attached to the detainee's property (see Figure D-1).

DOCUMENT HANDLING

DOCUMENTS FOUND ON ENEMY PRISONER OF WAR

D-17. The battalion S2 and subordinate unit commander ensure that CEDs found on detainees are handled as follows. The capturing unit will—

- Search each detainee.
- Return identification documents to detainees. It may be preferable to return only one identity document, to preclude the detainee from spreading extras around to cause confusion. The preferred ID document to return to the detainee is a picture ID (such as a military

or government ID card). If the detainee has several identification documents, the S2 returns the ID that most accurately reflects the detainee's official status. This might be a military ID for a soldier and a passport or government-issue ID for a civilian. If the detainee has several identification documents with different names, this may be an indicator of CI interest. The S2 notifies the nearest CI unit.

- Write the following on the top and bottom half of the EPW capture tag: Number of documents taken, date and time, location and circumstances of capture, capturing unit's designation.
- Put CEDs in a waterproof bag, one per detainee.
- Affix Part C of the capture tag to the bag.
- Give CEDs to the senior escort.
- Direct the senior escort to evacuate CEDs with the detainee.

DOCUMENTS FOUND IN THE AO

D-18. An example of CEDs found in the AO is paperwork discovered in an overrun CP, but not on a detainee. The capturing unit will—

- Put CEDs in a waterproof bag.
- Follow the same procedures described above, and tag the bag.
- Evacuate the CEDs to the battalion S2.
- Evacuate all CEDs as dictated by Annex B of the OPORD. This is normally through the MI chain (for example, from Battalion S2 to Brigade S2, to the first HUMINT collection or DOCEX unit in the MI chain). The S2 normally coordinates with the S4 for the use of supply vehicles returning empty to the rear for the transportation of large numbers of documents.

INITIAL DOCUMENT EXPLOITATION

D-19. A combat unit without language-qualified personnel can perform limited battlefield DOCEX, mainly on maps and overlays. The unit S2 is normally responsible for any initial exploitation by the capturing unit. The S2 safeguards the items pending disposition. At the same time he—

- Looks over the document.
- Does not mark or harm it in anyway.
- Uses whatever resources are available to decipher it; for example, dictionaries and enemy map symbol guides.
- Looks for information that has a direct bearing on his current mission.

D-20. The S2 extracts the combat information and uses the SALUTE format as a template to organize the information.

EQUIPMENT HANDLING PROCEDURES

D-21. CEE includes all types of foreign materiel found on a detainee or in the AO that may have military application. The capturing unit—

- Always permits the detainee to keep protective equipment and equipment for his personal well being unless this gear is replaced by equivalent items by the capturing unit. This equipment includes helmet, NBC gear, mess gear (excluding knife and fork).

- Disposes of equipment in accordance with unit SOPs and instructions in Annex B of the OPORD. Most routine equipment is normally destroyed in place. Unusual or new equipment or equipment identified as being of TECHINT interest is tagged with a CEE tag (Part C of DD Form 2745) and evacuated to the nearest TECHINT unit. Communications equipment is also tagged and evacuated to the nearest SIGINT unit.

- Identifies equipment that cannot be easily evacuated; its location is passed through intelligence channels to the nearest unit that will be involved in its exploitation.

FIELD-EXPEDIENT TAGGING PROCEDURES

D-22. When no standard tag forms are available, the following field-expedient methods may be used:

- Use meals, ready-to-eat (MRE) cardboard or other type of paper.
- Write the capturing unit's designation.
- Write data and time of capture.
- Write POC coordinates.
- Write circumstances of capture.
- Identify EPW, captured document, or equipment captured.
- Put tag, without damaging the CED, in a waterproof bag.
- Attach EPW and CEE tags so they will not come off.

MEDICAL CARE

D-23. Medical equipment and supplies to permit the administering of emergency first aid should be available at each EPW collecting point and EPW holding area. A qualified medical retained person, if available, may administer first aid to other detainees. All detainees suspected of having communicable diseases are isolated for examination by a medical officer. Wounded detainees may be questioned by intelligence personnel once the detainees are cleared by competent medical authority for questioning.

D-24. For evacuation purposes, detainees may be classified as walking wounded or sick, or as non-walking wounded or sick. Walking wounded detainees are evacuated through MP EPW evacuation channels. Non-walking wounded are delivered to the nearest medical aid station and evacuated through medical channels.

1. DATE AND TIME OF CAPTURE	2. SERIAL NO. A

3. NAME   **4. DATE OF BIRTH**

5. RANK   **6. SERVICE NO.**

7. UNIT OF EPW   **8. CAPTURING UNIT**

9. LOCATION OF CAPTURE *(Grid coordinates)*

10. CIRCUMSTANCES OF CAPTURE   **11. PHYSICAL CONDITION OF EPW**   **12. WEAPONS, EQUIPMENT, DOCUMENTS**

DD Form 2745, May 96   Replaces DA Form 5976, JAN 91, Usable until exhausted.

ENEMY PRISONER OF WAR (EPW) CAPTURE TAG (PART A)

For use of this form, see AR 190-8. The proponent agency is DCSOPS.

Attach this part of tag to EPW. *(Do not remove from EPW.)*

1. **Search** – For weapons, military documents, or special equipment.
2. **Silence** – Prohibit talking among EPWs for ease of control.
3. **Segregate** – By rank, sex, and nationality.
4. **Safeguard** – To prevent harm or escape.
5. **Speed** – Evacuate from the combat zone.
6. **Tag** – Prisoners and documents or special equipment.

DD Form 2745 (BACK), May 96

B — Part B: **UNIT RECORD CARD (PART B)** — Forward to Unit. *(Capturing unit retains for records.)* Use string, wire, or other durable material to attach the appropriate section of this form to the EPW's equipment or property.

Part C: **DOCUMENT/SPECIAL EQUIPMENT WEAPONS CARD (PART C)** — Attach this part of tag to property taken. (Do not remove from property.)

As a minimum, the tag must include the following information:
Item 1. Date and time of capture (YYYYMMDD).
Item 8. Capturing Unit.
Item 9. Place of capture *(grid coordinates)*.
Item 10. Circumstances of capture *(how the EPW was captured)*.

Part C also includes: 10. DESCRIPTION OF WEAPONS, SPECIAL EQUIPMENT, DOCUMENTS.

DD Form 2745, May 96 / DD Form 2745 (BACK), May 96

Figure D-1. DD Form 2745 (Enemy Prisoner of War Capture Tag).

Appendix E

Extracts from Allied Joint Publication (AJP)-2.5

Following are excerpts from Allied Joint Publication-2.5, Handling of Captured Personnel, Equipment and Documents. AJP-2.5 is primarily an amalgamation of procedures outlined in STANAG 2033, STANAG 2044, and STANAG 2084, and will be used as the authoritative source for matters governed by those STANAGs. S2s and HUMINT collectors should maintain a copy on hand.

THE GENEVA CONVENTIONS

E-1. Treatment of prisoners of war (PWs) and other detainees as well as the handling of personal possessions including personal documents belonging to them will at all times be in accordance with the 1949 Geneva Conventions and, if ratified by the nations concerned, with the 1977 Protocols.

THE DETAINING POWER

E-2. The responsibilities of the "Detaining Power" as set out in GC3 shall rest with the nation or the deployed NATO command which initiated the capture or detention of the person or persons in question.

E-3. Transfers of PWs between NATO nations must be in accordance with GC3, Article 12, as interpreted but not superseded by national agreements.

KNOWLEDGE OF THE GENEVA CONVENTIONS AND THE 1977 PROTOCOLS

E-4. The Geneva Conventions require the signatories in time of both peace and war to circulate the text of the Conventions as widely as possible within their countries. They are especially required to ensure that the provisions and implications of the Conventions are clearly understood by the members of their armed forces and by the civilians attached to them. In order to assist this process, it is suggested in the Conventions that instruction in them is included in the syllabus of appropriate military and civilian courses of instruction.

E-5. The conventions also state that any military or other authority assuming responsibility for dealing with PWs in time of war or armed conflict must be in possession of the text of the Conventions and that its personnel must be instructed in how the Conventions affect them in the execution of their duties with PWs.

E-6. If ratified by the nations concerned, these requirements are, by inference, also applicable to the 1977 Protocols.

PRISONER-OF-WAR STATUS

E-7. Captured personnel (CPERS) can be divided into two categories: Captured personnel who are PWs and other detainees.

E-8. Prisoner-of-war status is conferred on captured personnel who fall within the categories listed in Article 4 of GC3, which is reproduced in Annex A, or, if ratified by the nations concerned, those who meet the expanded definition of a PW as stated in Protocol I of the 1977 Protocols.

E-9. Other detainees are personnel being held by NATO forces until their status can be determined and their further disposition can be decided.

E-10. Furthermore, when NATO forces are engaged in Peacetime Support Operations (PSOs), the operational plan should contain specific instructions as to which individuals will have PW status. Directions for handling other detainees, including those suspected of crimes against humanity and war crimes, are also to be included in the operational plan or in the Standing Operating Procedures used in the operation.

PERSONNEL

E-11. Conditions allowing, the following procedures should be followed by the capturing unit:

a. CPERS should be disarmed immediately, and all documents and effects of military or investigative interest except for necessary clothing, identity documents and protective equipment (Geneva Convention Relative to the Treatment of Prisoners of War (GC3), Article 18) should be removed. CPERS should then be tagged in accordance with the procedures outlined at Annex B. A Common Capture Report should also be completed and forwarded in accordance with the procedure set out in Annex C. It is important that the documents, equipment, maps, etc., taken from a CPERS accompany him to the next receiving unit. Valuable information may be lost by not having these items available during processing and interrogation.

b. Within the confines of the tactical situation, CPERS are to be segregated according to rank, grade, service, sex and nationality or ethnic group/warring faction to minimize the opportunity to prepare counter-interrogation measures. Furthermore, deserters, civilians and political indoctrination personnel will be individually segregated from other CPERS. Such segregation shall be undertaken in a manner which does not violate GC3, Article 16.

c. Talking or fraternization between CPERS is to be prevented in order not to prejudice future intelligence collection operations. CPERS will be allowed no opportunity to exchange information between themselves, to exchange identities or to dispose of articles of intelligence interest.

d. Interrogation operations must not be compromised by contact between CPERS and personnel not concerned with interrogation duties.

e. CPERS will also be prevented from observing sensitive and critical activities, equipment and procedures involving NATO, national or allied forces.

f. CPERS are to be guarded in a manner which shall deny them the opportunity for escape or sabotage.

g. Defectors and political refugees should also be segregated from other CPERS wherever possible. These personnel shall be screened by the nearest Interrogation Unit (IU), which will decide on their value to the intelligence organization and consequent future movements. In all cases, defectors are to be treated in accordance with the Geneva Convention Relative to the Protection of Civilians in Time of War (GC4). National policy may provide defectors treatment in accordance with GC3 where such treatment provides greater protection than GC4.

h. Personnel claiming to be agents of an allied power shall also be segregated from other CPERS. The intelligence organization (G2 or CJ2) is to be informed of all such individuals as soon as possible and will arrange for their screening to determine their future disposition.

i. CPERS suspected of crimes against humanity and war crimes will also be segregated from other prisoners. Legal authorities and the intelligence organization are to be informed of such suspects as soon as possible. They will be taken into custody by law enforcement personnel. Intelligence exploitation should be undertaken in cooperation with the legal authorities.

j. All CPERS are to be treated humanely.

k. Naval and Air Force personnel are to be identified and the intelligence organization is to be notified in order that interrogation by naval/air force personnel may take place at the earliest opportunity.

l. CPERS are to be escorted to the nearest Collecting Point or Holding Area as quickly as possible.

m. Should any doubt arise as to whether any persons, including those appearing to be PWs, having committed a belligerent act and having fallen into Allied hands, belong to any of the categories of persons entitled to PW status pursuant to Article 4 of GC3, such persons shall enjoy the protection of GC3 until such time as their status has been determined by a competent tribunal. If such a tribunal determines that an individual does not qualify for PW status, then the detaining commander must determine whether the detainee qualifies as a "protected person" pursuant to GC4, and obtain legal advice relative to the proper course of action for dealing with such detainees.

DOCUMENTS

E-12. Captured documents (CDOC) considered of intelligence interest will be handled by the capturing unit in the following manner:

a. The capturing unit will conduct a preliminary screening to obtain information of immediate technical or tactical value.

b. An intelligence report (INTREP) identifying the CDOC and its disposition as well as giving information considered to be of immediate tactical value will be prepared and submitted by the capturing unit. (See STANAG 2022.)

c. The capturing unit will tag or otherwise mark the CDOC as follows:

- National identifying letters of capturing unit as prescribed in STANAG 1059.
- Designation of capturing unit including service.
- Serial number of the CDOC. This will consist of a number allocated sequentially by the capturing unit.
- DTG of capture.
- Location of capture (geographic coordinates or UTM grid reference including grid zone designation and 100,000-meter square identification).
- Captured from Unit (enemy or warring faction) (including national identifying letters in accordance with STANAG 1059).
- Summary of circumstances under which the CDOC was obtained. Interrogation serial number of any associated CPERS, if appropriate or known.
- CED associated with a captured person should be marked with part C of the Standardized EPW Capture and Personal Equipment Tag.

d. CED to be used as evidence in legal proceedings against CPERS suspected of crimes against humanity and war crimes will be kept under guard or in a secure area separate from other CED.

EQUIPMENT

E-13. Captured Equipment (CE) and Associated Technical Documents (ATDs) considered of intelligence interest will be handled by the capturing unit in the following manner:

a. A Capture Report as set out in Annex C of the OPORD will be prepared by the capturing unit upon the discovery or capture of CE or ATD believed to be of intelligence interest. The reporting channels are from the capturing unit through the chain of command to the first TECHINT element.

b. The capturing unit will conduct a preliminary screening to obtain information of immediate technical or tactical value. A Preliminary Technical Report (PRETECHREP) as set out in Annex D of the OPORD will be prepared and submitted through established intelligence channels.

c. Intelligence Reports (INTREP) may, as circumstances dictate, be submitted in advance but not in lieu of the Capture Report and PRETECHREP. (See STANAG 2022.)

d. CE and ATD will be tagged or marked by the capturing unit as follows:

- National identifying letters of capturing unit as prescribed in STANAG 1059.
- Designation of capturing unit including service.
- DTG of capture.
- Location of capture (geographic coordinates or UTM grid reference including grid zone designation and 100,000-meter square identification).

- Captured from Unit (enemy or warring faction) (including national identifying letters in accordance with STANAG 1059).
- Summary of circumstances of capture.
- Associated CPERS.

e. CE and ATD to be used as evidence in legal proceedings against CPERS suspected of crimes against humanity and war crimes will be kept under guard separate from other CE and ATD.

Appendix F

NATO System of Allocating Interrogation Serial Numbers

1. Every captured person selected for interrogation will be given an Interrogation Serial Number. This number shall be allocated by the Interrogation Unit conducting the **first** interrogation of the captured person.

2. The number should not be confused with the Prisoner of War Internment Serial Number (ISN), which is to be used for administrative purposes only.

3. The purpose of the Interrogation Serial Number is to identify the source of information to ensure its proper evaluation, processing, and follow-up action. It will also identify the nationality and location of the interrogation unit.

4. The number shall be constituted as follows:

   a.   2 letters to indicate the Nationality of the captured person (see para 6).

   b.   2 letters to indicate the Service of the captured person (see para 7).

   c.   1 letter to indicate the Arm of Service of the captured person (see para 8).

   d.   4 numbers to indicate the Interrogation Sequence Number of the captured person (see para 9).

   e.   4 numbers to indicate the Day and Month of Capture.

   f.   A dash (-) to show a sequence break.

   g.   4 letters to indicate the Nationality and Service of the Interrogation Unit (see para 10).

   h.   4 letters to indicate the Interrogation Unit.

5. Each group shall be separated by a dash. The final number shall therefore appear as in the following examples:

   LS – NV – B – 0012 – 2105 – USNV – 0159

   Liechtenstein – Naval prisoner – Seaman – 12th captured person interrogated - captured 21 May – interrogated by US Navy – team 159

   AN – AF – H – 0357 – 0211 – GEAF – 0007

   Andorra – Air force prisoner – Intelligence – 357th captured person interrogated – captured 2 Nov – interrogated by GE Air Force – team 7

SM – AR – X – 0431 – 0707 – UKAR – 0019

San Marino – Army prisoner – One star or above – 431st captured person interrogated – captured 7 Jul – interrogated by UK Army – team 19

6. The two letters used for indicating the <u>Nationality of the captured person</u> will be in accordance with established NATO Country codes in STANAG 1059.

7. The following list of letters is to be used to indicate the <u>Service of the captured person</u>:

AR	- Army	NI	- Naval Infantry
NV	- Navy	AB	- Airborne Forces
NA	- Naval Air Arm	SF	- Special Purpose Forces
AF	- Air Force	PL	- Police
IR	- Irregular	CV	- Civilian (other than Police)

8. The following list of letters is to be used to indicate the <u>Arm of Service of the captured person</u>:

Navy (A)	Army (B)	Air Force (C)	Other/Partisan Forces (D)
A. Aircrew	Aircrew	Aircrew	Aircrew
B. Seamen	Infantry	Ground Crew	Merchant Marine
C. Communications	Signals	Communications	Radio Officers/Operators
D. Weapons/Electronic Engineer	Electrical/Electronic Engineer	Electrical/Electronic Engineer	
E. Mechanical/Marine/Engineers	Engineers	Mechanical/Air Frame/Engineers	
F. Gunnery	Artillery	Ordnance	Weapons/Ordnance Explosives
G. *HQ Staff	*HQ Staff	*HQ Staff	*HQ Staff
H. Intelligence	Intelligence	Intelligence	Intelligence
I. Marines	Airborne Forces	Airfield Defense	
J. Cooks/Stewards	Catering	Catering	Catering
K. Legal/Political	Legal/Political	Legal/Political	Legal/Political
L. Medical/Dental/Nursing	Medical/Dental/Nursing	Medical/Dental/Nursing	Medical/Dental/Nursing
M. Electronic Warfare	Electronic Warfare	Electronic Warfare	Electronic Warfare
N. Operations	Operations	Operations	Operations
O. Police	Police	Police	Police
P. Supply	Quartermaster	Supply	Supply
Q. Strategic Weapons	Strategic Weapons	Strategic Weapons	Strategic Weapons
R. Special Purpose Forces	Special Purpose Forces	Special Purpose Forces	Special Purpose Forces
S. Air Traffic Control	Armored	Air Traffic Control	Air Traffic Control
T. *Unknown	*Unknown	*Unknown	*Unknown

Navy (A)	Army (B)	Air Force (C)	Other/Partisan Forces (D)
U. Instructors	Education	Education	Instructors
V. Ministers of Religion	Ministers of Religion	Ministers of Religion	Ministers of Religion
W. *Submarines	Artillery Spotters	Forward Air Controllers	Couriers, etc.
X. *One Star +	*One Star +	*One Star +	*Officers in Command of Irregular Forces
Y. *Other	*Other	*Other	*Other

*Notes: G – Headquarters staffs below one star rank
      T – Captured Person's arm of service not known to report writer
      W – To be used in respect of all submarines regardless of arm or specialization
      X – To be used in respect of all ONE STAR or above officers regardless of army
      Y – PW arm of service not included in the appropriate list

9. The sequence number of the interrogated captured personnel shall be of four digits allocated in numerical order of interrogation by the team first interrogating the captured person.

10. Country, services and team codes.

    a. Country codes, as laid down in STANAG 1059, are to be used for indication of the Nationality of the Interrogation Team.

    b. The following lists of letters are to be used for indicating the Service of the Interrogation Team:

ARMY	- AR	AIR FORCE	- AF
NAVY	- NV	MARINE	- MR
COAST GUARD	- CG	PARA-MILITARY	- PM

    c. NATO teams will use the following abbreviations: (These will be determined as command changes are implemented and STANAG 1059 is revised).

SC EUROPE	RC SOUTH
RC NORTH	JHQ SOUTHWEST
JHQ NORTH	JHQ SOUTH
JHQ NORTHEAST	JHQ SOUTH CENTER
JHQ CENTER	JHQ SOUTHEAST
CC AIR NORTH	CC AIR SOUTH
CC NAV NORTH	CC NAV SOUTH
SC ATLANTIC	RC EAST
RC WEST	STRIK FLTLANT
RC SOUTHEAST	SUB ACLANT

11. Nations will allocate 4-digit serial numbers to their interrogation teams. NATO Commands will issue similar numbers to subordinate national interrogation teams under their command.

Appendix G

Questioning Quick Reference

This appendix offers a quick reference for the trained HUMINT collector. It is not meant to be all-inclusive, nor instructive in proper questioning technique, but lays out frequently used requirements grouped logically by OB factor. Proper formation of questions is covered in detail in Chapter 9.

Missions: Mission questioning consists of three areas: Time of Capture Mission, Future Mission, and Past Mission. Missions are questioned in that order, to ensure that the information is collected in the order of importance to a supported commander. Logical follow-up questioning may lead the collector into any of the OB factors at any time during questioning in order to provide complete information.

> **Offensive Missions:**
> When will the enemy attack?
> Where will they attack?
> What is the main objective of the attack?
> What units will participate in the attack?
> What tactics will be employed?
> What artillery, air, and other units will support the attack?
> **Defensive Missions:**
> Where will the enemy establish lines of defense?
> What units have been assigned to the defensive lines?
> What obstacles have been emplaced (mines, trenches, wire, etc.)?
> What artillery support is there for the defensive operation?
> **Retrograde Operations:**
> What units will take part in the retreat?
> What are the current positions of the retreating units?
> When will they start to retreat?
> What routes will be used?
> What is the planned destination of the retreating units?
> What units will cover the retreat?

Composition:
> What is the command and control element of (the target unit)?
> What types of units are directly subordinate to (the target unit)?
> What is the designation of (each of the subordinate units)?
> How many units of that type are directly subordinate to (target unit)?
> What units are attached? When? Why? What unit(s) are they detached from?
> What units are detached? When? Why? What unit(s) are they attached to now?

Weapons and Equipment Strength:
Individual Weapons:
What individual weapons are there in (target unit)?
How many?
What is the distribution of the weapons?
Crew-Served Weapons: What crew-served weapons are in (target unit)?
How many?
What is the distribution of the weapons?
Other Weapons: What other weapons are there in (target unit)?
What types?
How many?
How are they distributed?
Vehicles: What armored vehicles are in (unit)?
How many?
What nomenclature?
What other vehicles are in (unit)?

Dispositions:
Disposition questioning is ideally done with the aid of a map.
Where is (the target disposition)?
Show (on the map) the location of (the target disposition).
What enemy units, activities, or equipment are at that location?
What security measures are being employed at that location?
Additionally, specific types of dispositions require additional follow-up:
Area-Type Dispositions: (Staging Areas, Assembly Areas, Trains, etc.)
Show on the map (or describe) the perimeter of the disposition.
Where are units or activities located within it?
Where are the approaches/entrance?
What unit is in charge?
What vehicles or equipment are located within the disposition?
What is the date of information?
Line-Type Dispositions: (Line of Departure, Artillery emplacement, etc.)
Show on the map (or describe) the location of the disposition.
Define and locate both ends of the disposition.
What equipment is located there?
In the case of artillery, describe the placement and orientation of the guns,
ammunition, radar, and support vehicles.

Tactics:
Offensive: What offensive tactics are being employed by (unit)?
What other units are involved?
When did (unit) begin employing these tactics?
What are the major objectives?
Defensive: What defensive tactics are being employed by (unit)?
What other units are involved?
When did (unit) begin employing these tactics?
Special Operations: What special operations tactics are being employed by (unit)?
What are the designations of the units employing special operations tactics? When did
(unit) begin employing special operations tactics?
Where/Why are these tactics being employed?

Training:
> **Individual Training:** What individual training is being conducted by (unit)?
> Who is being trained?
> How effective is the training?
> Where is the training conducted?
> What are the training standards?
> **Unit Training:** What unit training is being conducted by (unit)?
> Who is being trained?
> How effective is the training?
> Where is the training conducted?
> What are the training standards?
> **Specialized Training:** What specialized training is being conducted by (unit)?
> Who is being trained?
> How effective is the training?
> Where is the training conducted?
> What are the training standards?
> What specialized equipment is used in the conduct of the training?

Combat Effectiveness:
> **Losses:**
> **Personnel:** What personnel losses have been incurred by (unit)?
> When? Where? How?
> What were the duty positions/ranks of the lost personnel?
> **Equipment:** What equipment losses have been sustained by (unit)?
> What type of equipment was lost?
> When? Where? How many?
> How were they lost?
> **Replacements (Personnel):**
> **Received:** What replacements have been received by (unit)?
> What replacements are available to (unit)?
> How many?
> From where were the replacements received?
> **Available:** What personnel replacements are available to (unit)?
> From where are replacements available?
> How many? What ranks?
> How long would it take for replacements to arrive once requested?
> **Replacements (Equipment):**
> **Received:** What equipment replacements have been received by (unit)?
> How many?
> From where were the replacements received?
> How does the quality of the replacement equipment compare to that of the original equipment?
> What is the level of confidence in the replacement equipment, compared to the original?
> **Available:** What equipment replacements are available to (unit)?
> From where are replacements available?
> How long would it take to receive replacement equipment once requested?
> How many of each type of equipment are available?

Reinforcements:

  Received: What reinforcements have been received by (unit)?
  What type was the reinforcing unit?
  What is the designation of the reinforcing unit?
  What equipment did the reinforcing unit bring?
  To which unit was the reinforcing unit further assigned?
  Why did (receiving unit) receive reinforcements?
  How long will the reinforcing unit be assigned as reinforcement to (unit)?
  Available: What reinforcements are available to (unit)?
  From where (parent unit/location) are reinforcements available?
  What types of units are available to reinforce (unit)?
  How long would it take for reinforcements to arrive, once requested?

Morale:

  Describe the morale of the members of (unit).
  How long has the morale been...?
  What is the unit leadership doing to (maintain/improve) the morale?
  What effect has (high/low) morale had on the behavior or performance of the unit
    members?

Logistics:

  Weapons: What is the condition of the (specific weapons) in (the unit)?
  Why are (the weapons) in that condition?
  What is being done to improve the condition of (the weapons)?
  How often is maintenance performed? By whom?
  Who inspects weapons? How often?
  Is there an increased emphasis on maintenance? If so, why?
  What is done to prevent/alleviate a shortage while weapons are being maintained?
  What spare parts are there for weapons in (unit)?
  What shortages of spare parts are there?
  What problems are there with spare parts (quality, fit, delays, etc)?
  Ammunition: What types of ammunition are available for the (weapon/weapon system)
    in the (unit)?
  What problems are being experienced with ammunition for (weapon or weapon system)
    in (unit)? Why? Since when?
  What is being done to correct the problem(s)?
  What shortages of ammunition for (weapon or weapon system) are there in (unit)?
  What is being done to correct the shortage?
  When was the last issue of ammunition for the (weapon or weapon system) in the (unit)?
  How much was issued?
  When is the next issue of ammunition planned?
  (For insurgents/irregulars - Where is funding obtained for ammunition/explosive
    purchases?
  Where are ammunition/explosives obtained?
  How are ammunition/explosives transported/delivered?
  Vehicles: What is the condition of the (specific vehicle) in (the unit)?
  Why are (the vehicles) in that condition?
  What is being done to improve the condition of (the vehicles)?
  How often is maintenance performed? By whom?
  Who inspects vehicles? How often?

Is there an increased emphasis on maintenance? If so, why?
What is done to prevent/alleviate a shortage while vehicles are being maintained?
What spare parts are there for vehicles in (unit)?
What shortages of spare parts are there?
What problems are there with spare parts (quality, fit, delays, etc)?

POL:
What problems have been experienced with the petroleum, oils, and lubricants (POL) for the (vehicle type) in the (unit)?
Why are there problems?
Since when have there been problems?
What is being done to correct the problems?
What shortages of POL are there? Why?
What is being done to alleviate the shortages?
When was the last resupply of POL in (unit)? Where? How much?
When is the next resupply of POL planned?

Communications Equipment:
What is the condition of the (specific radio set) in (the unit)?
Why are (the radios) in that condition?
What is being done to improve the condition of (the radios)?
How often is maintenance performed? By whom?
Who inspects communication equipment? How often?
Is there an increased emphasis on maintenance? If so, why?
What is done to prevent/alleviate a shortage while radios are being maintained? What spare parts are there for communication equipment in (unit)?
What shortages of spare parts are there?
What problems are there with spare parts (quality, fit, delays, etc)?

Medical:
Equipment: What individual medical equipment is in (unit)? How many?
What is the distribution?
What are the contents of individual medical kits in the (unit)?
What is the quality?
What shortages are there?
What problems are there with the individual medical equipment/supplies in (unit)?
What vehicular medical equipment is in (unit)? How many?
What is the distribution?
What are the contents of vehicular medical kits in (unit)?
What is the quality?
What shortages are there?
What problems are there with the vehicular medical equipment/supplies in (unit)?
Personnel: What medical personnel are assigned to (unit)?
What medical personnel are available to treat members of (unit)? How many? What are the duty positions of the medical personnel?
What level of medical care are the medical personnel able to provide to members of (unit)?
Facilities: What medical facilities are available to members of (unit)?
Where are the medical facilities?
What level of care is available there?
What higher echelons of medical care are available?

Medical Evacuation (MEDEVAC) Procedures: What MEDEVAC procedures are available to members of (unit)?

Where are MEDEVAC collection points?

What different MEDEVAC procedures are used for more seriously wounded personnel, compared to lightly wounded?

Food: What rations are members of (unit) eating?

What shortages are there of food?

What is the quality?

What problems have been encountered with the rations provided to (unit)?

When was the last issue of rations to members of (unit)?

When is the next issue?

What reserve stocks of field rations are there in (unit)?

Water: What is the source of drinking water for members of (unit)?

What problems have been encountered with quality, shortages, or contamination of drinking water?

What water purification methods are available to members of (unit)?

Under what circumstances will they be used?

If water purification tablets are used, what color are they?

Electronic Technical Data:

Radio Frequencies: What are the primary and alternate frequencies for radios in (each unit)?

Under what circumstances will alternate frequencies be used?

What networks operate on the various frequencies?

Call Signs: What is the call sign of (unit)?

What is the call sign of (each) person of authority in the unit?

When do call signs change?

What informal call signs are in use?

Miscellaneous:

Personalities: Question for name, rank, unit, duty position, and unit of key leaders.

Collect identifying characteristics such as age, height, weight, build, hair and eye color, writing hand, facial hair and teeth.

Report contact information such as work and home addresses, telephone numbers, fax number, and email address.

Complete biographical IIR format is in DIAM 58-12 (S//NF).

Code Names: What is the code name of (each unit)?

What code names are being used for specific operations?

Passwords: What is the current challenge/password for (unit)?

When did it go into effect?

When will it change?

What will the next challenge/password be?

What other (informal, "run in") passwords are in use?

Obstacles:

Enemy: What obstacles have the enemy forces emplaced?

Where, when, what type of obstacles?

What safe lanes are there through or around the obstacles?

(If mine fields, collect type, pattern, quantity, and method of emplacement of mines.)

How are the obstacles being covered (artillery, ambush, etc.)?

Why have the obstacles been emplaced where they are (denial of terrain, canalization into a fire sack, etc.)?

Friendly: What friendly obstacles have enemy forces encountered? Where? When?

Have those obstacles been breached or otherwise neutralized?

What effect have the friendly obstacles had on enemy maneuver?

PSYOP:

Enemy: What PSYOP are being conducted by (unit)?

What is the text of the messages?

Who is the target audience?

Where are the PSYOP materials prepared?

Where and how are they delivered? What is the hoped-for effect?

Friendly: What PSYOP have members of (unit) encountered? Where? When?

What form of PSYOP was encountered?

What effect have the PSYOP had on the officers/NCOs/enlisted of (unit)?

What is the leadership of (unit) doing to counter the effects of friendly PSYOP?

Appendix H

SALUTE Reporting

H-1. The SALUTE report format requires brief entries which require the collector to break information down into basic elements: who, what, where, when, why, and how. This allows for efficient reporting via electronic or hardcopy medium. It also allows the analyst to quickly scan multiple reports to find specific information.

H-2. Figure H-1 provides guidance and is not to be construed as strict requirements. SALUTE reports of combat activity may only contain a word or two in each entry, whereas Intelligence reports tend to include more detail.

TO: Usually the address of the supported S2/G2 (according to unit SOP)

FROM: Your unit or team designation, or your duty position, as appropriate

DTG: The date-time group of when the report is being submitted

Report Number: From local SOP

1. (S)ize/Who: Expressed as a quantity, and echelon, or size (e.g., 1 x BDE). If multiple echelons are involved in the activity being reported, there can be multiple entries (e.g.,1 x BDE; 2 x BN). Non-standard units are reported as such (e.g., bomb-making class; support staff).

2. (A)ctivity/What: This line is the focal point of the report and relates to the PIR or important non-PIR information being reported. It should be a concise bullet statement.

3. (L)ocation/Where: Generally a grid coordinate, including the 100,000 meter grid zone designator. The entry can also be an address, if appropriate, but still should include an 8-digit grid coordinate. City names will always be followed by the two-character country code. If the activity being reported involves movement (advance, withdrawal, etc.) the location entry will include "From" and "To" entries. The route used will be reported under "Equipment/How."

4. (U)nit/Who: This entry identifies who is performing the activity described in the "Activity/What" entry. Include the complete designation of a military unit, identification of a civilian or insurgent group, or the full name of an individual, as appropriate.

5. (T)ime/When: For a future event, this is when the activity will initiate. Past events are usually not the subject of SALUTE reports, but if a past event is to be reported, the Time/When entry will generally reflect when the event ended. Ongoing events are reported as such. Reports of composition of forces, morale, and Electronic Technical Data and other non-event topics are reported as ongoing. When reporting on a disposition, the "Time/When" entry is generally the last time the source was at the disposition.

6. (E)quipment/How: The information reported in this entry clarifies, completes, and/or expands upon information reported in any of the previous entries. It includes information concerning equipment involved, tactics used, and any essential elements of information not reported in the previous paragraphs.

7. Remarks: Use this entry to report the source of the information, whether a person, a CED, open-source media, or other source. Include the date of information and the PIR that the reported information addresses. Map data for coordinates given in the "Location/Where" entry are included, stating map series name, sheet number, scale and edition. If there are enclosures to the SALUTE report, such as sketches, they are annotated here.

Figure H-1. Example of a Written SALUTE Report.

Appendix I

Document Exploitation and Handling

I-1. DOCEX is a vital information source in the development of the all-source intelligence picture. Unless planned for and carefully monitored, the volume of CEDs in all operations can rapidly overwhelm a unit's capability to extract meaningful information.

DEFINITIONS

I-2. A document, as defined by AJP 2.5, is any piece of recorded information, regardless of form. Documents include printed material such as books, newspapers, pamphlets, OPORDs, and identity cards as well as handwritten materials such as letters, diaries, and notes. Documents also include electronically recorded media such as computer files, tape recordings, and video recordings and the electronic equipment which contains documents or other vitally important intelligence. Examples include hard drives, operating systems, and personal electronic devices, including phones, PDAs, and GPS devices. A CED may be needed by several collection or exploitation activities at the same time, requiring copies to be made. Collectors must have ready access to copying equipment. Documents often must be evacuated through two different channels for proper exploitation, which also makes copying necessary. Such documents and equipment require special handling to assure that they are returned to their owners.

I-3. DOCEX is the systematic extraction of information from threat documents for the purpose of producing intelligence or answering IRs. A threat document has been in the possession of the threat, written by the threat, or is directly related to a future threat situation. DOCEX can occur in conjunction with HUMINT collection activities or as a separate activity.

I-4. A CED is any document that was in the possession of an enemy force that subsequently comes into the hands of a friendly force, regardless of the origin of that document. There are three types of CEDs.

- Official - documents of government or military origin.

- Identity - personal items such as identification (ID) cards or books, passports, driver licenses.

- Personal - documents of a private nature such as diaries, letters, and photographs.

I-5. Open-source documents are documents that are available to the general public including but not limited to newspapers, books, videotapes, public records, and documents available on the Internet or other publicly available electronic media.

I-6.   Source-associated documents are documents that are encountered on or in immediate association with a human source. These may include both official and personal documents. Documents associated with human sources are normally exploited, at least initially, during the interrogation or debriefing of the source. Interrogators typically use these documents during planning and preparation for interrogation of the associated EPW. These personal documents and source identification documents are therefore evacuated in conjunction with the associated source and sent through prisoner, detainee, or refugee evacuation channels rather than through intelligence channels. If the duplication capability exists, collectors should copy personal documents that contain intelligence information and evacuate the copy through intelligence channels. The original personal document should be evacuated with the detainee but not on his person until the HUMINT collector has exploited it. Collectors evacuate official documents through intelligence channels after initial exploitation. If possible, the collector will copy official documents and evacuate the copy with, but not on, the source.

OPEN-SOURCE INFORMATION

I-7.   Open-source information is publicly available information appearing in print or electronic form. Open-source information may be transmitted via radio, television, newspapers, commercial databases, electronic mail networks, or other electronic media like CD-ROMs. Whatever form they take, open sources are not—

- Classified at their origin.

- Subject to proprietary constraints.

- The product of sensitive contacts with US or foreign persons.

I-8.   In all operations, open-source collection can be a valuable addition to the overall intelligence collection and each intelligence discipline's efforts. Open-source information supplements the HUMINT collection effort, and all types of open sources must be considered for exploitation.

I-9.   Open sources are evaluated and categorized as friendly, neutral, or hostile. Certain high-value, open-source information sources may be identified for continuous monitoring. Other open-source information sources may be identified to screen for the presence or lack of specific indicators. In addition, the information obtained from open sources is extremely helpful for the HCT members to be current with the latest developments in the AO, which enables them to establish rapport and effectively converse with their sources. Open-source documents are exploited in the same manner as CEDs.

OPEN-SOURCE DOCUMENT OPERATIONS

I-10. Open-source document operations are the systematic extraction of information from publicly available documents in response to command IRs. Open-source document operations can be separate operations or can be included as part of other ongoing operations. Open-source documents are significant in the planning of all operations, especially during the execution of stability and reconstruction operations and civil support operations. As well as hard data, open-source information can provide valuable background information on the opinions, values, cultural nuances, and other sociopolitical aspects in AOIs. In evaluating open-source documents, collectors and analysts must be careful to determine the origin of the document and the possibilities of inherent biases contained within the document.

CAPTURED DOCUMENT OPERATIONS

I-11. One of the significant characteristics of operations is the proliferation of recordkeeping and communications by digital methods (faxes, e-mails, typed, or computer-generated documents). The rapid and accurate extraction of information from these documents contributes significantly to the commander's accurate visualization of his battlefield. Documents may be captured on or in immediate association with EPWs and detainees, may be found on or turned in by refugees, line crossers, DPs or local civilians, or may be found in abandoned enemy positions or anywhere on the battlefield.

DOCUMENT EVACUATION AND HANDLING

I-12. The rapid evacuation and exploitation of documents is a shared responsibility. It originates with the capturing unit and continues to the complete extraction of pertinent information and the arrival of the document at a permanent repository, normally at the joint level, either within the theater of operations or outside of it. Documents captured in association with detainees and EPWs, with the exception of identity documents, are removed from the individual to ensure that documents of intelligence interest are not destroyed. These documents are evacuated through EPW evacuation channels with, but not on the person of, the detainee. With the exception of official documents, all documents are eventually returned to the detainee.

I-13. CEDs not associated with a detainee are evacuated through MI channels, generally starting with the capturing unit's S2. Depending on the type of documents, they may eventually be evacuated to the National Center for Document Exploitation. HUMINT collectors and translators can extract information of intelligence interest from CEDs at every echelon; they will make an attempt to exploit CEDs within their expertise and technical support constraints. Collectors evacuate CEDs to different elements based upon the information contained and the type of document concerned. For example, documents related to criminal activity may be evacuated to the nearest criminal investigative unit. Direct evacuation to an element outside the chain of command takes place at the lowest practical echelon but is normally done by the first MI unit in the chain of command. Document evacuation procedures are outlined in Annex B (Intelligence) of the unit's OPORD and SOPs.

Actions by the Capturing Unit

I-14. Document accountability begins at the time the document comes into US possession. Original documents must not be marked, altered, or defaced in any way. The capturing unit attaches a DD Form 2745 (Enemy Prisoner of War Capture Tag), Part C, to each document. Only in the case where a capturing unit does not have the time nor the manpower to mark each document due to ongoing combat operations should the capturing unit fill out one capture tag for a group of documents. In this case, the capturing unit should place the documents in a weatherproof container (box or plastic bag). The capturing unit should fill out two copies of the DD Form 2745, placing one copy inside the container and attaching one to the outside of the container. If these forms are not available, the capturing unit records the required data on any piece of paper. Figure I-1 shows an example of a field expedient tag. At a minimum, the capturing unit should record the information as follows:

- Time the document was captured as a DTG.

- Place document was captured, including an 8-digit coordinate, and description of the location. This should be as detailed as time allows. For example, if a terrorist safe house was searched, documents might be "bagged and tagged" based on what room of the house they were in, what file cabinet, what desk, and so forth.

- Identity of the capturing unit.

- Identity of the source from whom the document was taken, if applicable.

- Summary of the circumstances under which the document was found.

I-15. Document evacuation procedures are listed in Annex B (Intelligence) to the OPORD. If the capturing unit does not contain a supporting HCT, it forwards any CEDs found on the battlefield directly to its S2. The S2 extracts PIR information as practicable, ensures that the documents are properly tagged, and ensures speedy evacuation to the next higher echelon through intelligence channels. Normally, a capturing unit will use any available vehicle, and in particular empty returning supply vehicles, to evacuate documents. Documents captured on or in association with detainees, including EPWs, should be tagged and removed from the detainee. They are evacuated with (but not on) the detainees to an MP escort unit or an EPW holding facility.

I-16. When large numbers of documents are captured in a single location, it is often more expedient for the capturing unit to request a DOCEX team or HCT from the supporting MI unit be sent to the documents rather than attempting to evacuate all the documents. This reduces the burden on the capturing unit, facilitates the rapid extraction of information, and enables the priority evacuation of documents of importance to higher echelons. This method should only be used if the capturing unit can adequately secure the documents until the arrival of the DOCEX team and if the battlefield situation and MI resources permit the dispatch of a team. The capturing unit should include in its request the following:

- The identification of the capturing unit.

- Its location and the location of the documents.

```
CAPTURED DOCUMENT TAG

NATIONALITY OF CAPTURING FORCE:___US_____

_____

DATE/TIME CAPTURED:_____151310ZAUG2004_____

PLACE CAPTURED_____BH56321785, Smalltown, IZ_____

CAPTURING UNIT:_____1stPlt/B Trp/1-1 Cav_____

IDENTITY OF SOURCE (If Applicable):____MAJ, Republican Guard_____

_____

CIRCUMSTANCES OF CAPTURE:____Surrendered his company to_____

__a passing US cavalry patrol_____

DESCRIPTION OF WEAPON/DOCUMENT:____1 x PSYOP document_____

____produced by US 16th PSYOP Bn_____
```

Figure I-1. Example of a Field Expedient Capture Document Tag.

- The general description of the document site (such as an enemy brigade headquarters).

- The approximate number and type of documents.

- The presence of captured computers or similar equipment.

I-17. The MI unit dispatching the team should notify the requesting team as soon as possible to provide them an estimated time of arrival and to coordinate the arrival of the team. There is no set time for how long any particular echelon may keep a document for study. The primary aim of speedy evacuation to the rear for examination by qualified DOCEX elements remains. Each echelon is responsible to prevent recapture, loss, or destruction of the CEDs.

ACTIONS BY THE FIRST HUMINT COLLECTION OR DOCEX UNIT

I-18. The first HUMINT collection or DOCEX unit to receive CEDs should log, categorize, and exploit the documents to the best of its abilities based on METT-TC factors. They should rapidly identify documents requiring special handling or special expertise to exploit and evacuate those documents to the appropriate agencies. The MI unit SOP or OPORD should list special document evacuation requirements and priorities.

Accountability

I-19. The capturing unit and each higher echelon take steps to ensure that they maintain CED accountability during document evacuation. To establish accountability, the responsible element inventories all incoming CEDs. Anyone who captures, evacuates, processes, or handles CEDs must maintain accountability. All CEDs should have completed captured document tags. An incoming batch of documents should include a captured document transmittal. Figure I-2 shows this format. The exact format for a document transmittal is a matter of local SOP, but it should contain the information listed below:

- The identity of the element to which the CEDs are to be evacuated.

- The identity of the unit forwarding the CEDs.

- The identification number of the document transmittal.

- Whether or not CEDs in the package have been screened and the screening category. (If not screened, NA is circled.) Document handlers should package documents that have been screened separately, by category.

- A list of the document serial numbers of the CEDs in the package.

```
TO:_____          DTG:_____
FROM:_____          TRANSMITTAL NO:_____

SCREENED: YES / NO                   CATEGORY: A B C D NA
CED SERIAL NUMBERS:

_____                        _____

_____                        _____

_____                        _____

_____                        _____

_____                        _____
```

Figure I-2. Example of a Captured Document Transmittal Sheet.

I-20. When a batch is received without a transmittal, the HUMINT collection element contacts the forwarding units and obtains a list of document serial numbers (if possible). The HUMINT collection element records all trace actions in its journal. Accountability includes—

- Inventorying the CEDs as they arrive.

- Initiating necessary trace actions.

- Maintaining the captured document log. (See Figure I-3.)

I-21. When a collector includes intelligence derived from a CED in an intelligence report, he references the identification letters and number of the document concerned to avoid false confirmation.

Inventory

I-22. The receiving element conducts an initial inventory of incoming CEDs by comparing the CED to the captured document tag and accompanying transmittal documents. This comparison identifies—

- Transmittals that list missing CEDs.
- Document tags not attached to CEDs.
- CEDs not attached to document tags.
- CEDs not listed on the accompanying transmittal documents.

UNIT:_____

FILE NUMBER	RECEIVED DOCUMENT DTG	SERIAL #	INCOMING TRANSMITTAL#	FORWARDING UNIT	RECEIVED BY	DTG AND PLACE OF CAPTURE
1501	150830AUG99	0102368	1T08	1/82ᵈ Abn Div	SSG KIM	150500AUG99/EK030949
1502	150930AUG99	0110443	2T11	2/82ᵈ Abn Div	SSG KIM	150620AUG99/EK045860
1503	150930AUG99	1039964	2T11	2/82ᵈ Abn Div	SSG KIM	150725AUG99/EK058383
1504	150930AUG99	1192583	2T11	2/82ᵈ Abn Div	SSG KIM	150725AUG99/EK058383

CAPTURING UNIT	SCREENING CATEGORY	DESCRIPTION OF DOCUMENT	DESTINATION/ TRANSMITTAL #	REMARKS
Co A, 1/504ᵗʰ, 1 Bde, 82ᵈ	A	Letter of promotion, KP, 1Pg	JDEC, 15T01	SALUTE written
Co B, 2/504ᵗʰ, 2 Bde, 82ᵈ	A	Letter describing attack, RU, 3 Pg	JDEC, 15T01	SALUTE written
Co B, 2/504ᵗʰ, 2 Bde, 82ᵈ	B	List of call signs, RU, 1Pg	JDEC, 15T03	None
Co B, 2/504ᵗʰ, 2 Bde, 82ᵈ	C	Personal letter, KP, 2 Pg	JDEC 15T02	Translation end

Figure I-3. Example of a Captured Document Log.

Trace Actions

I-23. The receiving unit initiates trace actions on all missing CEDs, missing captured document tags, and all information missing from the captured document tags. They initiate trace actions by contacting elements from which the documents were received. The receiving unit can complete this corrective action swiftly if that unit's captured document log was filled out completely. If necessary, the trace action continues to other elements that have handled the document. If a captured document tag is unavailable from elements that have previously handled the CED, the document examiner fills out a captured document tag for the document using whatever information is available. Attempts to obtain missing CEDs are critical because of the information those CEDs might contain.

Document Logs

I-24. The captured document log is a record of what an element knows about a CED. After trace actions are initiated for any missing documents, the CEDs are entered in the REMARKS section of the captured document log. This log must contain the following:

- Name of capturing unit.

- File number (a sequential number to identify the order of entry).

- DTG the CED was received at this element.

- Document serial number of the captured document tag.

- Identification number of the transmittal document accompanying the CED.

- Complete designation of the unit that forwarded the CED.

- Name and rank of individual that received the CED.

- DTG and place of capture (as listed on the captured document tag).

- Identity of the capturing units (as listed on the captured document tag).

- Document category (after screening).

- Description of the CED. (At a minimum, the description includes the original language; number of pages; type of document such as a map, letter, or photograph; and the enemy's identification number for the CED, if available.)

- Destination and identification number of the outgoing transmittal.

- Remarks to include any other information that can assist the unit in identifying the CED including processing codes. These are set up by local SOPs to denote all actions taken with the document while at the element, including intelligence reports, translations, reproductions, or return of the CED to the source from whom it was taken.

DOCUMENT SCREENING

I-25. Document screening is the rapid but systematic evaluation of documents to determine which documents contain priority information. Selected priority documents will be exploited immediately for PIR information and evacuated expeditiously (often electronically) to a DOCEX facility. Document screening can be done manually (requiring a linguist who is well versed in the current collection requirements) or through the use of scanning devices with key word identification capability. Document processing does not require the complete translation of a document but requires sufficient translation to determine the significance of the document. A non-linguist may be able to do a degree of preliminary screening based on document format and the location where the document was found.

I-26. As screeners screen each document, they assign one of four category designations. The assigned category determines the document's priority for exploitation and evacuation. Document screening requires that the screening units receive the most current PIR and intelligence requirements; current friendly and enemy situation update; and relevant OB information. Screeners at higher echelons can re-categorize CEDs, to more accurately reflect the requirements at that level or information that has past its LTIOV.

DOCUMENT CATEGORIES

I-27. Documents are divided into categories to prioritize their evacuation and the extraction of information from them for intelligence purposes. Document categories are discussed below.

Category A

I-28. Category A documents are those that require priority evacuation and/or special handling because of their special intelligence value. They contain SALUTE reportable information. Category A documents also include those that are of interest to another command, collecting agency, or other agency (for example, TECHINT, Air Force, Navy, PSYOP, Cryptography).

I-29. What determines if a document is a Category A document changes according to the operational environment and will be set forth in each DOCEX element's SOP and in Annex B (Intelligence) of the unit's OPORD. Documents that are evidence in legal proceedings against captured personnel suspected of crimes against humanity and war crimes will be handled as Category A documents. All Category A documents are handled as SECRET. Standard Category A documents include but are not limited to—

- Unmarked maps.

- Maps and charts containing any operational graphics, which are sent to the G2/S2. From G2/S2, they would be evacuated to the all-source analysis center.

- Air Force related documents, which should go to the nearest Air Force headquarters.

- Navy related documents, which should be sent to the nearest Navy headquarters.

- TECHINT-related documents, which are evacuated to the nearest TECHINT unit.

- Cryptographic and communications-related documents, which are evacuated to the nearest SIGINT analysis unit.

- Documents constituting evidence to be used in legal proceedings against persons suspected of crimes against humanity and war crimes,

which will be marked "CRIMINAL EVIDENCE." Such documents will
be kept separate from other documents and will be stored under guard
or in a secure area until turned over to a war crimes investigative unit.
SJA should be consulted concerning chain of custody requirements.

Category B

I-30. Category B documents contain information of intelligence interest to the
supported command. The lowest echelon possible exploits the documents and
evacuates them through intelligence channels. Category B documents are
handled as SECRET.

Category C

I-31. Category C documents and items contain no information of intelligence
interest but still require special administrative accountability (for example,
currency, works of art, narcotics). Currency is accounted for on DA Form
4137 (Evidence/Property Custody Document).

Category D

I-32. Category D documents contain no information of intelligence value.
Only the theater or higher document repository can categorize documents as
Category D.

GROUP DOCUMENTS

I-33. CEDs are first grouped according to their assigned screening category.
Personnel must be careful when sorting CEDs to ensure no CED is separated
from its associated documents. These large groupings can be broken down
into smaller groups. Each of these smaller groupings consists of CEDs that
were—

- Captured by the same unit.

- Captured in the same place.

- Captured on the same day at the same time.

- Received at the DOCEX element at the same time.

TRANSMITTAL OF CEDS FROM FIRST AND SUBSEQUENT MI UNITS

I-34. Unless they have an HCT in DS, most units that capture or find
documents normally have no way of evaluating, categorizing, or otherwise
differentiating documents. They are all tagged and evacuated together by the
most expedient means through MI channels. Once these documents arrive at
a HUMINT collection or DOCEX unit, the unit can screen, categorize, and
extract information from the documents. The degree that documents are
exploited at each echelon is dependent on mission priorities and available
resources. Document handlers must note any attempts to exploit CEDs on
the transmittal documents to prevent unnecessary duplication of effort by
higher echelons.

I-35. When transportation assets are limited, CEDs are evacuated according to priority based on document categorization. All Category A CEDs will be evacuated first, followed in order by Categories B, C, and D. Documents that have not yet been screened will be evacuated as Category C documents, but the transmittal slip will clearly indicate that the documents have not been screened.

I-36. Documents will be evacuated in accordance with unit SOP and Annex B (Intelligence) in the unit OPORD. Lower priority CEDs, no matter how old, are never evacuated ahead of those with higher priority. CEDs are packaged so that a package of documents contains CEDs of only one category. If the CED cannot be screened because of time or language constraints, it should be treated as a Category C, but kept separate from screened Category C CEDs.

I-37. When CEDs are evacuated from any echelon, a document transmittal sheet is used (Figure I-2). A separate transmittal document is prepared for each group of CEDs to be evacuated. The sending unit prepares a separate transmittal document for each separate addressee. The transmittal identification number is recorded in the captured document log (Figure I-3) as part of the entry for each captured document. Copies of all translations should accompany the documents to avoid duplication of effort. If the sending unit submitted intelligence reports electronically, it should note the report number or include a copy of the report with the document to avoid duplicate reporting.

I-38. All CEDs being evacuated must be accompanied with the appropriate—

- Technical document (TECHDOC) cover sheet.

- SECRET cover sheet on Categories A and B documents.

- Translation reports and hardcopy reports accompanying translated documents.

- Captured document tags.

JOINT DOCUMENT EXPLOITATION FACILITY

I-39. The Theater MI brigade or group is normally tasked with the establishment of the theater JDEF. The JDEF is staffed by Army linguists, supported by technical experts from the Army and from the other services, and supplemented as required by military and civilian contract translators. The JDEF will normally contain MI experts from SIGINT, CI, TECHINT, and other areas as required to identify and exploit documents of interest to these specialties.

I-40. Military and civilian translators must have security clearances appropriate to their mission requirements. This normally equates to at least a Secret clearance since the translators must be made aware of US collection requirements to facilitate their work. The JDEF performs a final examination of all documents of possible theater intelligence value before storing or evacuating them. The DIA sets procedures for exploitation of documents above theater Army level.

DOCUMENT PROCESSING (RECOVERY AND TRANSLATION)

I-41. Units must normally process documents prior to exploiting them. Document processing includes the translation of foreign language documents into English, the recovery of damaged documents, the decryption of encrypted documents, and the extraction of documents from electronic media such as the extraction or downloading of files from a computer disc or hard drive. This need for processing frequently limits the amount of DOCEX that can be done outside a DOCEX facility.

DOCUMENT RECOVERY

I-42. At a minimum, the JDEF manning includes teams trained in extracting and downloading information from electronic media such as computer hard drives. These individuals work in conjunction with TECHINT personnel responsible for the evaluation of captured computer hardware and software. These teams are prepared to deploy forward as necessary to accomplish their mission.

DOCUMENT TRANSLATION

I-43. Translations are not intelligence information reports. They are, however, often a precondition for DOCEX. Once translated, intelligence information can be extracted and reported on an IIR, SALUTE, or similar report. A translation should accompany the original foreign language document; a copy of the translation should accompany any copies of the original foreign language document and, as required, the intelligence reports. A translation report should contain the following information:

- Where the report will be sent.

- Which element prepared the report.

- DTG of the document translation.

- Report number as designated by local SOPs.

- Document number taken from the captured document tag.

- Document description including type of document, number of pages, physical construction of document, and enemy identification number, if applicable.

- Original captured document language.

- DTG document was received at element preparing the report.

- DTG document was captured.

- Place document was captured.

- Identity of capturing unit.

- Circumstances under which document was captured.

- Name of translator.

- Type of translation: full, extract, or summary.

- Remarks for clarification or explanation, including the identification of the portions of the document translated in an extract translation.

- Classification and downgrading instructions in accordance with AR 380-5.

TYPES OF TRANSLATION

I-44. There are three types of translations:

- Full—one in which the entire document is translated. This is both time and manpower intensive, especially for lengthy or highly technical documents. Normally only a DOCEX facility at theater or national level will do full translations, and then only when the value of the information, technical complexity, or political sensitivity of the document requires a full translation. Even when dealing with Category A documents, it may not be necessary to translate the entire document to gain the relevant information it contains.

- Extract—one in which only a portion of the document is translated. For instance, a TECHINT analyst may decide that only a few paragraphs in the middle of a 600-page helicopter maintenance manual merit translation, and that a full translation is not necessary. The analyst would request only what he needed.

- Summary—one in which a translator begins by reading the entire document. He then summarizes the main points of information instead of rendering a full or extract translation. A summary translation is normally written, but may be presented orally, particularly at the tactical level. Summary translations may be done as part of the document screening process. A summary translation requires a translator have more analytical abilities. The translator must balance the need for complete exploitation of the document against time available in combat operations. Translators working in languages of which they have a limited working knowledge may also use a summary translation. For instance, a Russian linguist may not be able to accurately deliver a full translation of a Bulgarian language document. However, he can probably render a usable summary of its content.

TRANSLATOR SUPPORT TO DOCEX

I-45. When HUMINT collectors are not available due to shortages or other mission requirements, DOCEX can be performed by military or civilian linguists under the management of a cadre of HUMINT collectors.

SECURITY REQUIREMENTS

I-46. Document translators will usually need to have a clearance in order to conduct document translation. An exception is that open-source document translation does not require a security clearance unless the information collected from the open-source documents is linked to specific US collection requirements, plans, or operations. In this case the translator will need a clearance appropriate to the level of the particular contingency plan or operation to which the information is linked. Individuals without a security clearance should not be used in the exploitation of a closed source or CEDs. By their very nature, the translation of these documents gives keys into the level of US knowledge and the direction of US planning and intentions that precludes their translation by individuals without a security clearance.

SOURCES OF TRANSLATORS

I-47. There are various sources that a commander can use to obtain the linguists necessary to support DOCEX operations.

- RC and AC MI linguists. Dependent on their mission requirements, any MI soldiers with the required language qualification may be used as document translators. The advantage of using MI soldiers is that they have the appropriate security clearance and have a firm grasp of collection requirements. The DOCEX operation will usually require MI officers to manage the administrative portions.

- Other linguists. Non-MI Army linguists include numerous soldiers who have proficiency in a foreign language, regardless of MOS. US civilians can be contracted to provide translator support as can local nationals. Local national hires will provide the bulk of the translator support.

REQUIREMENTS FOR A DOCEX OPERATION

I-48. The number of personnel required to conduct DOCEX varies with the echelon and with the volume of documents. Regardless of the size of the operation, certain basic functions must occur:

- Supervision and Administration. These are the C2 and logistical aspects of the DOCEX operation that ensure that operations are smooth and uninterrupted.

- Accountability. Accountability includes logging documents in and out, copying documents as required, storing documents, receiving and transmitting documents, maintaining files, and other routine activities. This ranges from a parttime task for one individual at a low echelon, to warehouse-sized operations involving tons of documents at an EAC document repository.

- Screening. Screening involves the review and categorization of all documents, the prioritization for translation and exploitation, and the identification of documents for special handling and immediate transmittal to specialized units for exploitation. Screening requires senior, experienced individuals well versed in the target language and the collection requirements, capable of making rapid decisions based on minimal information. The number of screeners required depends on the document flow that may range from only a few per day at a low echelon, to literally thousands of documents a day at a theater-level activity.

• Security Requirements. Security requirements include ensuring that the personnel dealing with the documents have the appropriate security level and that they do not pose a security risk to the US. This is particularly important when dealing with non-US translators. Security also evolves ensuring that the documents are marked in accordance with regulation and that proper security measures are in place to prevent compromise of information. At higher echelons, dealing with large numbers of non-US translators normally requires a designated CI team conducting recurring personnel security evaluations.

• Translation. This function involves the directed translation of documents from the target language to English. It requires personnel with the appropriate clearance level who have a working idiomatic knowledge of the target language and English. Higher echelon activities, such as the EAC document repository, group their translation efforts by subject area. For example, one person or group could translate all medical-related documents. At lower echelons, the translators require a more general knowledge. At lower echelons, the same person may translate a document, extract the pertinent information, and report that information. At higher echelons, these are often separate functions.

• Exploitation and Reporting. This task is the identification and extraction of information in response to collection requirements and requires a high level of expertise. The individual must be totally knowledgeable of collection requirements and must be able to readily identify indicators of activity or identify the significance of minute pieces of information that could contribute to answering requirements. Reporting involves placing that extracted information into a coherent, properly formatted report so that the all-source analyst can add it to the intelligence picture.

• Quality Control. This aspect includes ensuring that all aspects of the DOCEX operation—including accounting for, screening, translating, exploiting, and reporting—are functioning correctly.

REPORTING

I-49. Information collected from documents is normally reported in a SALUTE report or an IIR. Reporting through other reporting formats is discouraged. Intelligence reports are normally forwarded electronically or as otherwise directed by SOPs and operational instructions. Normally an electronic or hardcopy file of each report is maintained at the unit of origin; one electronic or hardcopy is submitted through intelligence reporting channels; and one is forwarded with evacuated documents to the next unit to receive the document to prevent redundant reporting. In the event that the document itself cannot be evacuated in a timely manner, a verified copy of a translation report can be forwarded separately from the original document to an exploitation agency.

DOCUMENT EXPLOITATION IN SUPPORT OF HUMINT COLLECTION

I-50. Documents found on detainees, including EPWs—that can be exploited more efficiently when combined with HUMINT collection—are forwarded with the detainee to the next echelon in the EPW/detainee evacuation channel. In exceptional cases, documents may be evacuated ahead of the EPW or other detainee for advance study by intelligence units. A notation should be made on the EPW's capture tag or accompanying administrative papers about the existence of such documents and their location if they become separated from the detainee.

SOURCE-ASSOCIATED DOCUMENTS

I-51. Documents captured on or in association with a human source play an important role in the HUMINT collection process. These documents may contain reportable information the same as with any other CED. The information is immediately extracted from the documents and forwarded in the appropriate intelligence report. In addition to reportable information, documents (in particular personal documents) may provide valuable insight into the attitude and motivation of the source and can be effectively used by the HUMINT collector in the approach process (see Chapter 8). Guidelines for the disposition of the detainee's documents and valuables are set by international agreements and discussed in more detail in AR 190-8 and FM 19-4.

I-52. The capturing unit removes all documents, with the exception of the source's primary identification document, from an EPW or other detainee to prevent their destruction. These are placed in a waterproof container (usually a plastic bag) and Part C of the capture tag is placed in the bag. Documents from each source should be placed in a separate bag. These documents are turned over to the first MP EPW handling unit in the chain of command. The MPs will inventory all documents and prepare a handreceipt and provide a copy to the EPW or detainee.

I-53. To ensure proper handling and expeditious disposition of these documents, the first HUMINT collection element to see the detainee should review the documents as part of the source screening process. If an official document is confiscated and evacuated through MI channels, the HUMINT collector must obtain a receipt for that document from the MPs. If possible, the HUMINT collection unit copies any documents that contain information of intelligence interest and forwards the copies through MI channels. With the exception of an identification document, documents are normally kept separate from the detainee until the detainee arrives at a permanent confinement facility, at which time documents are returned to them per AR 190-8.

I-54. Three possible actions may be taken with documents captured with a source. The documents may be confiscated, impounded, or returned to the source.

Confiscated

I-55. Documents confiscated from a source are taken away with no intention of returning them. Official documents, except identification documents, are confiscated and appropriately evacuated. The intelligence value of the document should be weighed against the document's support in the HUMINT collection of the source. The HUMINT collector must comply with the accounting procedures established for CEDs by the MPs in accordance with AR 190-8.

Impounded

I-56. Some CEDs will contain information that must be exploited at higher echelons. These documents may be impounded by the HUMINT collector and evacuated through intelligence channels. The HUMINT collector must issue a receipt to the source for any personal documents that he impounds. He must comply with the accounting procedures established for CEDs impounded by the MPs in accordance with AR 190-8. When a CED is impounded, it is taken with the intent of eventual return. Personal documents with military information will be impounded if the military value is greater than the sentimental value. An example of a personal document whose military value might outweigh the sentimental value could be a personal photograph that includes military installations or equipment.

I-57. When a CED is impounded, it must be receipted. The receipt will include an itemized list of all the items taken from the prisoner, and the name, rank, and unit of the person issuing the receipt. Items of high value may be impounded for security reasons. For instance, an EPW or detainee apprehended with an unusually large amount of money would have the money impounded and receipted. The MPs will establish and maintain a DA Form 4237-R (Detainee Personnel Record) for impounded items. The register will identify the owner of the impounded items and provide a detailed description of the items impounded. A receipt will be given to anyone who has items impounded. Also, the OIC or authorized representative will complete and sign DA Form 1132-R (Prisoner's Personal Property List – Personal Deposit Fund). A copy will be provided the source. See AR 190-8 for procedures on handling personal effects.

Returned

I-58. Returned CEDs are usually personal in nature. They are taken only to be inspected for information of interest and are given back to the source. Personal documents belonging to a source will be returned to the source after examination in accordance with the GPW. These documents are CEDs whose sentimental value outweighs their military value and may be returned to the source. Copies of these documents may be made and forwarded if deemed necessary. Except for an identification document (which is always returned to the source), documents are evacuated with the source, rather than on the source, until the source reaches a permanent confinement facility at EAC.

Appendix J

References

The following references are provided to assist HUMINT collectors, commanders, and trainers in obtaining regulatory guidance for HUMINT collection operations. These and other references are in the bibliography.

1. AR 350-1. *Army Training and Education.* 9 April 2003.

2. AR 380-5. *Department of the Army Information Security Program.* September 2000.

3. AR 380-10. *Technology Transfer, Disclosure of Information, and Contacts with Foreign Representatives.* 15 February 2001.

4. AR 380-15. *(C) Safeguarding NATO Classified Information.* 1 March 1984.

5. AR 380-28. *Department of the Army Special Security System.* 12 December 1991.

6. AR 380-40. *Policy for Controlling and Safeguarding Communications Security (COMSEC) Material.* 22 October 1990.

7. AR 380-49. *Industrial Security.* 15 April 1982.

8. AR 380-53. *Telecommunications Security Monitoring.* 29 April 1998.

9. AR 380-67. *Personnel Security Program.* 9 September 1988.

10. AR 381-1. *Security Controls on Dissemination of Intelligence Information.* 12 February 1990.

11. AR 381-10. *U.S. Army Intelligence Activities.* 1 July 1984.

12. AR 381-12. *Subversion and Espionage Directed Against the US Army (SAEDA).* 15 January 1993.

13. AR 381-14. *Technical Counterintelligence (TCI).* 30 September 2002.

14. AR 381-20. *(U) US Army Counterintelligence Activities (S).* 26 September 1986.

15. AR 381-45. *Investigative Records Repository.* 25 August 1989.

16. AR 381-47. *(U) US Army Offensive Counterespionage Activities (S).* 30 July 1990.

17. AR 381-100. *(U) Army Human Intelligence Collection Program (S).* 15 May 1988.

18. AR 381-102. *(U) US Army Cover Support Program (S).* 10 January 1991.

19. AR 381-143. *Military Intelligence Nonstandard Material Policies and Procedures.* 1 December 1986.

20. AR 381-172. *Counterintelligence Force Protection Source Operations and Low-Level Source Operations.* 30 December 1994.

21. AR 614-115. *Military Intelligence Excepted Career Program.* 12 April 2004.

22. AR 614-200. *Military Intelligence Civilian Excepted Career Program.* 30 September 2004.

23. DA Pam 381-15. *Foreign Military Intelligence Collection Activities Program.* 1 June 1988.

24. Executive Order 12333. *United States Intelligence Activities.* 4 December 1981.

25. DOD Directive 2310.1. *DOD Program for Enemy Prisoners of War (EPOW) and Detainees (Short Title: DOD Enemy POW Detainee Program).* 18 August 1994.

26. DOD Directive 3115.09. *DOD Intelligence Interrogations, Detainee Debriefings, and Tactical Questioning.* 3 November 2005.

27. DOD Directive 5100.77. *DOD Law of War Program.* 9 December 1998.

28. DOD Directive 5240.1. *DOD Intelligence Activities.* 3 December 1982.

29. DOD Instruction 3020.41. *Contractor Personnel Authorized to Accompany the U.S. Armed Forces.* 3 October 2005.

30. The Under Secretary of Defense for Intelligence (USD(I) Memorandum, *Guidance for the Conduct and Oversight of Defense Human Intelligence (HUMINT) (U).* 14 December 2004.

31. USD(I) Memorandum, Implementation Instructions for USD(I) Memorandum, *Guidance for the Conduct and Oversight of Defense Human Intelligence HUMINT (U).* 7 September 2005.

32. DOD Regulation 5240.0-R. *Procedures Governing the Activities of DOD Intelligence Components That Affect United States Persons.* December 1982.

33. Detainee Treatment Act of 2005, Public Law No. 109-163, Title XIV.

Further information and links to many of the above publications can be found at: http://www.dami.army.pentagon.mil/offices/dami-cd/

ARTEPS, MTPs, and other intelligence training products are available at the Reimer Digital Library http://www.adtdl.army.mil.

Appendix K

Contract Interrogators

INTRODUCTION

K-1. Contractors are used increasingly to augment existing capabilities and bridge gaps in the deployed force structure. With the increased use of contractors comes the need to identify the doctrine and procedures affecting their employment. Leaders and those working with contractors must understand that contractors are civilians authorized to accompany the force in the field and should be provided with an ID card as proof of their authorization. In accordance with GPW Article 4, contractors are to be accorded POW status if captured.

KEY TERMS

K-2. **Contract Interrogator.** A contract interrogator is a contractor who is specifically trained and DOD certified for, tasked with, and engages in the collection of information from individuals (HUMINT sources) for the purpose of answering intelligence information requirements. Their operations must be conducted in accordance with all applicable law and policy. Applicable law and policy include US law; the law of war; relevant international law; relevant directives including DOD Directive 3115.09, "DOD Intelligence Interrogations, Detainee Debriefings, and Tactical Questioning"; DOD Directive 2310.1E, "The Department of Defense Detainee Program"; DOD instructions; and military execute orders including FRAGOs. Contract interrogators will operate only in fixed facilities and not in tactical operations. See DOD Instruction 3020.41 for additional information.

K-3. **External Support Contractor.** Contract interrogators fall into the category of External Support Contractor. They work under contracts awarded by contracting officers serving under the command and procurement authority of supporting headquarters outside the theater. Their support augments the commander's organic capability.

K-4. **Contracting Officer Representative (COR).** A COR is the contracting officer's designated representative who assists in the technical monitoring and administration of a contract. Typically, a COR is responsible for assisting the contracting officer in contractor-employee day-to-day management of issues that may affect contractor or unit requirements.

K-5. Statement of Work (SOW) or Performance Work Statement:

- Contractor roles and functional requirements, as well as security issues and the relationship to the military chain of command, must be accurately and adequately defined in the SOW. The SOW should include a description of the essential service and require the contractor to prepare contingency plans to reasonably ensure continuation. Contractors are required to perform all tasks identified within the SOW and all other provisions defined within the contract. The SOW may also stipulate the appointment of a senior contractor at each echelon or facility to be the point of contact for the commander and the COR for resolving contract performance and scheduling challenges.

- The contract must stipulate whether the duty of a senior contractor is an additional duty or the full-time job of the selected contractor. Considering that contractors are "managed" rather than "commanded," having a senior contractor will assist the commander in managing the contract. Without an onsite contractor supervisor, the default chain of command for contract interrogators might otherwise reside in CONUS, at the contractor's headquarters.

CIVILIAN STATUS

K-6. A contract employee cannot be made to engage in any activity inconsistent with his civilian status such as serving as a crew member on a weapon system. Interrogations are presumptively consistent with civilian contractor status, but other tasks should be vetted with the command's legal advisor to ensure they are legally permissible.

COMMAND CONSIDERATIONS

K-7. Theaters in which large-scale operations are conducted are very likely to employ contract interrogators, due to limited numbers of Army HUMINT collectors available. The presence of contractors in the AO adds an additional dimension to the commander's planning process. Force protection is a critical issue. Even in "permissive" environments such as in the case of humanitarian operations, contractors may be placed in danger. The commander must protect his contractors since they have limited capacity to protect themselves.

K-8. The commander has no command authority over contractor personnel. Their relationship with the Government is governed by the terms and conditions of their contract. In short, the commander must "manage" contractor personnel through the contracting process. He has no authority to command or discipline them unless it is a declared war, at which time contractors may fall under the provisions of the UCMJ. This distinction between command and management does not prohibit the commander from directing contractors to carry out essential activities, such as activities related to security and safety, consistent with the terms of their contracts.

K-9. The terms and conditions of any contract must be constructed so as to include provisions requiring contractor personnel to abide by all guidance and obey all instructions and general orders applicable to US Armed Forces and DOD civilians including those issued by the Theater Commander. The contractor can be required to "take reasonable steps" to ensure his personnel comply with the above and to ensure "the good conduct" of his employees. Furthermore, the contractor can be required to promptly resolve, to the satisfaction of the COR, all contractor employee performance and conduct problems identified by the COR. The COR can direct the contractor, at the contractor's expense (for example, a non-allowable charge to the contract) to remove and replace any contractor employee failing to comply with the above. This provides a significant tool to aid in achieving good order and discipline within an AO. The commander also has the authority to initiate proceedings that could lead to charges under Military Extraterritorial Jurisdiction Act (MEJA) or the War Crimes Act.

K-10. Contractors will be subject to the extraterritorial jurisdiction of the United States and will comply with all applicable law and policy. Applicable law and policy include US law; the law of war; relevant international law; relevant directives including DOD Directive 3115.09, "DOD Intelligence Interrogations, Detainee Debriefings, and Tactical Questioning"; DOD Directive 2310.1E, "The Department of Defense Detainee Program"; DOD instructions; and military execute orders including FRAGOs. Contractor misconduct may be subject to prosecution under federal jurisdiction pursuant to the MEJA or the War Crimes Act, or both. Procedures for initiating investigation into such misconduct will normally be established by the operational command.

K-11. The operational command will also establish procedures for referral to federal civilian authorities and necessary pre-trial confinement. Contractors normally will also be subject to the criminal jurisdiction of the HN unless granted immunity from jurisdiction through the provisions of a SOFA or equivalent agreement, or unless the HN waives jurisdiction. As a general rule, contractors are not subject to the UCMJ (with the possible exception for periods of formally declared war).

K-12. The commander should expect contractors to use all means at their disposal to continue to provide essential services, in accordance with the terms and conditions of contracts, until the military releases them. The combatant commander determines when to execute contingency plans for essential services and when to release a contractor.

INTERROGATOR CONTRACT CONSIDERATIONS

K-13. Contracts should be written with the following principles and considerations:

- Contractors will be deployable under all operational scenarios. They will be subject to the same time-phased force deployment data requirements as deploying military units.

- The contractor information system will interface with the Standard Army Management Information System at both retail and wholesale levels. Army units should not have to contend with two separate information systems.

- The contractors will provide interrogation support at fixed facilities as needed. During deployments, the commander (subject to contract terms and conditions) will determine where contractors operate in their AO.

- The contractors will not replace force structure. They will augment Army capabilities and provide an additional option for meeting support requirements.

- Force protection, including the protection of contractors, is the responsibility of commanders.
- The unit must integrate contractor support into the overall support plan. Transportation of contractors on the battlefield must be planned.
- Command and control of contract personnel is dependent upon terms and conditions of the contract. The contracting officer or his designated representative is the appointed liaison for monitoring contractor performance requirements and will ensure that contractors move material and personnel according to the combatant commander's plan. A good technique is for the command to designate and train a unit COR or Contracting Officer's Technical Representative (COTR) at each facility where contract interrogators will be working. Because international and domestic law, including SOFAs, affect the relationship between commanders and contract personnel, commanders and CORs should seek legal advice concerning issues arising during operations.
- The unit must establish a contractor personnel reporting and accountability system. Again, a good technique is for the command to designate and train a unit COR or TCOR at each facility where contract interrogators will be working.
- During deployments, contractors will live and work in field conditions comparable to those for the supported ARFOR. Living arrangements, transportation requirements, food, medical, and other support services will be provided according to the contract. These services may include but are not limited to—
  - Non-routine medical and dental care.
  - Mess.
  - Quarters.
  - Special clothing, equipment, weapons, or training mandated by the applicable commander.
  - Mail.
  - Emergency notification.
- Planning must be accomplished to ensure agreed upon support to contractors is available to the responsible commander.

CONTRACT INTERROGATOR REQUIREMENTS

SELECTION AND EMPLOYMENT CRITERIA

K-14. Contract personnel must meet certain minimum requirements to be qualified to work as contract interrogators. It is the responsibility of the hiring contracting company to ensure that these criteria, set by the Government, are met by the personnel they hire.

K-15. Policy will dictate employment criteria such as minimum education, military service, security clearance, and previous training. Certain civilian experience and training may also meet the policy requirements for contract interrogator qualifications. Commanders and CORs should refer to the current DA policy on contract interrogators for appropriate guidance.

TRAINING REQUIREMENTS

K-16. All contractor interrogators must successfully complete a training program approved by the United States Army Intelligence Center and Fort Huachuca, or the Defense HUMINT Management Office, which will serve as validation to perform MI interrogations (see DODD 3115.09). The organization Commander or Director initiating the contract must certify that the training has been completed. The training program will ensure that contract interrogators are trained and certified on—

* The most current TTP of interrogation as promulgated by DOD.

* The applicable law and policy related to the treatment of detainees. Applicable law and policy include US law; the law of war; relevant international law; relevant directives including DOD Directive 3115.09, "DOD Intelligence Interrogations, Detainee Debriefings, and Tactical Questioning"; DOD Directive 2310.1E, "The Department of Defense Detainee Program"; DOD instructions; and military execute orders including FRAGOs.

K-17. Contract interrogators must also receive training on the supported unit's mission and Interrogation Counter-Resistance Policy pertinent to the AOR they are supporting. This training will be conducted in-theater by the gaining unit.

EQUIPMENT

K-18. Contractors must be issued personal protection gear appropriate for the threat environment. This may include ballistic helmet, personal body armor, NBC protective gear, and in certain circumstances a personal weapon. Contract interrogators must have access to automation equipment to support their mission of intelligence collection and reporting.

UTILIZATION

K-19. Only trained and certified contractors working under the supervision of MI personnel are authorized to conduct interrogations. Contract interrogators may conduct interrogations with an interpreter in the room; however, an OIC or NCO will monitor the interrogation by personal presence or by means of live video camera feed.

K-20. Contract interrogators—

* Will not supervise any military personnel or federal government employees, nor will they be in charge of interrogation facilities.

* Will operate only in fixed facilities. This requirement provides a measure of force protection to the contract interrogator and frees up Army HUMINT collectors for tactical missions.

* Must submit a written interrogation plan to the OIC or NCOIC, and receive approval for the plan, in advance of each interrogation. The plan will specify the information desired and identify what techniques

and approaches for obtaining information will be used to conduct the interrogations.

WORK LOCATIONS

K-21. The commander may position contract interrogators in fixed detention facilities anywhere in the theater, in accordance with the terms and conditions of their contract.

ATTIRE

K-22. Contractors accompanying the force should be visibly distinct from soldiers so as not to jeopardize their status. The JTF or combatant commander may direct contract personnel to wear civilian clothing or a uniform that says "civilian." Specific items of military attire required for safety or security may also be worn, such as chemical defense equipment, cold weather equipment, or mission specific safety equipment.

FIRE ARMS

K-23. A contractor authorized to accompany the force in the field is *not entitled* to be issued a firearm. However, a contract employee may be armed only if **all** of following conditions are met:

- The theater commander grants his approval.

- The employee's corporate policy allows it.

- The employee agrees.

K-24. If all three conditions are met, then the employee must pass proper military side-arm training and will be issued a military specification weapon and ammunition (generally, a 9-millimeter pistol) for *personal protection*. The contractor must also comply with all applicable DOD, service, and local command policies on weapons.

RECOMMENDED READING

DOD Instruction 3020.41. *Contractor Personnel Authorized to Accompany the U.S. Armed Forces.* 3 October 2005.

AR 715-9. *Contractors Accompanying the Force.* 29 October 1999.

DA Pamphlet 715-16. *Contractor Deployment Guide.* 27 February 1998.

FM 3-100.21. *Contractors on the Battlefield.* 3 January 2003.

FM 63-11. *Logistics Support Element Tactics, Techniques, and Procedures.* 8 October 1996.

AMC Pamphlet 715-18. *AMC Contracts and Contractors Supporting Military Operations.* June 2000.

Appendix L

Sample Equipment for HCT Operations

L-1. This materiel and equipment list is provided as a sample of what an HCT may require to support the commander's intelligence requirements. Some of the equipment that is intended to be given to a source should be considered expendable.

L-2. HCT Kit Bag - Assumes four-person configuration for each echelon. Regardless of support relationship (Organic/DS/GS/GSR), HCTs inherently require the following.

L-3. Movement/Survivability/Security:

- Two M998 1-¼ ton vehicle (or non-tactical vehicle as appropriate for mission) per team.
- One crew/squad served weapon per team.
- One M16A2 per team.
- Three M4s per team.
- Four 9mms per team.
- M68 Aim-point System.
- One x body armor with protective plate per team member.

L-4. Collection and Reporting System - Hardware 1 x System per HCT:

- Scaleable collection and reporting tool that changes configuration depending on where it is on the battlefield.
- Should include an individual collection and reporting tool.
- Should include a tool for stand-alone SIPRNET connectivity via satellite or other.

L-5. Collection and Reporting System - Software 1 x per System:

- Simple, intuitive Graphic User Interface (GUI).
- Standardized Reports - SALUTE, IIR, Tactical Interrogation Report, Contact Report, BSD Report. (Policy states that once filled out, the BSD becomes classified; therefore, change policy or make the collection or reporting tool classified as well.)
- Source Management Tool.
- CI Investigations Reports.
- Link Analysis (capable of interface with coalition systems—in this case, Analyst Notebook).
- Mapping - Single, standardized tool (down to 1:12,500 scale maps, operational graphics, GPS interface).

- Biometrics Integration/Biometrics Enrollment Tools (Integrated Automated Fingerprint Identification System [IAFIS]-compliant 10-print fingerprint scanners, iris scanners, photographing station).
- Basic DOCEX application.
- Foreign Language Translation.
- Mission Planning Software.
- Query Tools: basic, advanced, multi-entity, multi-media, save user-defined queries.

L-6. Collection and Reporting System - Peripherals 1 x HCT:

- Digital video/still.
- Printer with print and scanner head (photo quality with ports for flash cards/memory sticks).
- Separate collection kit for biometrics (ink/card packets and/or IAFIS-compliant live scanner, iris scanner). Must be FBI compliant and portable).
- GPS.
- Digital voice recorder (Universal Serial Bus [USB] interface).

L-7. Communications - Requires organic communications systems to higher and laterally (non-line of sight and line of sight):

- Intra-team communications - 1 x individual:
  - Secure or nonsecure (prefer secure).
  - Hands-free and/or handheld.
- Team to all - 1 x HCT:
  - Secure.
  - FM/UHF/Microwave.
  - Line of sight.
  - Non-line of sight.
  - Voice.
  - Digital.
  - Cellular telephone:
    - Voice.
    - Backup for transmitting data.
    - GPS enabled.
  - Friendly force identification and tracking system.
- HCT to Source - 2 sets x HCT:
  - Phone cards.
  - Cell phones.
  - Radios.
  - Email or "Blackberry-like" communications.
  - One-way pager.
  - Clandestine signaling.

L-8. Source Support - 2 sets x HCT:

- Source tracking (blue-force tracker-like capability).
- Digital Voice Recorder (micro, USB interface).
- Digital video or still camera, with telephoto lenses.
- GPS.
- 1 x 1Gb Thumb Drive.

L-9. Data Storage - 1 x HCT:

- 2 x hard drives (one for UNCLASS, one for SECRET).
- 2 x Micro/Thumb, 1Gb drive (one for UNCLASS, one for SECRET).

L-10. Power Generation - 1 x liquid fuel generator or high-capacity (12-hour) (battery - silent, vehicle recharge), power source - 1 x HCT.

L-11. 1 x Universal Power Conversion Kit per HCT and OMT.

L-12. Vision Enhancement:

- 2 x night vision goggles (NVG) per HCT.
- 1 x thermal sensor per HCT.
- 1 x binoculars per individual (4 each).
- 1 x laser range finder per HCT.

Appendix M

Restricted Interrogation Technique – Separation

INTRODUCTION

M-1. As part of the Army's efforts to gain actionable intelligence in the war on terrorism, HUMINT collectors may be authorized, in accordance with this appendix, to employ the separation interrogation technique, by exception, to meet unique and critical operational requirements. The purpose of separation is to deny the detainee the opportunity to communicate with other detainees in order to keep him from learning counter-resistance techniques or gathering new information to support a cover story; decreasing the detainee's resistance to interrogation. Separation, further described in paragraphs M-2 and M-28, is the only restricted interrogation technique that may be authorized for use. Separation will only be used during the interrogation of specific unlawful enemy combatants for whom proper approvals have been granted in accordance with this appendix. However, separation may not be employed on detainees covered by Geneva Convention Relative to the Treatment of Prisoners of War (GPW), primarily enemy prisoners of war (EPWs). The separation technique will be used only at COCOM-approved locations. Separation may be employed in combination with authorized interrogation approaches—

- On specific unlawful enemy combatants.
- To help overcome resistance and gain actionable intelligence.
- To safeguard US and coalition forces.
- To protect US interests.

GENERAL

M-2. This appendix provides doctrinal guidance for the use of separation as an interrogation technique. Separation involves removing the detainee from other detainees and their environment, while still complying with the basic standards of humane treatment and prohibitions against torture or cruel, inhuman, or degrading treatment or punishment, as defined in the Detainee Treatment Act of 2005 and addressed in GPW Article 3 (Common Article III). Separation is to be distinguished from segregation, which refers to removing a detainee from other detainees and their environment for legitimate purposes unrelated to interrogation, such as when necessary for the movement, health, safety and/or security of the detainee, or the detention facility or its personnel. This appendix—

- Will be reviewed annually and may be amended or updated from time to time to account for changes in doctrine, policy, or law, and to address lessons learned.
- Is not a stand-alone doctrinal product and must be used in conjunction with the main portion of this manual.

M-3. Careful consideration should be given prior to employing separation as an interrogation technique in order to mitigate the risks associated with its use. The use of separation should not be confused with the detainee-handling techniques approved in Appendix D. Specifically, the use of segregation during prisoner handling (Search, Silence, Segregate, Speed, Safeguard, and Tag [5 S's and a T]) should not be confused with the use of separation as a restricted interrogation technique.

M-4. Members of all DOD Components are required to comply with the law of war during all armed conflicts, however such conflicts are characterized, and in all other military operations. Proper application of separation as a restricted interrogation technique in selective cases involving specific unlawful enemy combatants and in accordance with the safeguards outlined in this manual is consistent with the minimum humane standards of treatment required by US law, the law of war; and does not constitute cruel, inhuman, or degrading treatment or punishment as defined in the Detainee Treatment Act of 2005 and addressed in GPW Common Article III.

M-5. Use of separation for interrogation is authorized by exception. Separation will be applied on a case-by-case basis when there is a good basis to believe that the detainee is likely to possess important intelligence and the interrogation approach techniques provided in Chapter 8 are insufficient. Separation should be used as part of a well-orchestrated strategy involving the innovative application of unrestricted approach techniques. Separation requires special approval, judicious execution, special control measures, and rigorous oversight.

M-6. Additionally, the use of separation as a restricted interrogation technique shall be conducted humanely in accordance with applicable law and policy. Applicable law and policy for purposes of this appendix include US law; the law of war; relevant international law; relevant directives including DOD Directive 3115.09, "DOD Intelligence Interrogations, Detainee Debriefings, and Tactical Questioning"; DOD Directive 2310.1E, "The Department of Defense Detainee Program"; DOD instructions; and military execute orders including FRAGOs.

M-7. More stringent than normal safeguards must be applied when using the separation technique. Use of separation is subject to USD(I) oversight. Compared to approach techniques, there are two additional steps in the approval process (see Figure M-l) for the use of the separation technique:

- First, the COCOM Commander approves (after SJA review) use of the separation technique in theater.
- Second, following the COCOM Commander's approval, the first General Officer/Flag Officer (GO/FO) in an interrogator's chain of command approves each specific use of separation and the interrogation plan that implements separation (this is non-delegable). Interrogation supervisors shall have their servicing SJA review the interrogation plan before submitting it to the GO/FO for approval.

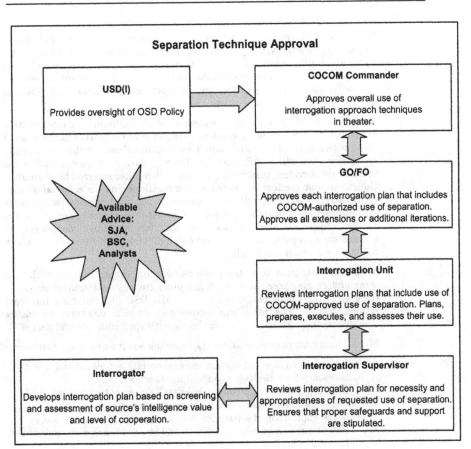

Figure M-1. Separation Approval Process.

M-8. The employment of separation requires notification, acknowledgment, and periodic review, in accordance with USD(I) Memorandum, "(S//NF) Guidance for the Conduct and Oversight of Defense Human Intelligence (HUMINT) (U)," dated 14 December 2004. This means that after the separation is approved for use by COCOMs, the I&WS must be notified as soon as practical. The Office of the Secretary of Defense will review these activities periodically in accordance with DOD Directive 3115.09.

M-9. The planning process for the employment of standard interrogation operations also applies to the employment of the separation technique (see Chapter 4).

RESPONSIBILITIES

M-10. Commanders of interrogation or detention facilities and forces employing the approved separation technique are responsible for compliance with applicable law and policy. Commanders must understand that separation poses a higher risk to the detainee than do standard techniques, and so require strenuous oversight to avoid misapplication and potential abuse.

M-11. The interrogation chain of command must coordinate the interrogation plan with the detention operations commander. Close coordination must occur between intelligence personnel and personnel responsible for detainee operations, including MP, security forces, Master at Arms, and other individuals providing security for detainees (hereafter referred to as guards). Guards do not conduct intelligence interrogations and, in accordance with DOD Directive 3115.09, will not set the conditions for interrogations. Guards may support interrogators as requested for detainee custody, control, escort, and/or additional security (for example, for combative detainees), in accordance with paragraphs 5-57 through 5-66 and FM 3-19.40, JP 3-63, and the approved interrogation plan.

M-12. The detention operations commander (in conjunction with the MI commander) may convene a multidiscipline custody and control oversight team including, but not limited to, MP, MI, BSC (if available), and legal representatives. The team can advise and provide measures to ensure effective custody and control in compliance with applicable law and policy.

M-13. Commanders must consider the following when employing separation:

- Is separation directed against the appropriate individual and is it necessary for collecting important intelligence?
- Does separation complement the overall interrogation strategy and interrogation approach technique or techniques?
- Is the application of separation with the specific detainee at issue consistent with humane treatment and in accordance with applicable law and policy?

M-14. Planning for the use of separation requires coordination with staff elements that provide support to interrogation operations. Staff elements that support interrogation facilities and forces employing separation will comply with paragraphs 4-59 and 4-60 and all controls and safeguards identified in paragraphs M-17 through M-26.

GENERAL CONTROLS AND SAFEGUARDS

HUMANE TREATMENT

M-15. All captured or detained personnel shall be treated humanely at all times and in accordance with DOD Directive 3115.09, "DOD Intelligence Interrogations, Detainee Debriefings, and Tactical Questioning"; DOD Directive 2310.1E, "Department of Defense Detainee Program," and no person in the custody or under the control of the DOD, regardless of nationality or physical location, shall be subject to cruel, inhuman, or degrading treatment or punishment as defined in US law, including the

Detainee Treatment Act of 2005. All intelligence interrogations, debriefings, or tactical questioning to gain intelligence from captured or detained personnel shall be conducted in accordance with applicable law and policy.

M-16. Any inhumane treatment—including abusive practices, torture, or cruel, inhuman, or degrading treatment or punishment as defined in US law, including the Detainee Treatment Act of 2005—is prohibited and all instances of such treatment will be reported immediately in accordance with paragraph 5-69 thru 5-72. Beyond being impermissible, these unlawful and unauthorized forms of treatment are unproductive because they may yield unreliable results, damage subsequent collection efforts, and result in extremely negative consequences at national and international levels. Review by the servicing SJA is required prior to using separation. Each interrogation plan must include specific safeguards to be followed: limits on duration, interval between applications, and termination criteria. Medical personnel will be available to respond in the event a medical emergency occurs.

FM 2-22.3 REQUIREMENTS

M-17. Separation must be employed in accordance with the standards in this manual. These standards include the following:

- Prohibitions against abusive and unlawful actions (see para 5-75) and against the employment of military working dogs in the conduct of interrogation (see paras 5-59 and 8-2).

- Requirement for non-DOD agencies to observe the same standards for the conduct of interrogation operations and treatment of detainees when in DOD facilities (see para 5-55).

- Prohibition on guards conducting intelligence interrogations or taking any actions to set the conditions for interrogations. Humane treatment, evacuation, custody and control (reception, processing, administration, internment, and safety) of detainees; force protection; and the operation of the internment facility are discussed in paragraphs 5-57 through 5-66. (FM 3-19 .40 and JP 3-63 also thoroughly discuss detainee operations.)

- Assignment of ISNs to all detainees in DOD control, whether or not interrogation has commenced, as soon as possible; normally within 14 days of capture. (See AR 190-8 and Secretary of Defense Memorandum dated 20 September 2005, "(S//NF) Policy on Assigning Detainee Internment Serial Numbers (ISN)(U)."

- Access to detainees by the ICRC.

REPORTING OF ABUSES AND SUSPECTED ABUSES

M-18. As an interrogation technique, separation is particularly sensitive due to the possibility that it could be perceived as an impermissible act. Interrogators applying the separation technique and the chain of command supervising must be acutely sensitive to the application of the technique to ensure that the line between permissible or lawful actions and impermissible or unlawful actions is distinct and maintained. Therefore, HUMINT collectors should have heightened awareness and understanding of the risks, control measures, and safeguards associated with the use of separation. Any interrogation technique that appears to be cruel, inhuman, or degrading as defined in US law; or impermissibly coercive, or is not listed in this manual, is prohibited and should be reported immediately to the chain of command or other appropriate channels for resolution. Orders given to treat detainees inhumanely are unlawful. Every interrogator must know how to respond to orders that the individual interrogator perceives to be unlawful (see paras 5-80 through 5-82).

M-19. If the order is a lawful order, it should be obeyed. Failure to obey a lawful order is an offense under the UCMJ.

COMMAND POLICY AND OPERATION ORDERS

M-20. The provisions of this appendix must be written into COCOM policy and/or OPORDs when using the restricted interrogation technique of separation.

MEDICAL

M-21. Commanders are responsible to ensure that detainees undergoing separation during interrogation receive adequate health care as described in greater detail in paragraph 5-91.

TRAINING AND CERTIFICATION

M-22. Only those DOD interrogators who have been trained and certified by the United States Army Intelligence Center (USAIC), or other Defense HUMINT Management Office (DHMO) designated agency, in accordance with guidance established by USD(I) to use separation, are authorized to employ this technique. The training must meet certification standards established by the Defense HUMINT Manager in coordination with applicable DOD components. Properly trained and certified contract interrogators are authorized to initiate interrogation plans that request the use separation, and, once the plan is approved, to employ the technique in accordance with the provisions of this appendix and Appendix K. Contract interrogators will always be utilized under the supervision and control of US government or military personnel (see para K-19 and K- 20). Non-DOD personnel conducting interrogations in DOD facilities must be certified by their agency and separately gain approval (through their agency's chain of command) for the additional technique described in this appendix. They must present this written certification and agency approval to the COCOM before use is permitted (see para 5-55).

PLANNING

M-23. Planning for the use of separation must include—

- An interrogation plan that addresses safeguards, limits of duration, interval between applications, termination criteria, and presence of qualified medical personnel for emergencies (see Figure M-2).
- A provision for detainees to be checked periodically in accordance with command health care directives, guidance, and SOPs applicable to all detainees.
- A legal review.

Page _____ of _____

INTERROGATION PLAN
FOR USE OF
RESTRICTED SEPARATION TECHNIQUE

THIS FORM IS TO BE COMPLETED IN CONJUNCTION WITH, AND MAINTAINED WITH, THE BASE INTERROGATION PLAN, FIGURE 10-3, FM 2-22.3. USE ADDITIONAL FORMS AS NEDED.

COCOM/SECDEF ORDER OR PLAN #:_____

RESTRICTED TECHNIQUE STRATEGY:
  JUSTIFICATION:_____
  _____
  _____
  _____

USED IN CONJUNCTION WITH THE FOLLOWING APPROACH STRATEGIES:_____
  _____

SPECIFIC DESCRIPTION OF THE SEPARATION TECHNIQUE EMPLOYMENT STRATEGY:_____
  _____
  _____
  _____
  _____

PROPOSED DURATION:_____ REQUIRED BREAK:_____

SPECIFIC SAFEGUARDS AND OVERSIGHT TO BE EMPLOYED:
  GUARDS:_____
  _____
  _____
  INTERROGATORS:_____
  _____
  OTHER:_____
  _____

DOCUMENTATION OF USE: (PHOTOS, VIDEO, NOTES)_____
  _____

TERMINATION CRITERIA TECHNIQUE:_____

REVIEW: Interrogation Supervisor:_____ MI Unit Cdr:_____ GO/OF SJA_____

APPROVAL AUTHORITY: PRINTED NAME   DTG OF APPROVAL
GENERAL OFFICER/FLAG OFFICER   _____   _____
  APPROVED FOR _____DURATION

MI UNIT COMMANDER   _____   _____

INTERROGATION SUPERVISOR   _____   _____

Figure M-2. Installation Plan for Restricted Separation Techniques.

- Documentation of the use of separation, including photographs and/or videotaping, if appropriate and available (see para 5-54).

M-24. Separation is only authorized for use in interrogation operations, not for other Military Source Operations. Separation may be approved for use in combination with authorized approach techniques. General controls and safeguards contained in this manual must be applied during the use separation, in conjunction with the safeguards specific to the separation technique. **Planning must consider the possible cumulative effect of using multiple techniques and take into account the age, sex, and health of detainees, as appropriate.**

TECHNICAL CONTROL

M-25. Requests for approval of separation will be forwarded (for information purposes only) via secure means through intelligence technical channels at the same time as they are sent through command channels. Intelligence technical channels are those used for forwarding of source information and technical parameters of collection operations from lower to higher and passing tasking specifics, source information, technical control measures, and other sensitive information from higher to lower. The technical chain extends from the HCT through the OMT and Operations Section (if one exists) to the C/J/G/S2X.

APPLICATION OF SEPARATION TECHNIQUE

M-26. The purpose of separation is to deny the detainee the opportunity to communicate with other detainees in order to keep him from learning counter-resistance techniques or gathering new information to support a cover story, decreasing the detainee's resistance to interrogation. Separation does not constitute sensory deprivation, which is prohibited. For the purposes of this manual, sensory deprivation is defined as an arranged situation causing significant psychological distress due to a prolonged absence, or significant reduction, of the usual external stimuli and perceptual opportunities. Sensory deprivation may result in extreme anxiety, hallucinations, bizarre thoughts, depression, and anti-social behavior. Detainees will not be subjected to sensory deprivation.

M-27. Physical separation is the best and preferred method of separation. As a last resort, when physical separation of detainees is not feasible, goggles or blindfolds and earmuffs may be utilized as a field expedient method to generate a perception of separation.

M-28. Objectives:

- Physical Separation: Prevent the detainee from communicating with other detainees (which might increase the detainee's resistance to interrogation) and foster a feeling of futility.

- Field Expedient Separation: Prolong the shock of capture. Prevent the detainee from communicating with other detainees (which might increase the detainee's resistance to interrogation) and foster a feeling of futility.

M-29. Safeguards:

- **Duration:** Self-explanatory.
- **Physical Separation:** Limited to 30 days of initial duration.
- **Field Expedient Separation:** Limited to 12 hours of initial duration at the initial interrogation site. This limit on duration does not include the time that goggles or blindfolds and earmuffs are used on detainees for security purposes during transit and evacuation.
- **Oversight Considerations for Field Expedient Separation:**
  - The intended use of field expedient means of separation must be specified on the interrogation plan that is submitted to the GO/FO for approval.
  - Detainees must be protected from self-injury when field expedient means of separation are used. The effect of the application of field expedient separation means on the detainee must be monitored to detect any possible health concerns.

M-30. The following safeguards apply to both Physical Separation and Field Expedient Separation.

- **Break:** Additional periods of separation will not be applied without the approving GO/FO's determination of the length of a break between iterations.

- **Extension:**
  - **Physical Separation Method:** Extensions of this technique past 30 days must be reviewed by the servicing SJA and approved by the original approving GO/FO or his replacement in that position.
  - **Field Expedient Method:** Extensions past 12 hours of initial duration at the initial interrogation site must be reviewed by the servicing SJA and approved by the original approving/replacement GO/FO.
  - **Medical:** Detainees will be checked periodically in accordance with command health care directives, guidance, and SOPs applicable to all detainees.
  - **Custody and Control:** The interrogation chain of command must coordinate the interrogation plan with the Detention Operations Commander. The Detention Operations Commander (in conjunction with the MI commander) may convene a multidiscipline custody and control oversight team including, but not limited to, MP, MI, BSC (if available), and legal representatives. The team can advise and provide measures to ensure effective custody and control in compliance with the requirements of applicable law and policy.

- **Oversight Considerations:**
  - Use of hoods (sacks) over the head, or of duct tape or adhesive tape over the eyes, as a separation method is prohibited.
  - If separation has been approved, and the interrogator subsequently determines that there may be a problem, the interrogator should seek further guidance through the chain of command before applying the technique.

- Care should be taken to protect the detainee from exposure (in accordance with all appropriate standards addressing excessive or inadequate environmental conditions) to—
  - Excessive noise.
  - Excessive dampness.
  - Excessive or inadequate heat, light, or ventilation.
  - Inadequate bedding and blankets.
  - Interrogation activity leadership will periodically monitor the application of this technique.
- Use of separation must not preclude the detainee getting four hours of continuous sleep every 24 hours.
- Oversight should account for moving a detainee from one environment to another (thus a different location) or arrangements to modify the environment within the same location in accordance with the approved interrogation plan.

M-31. Suggested Approach Combinations:

- Futility.
- Incentive.
- Fear Up.

Glossary

The glossary lists acronyms and terms with Army or joint definitions, and other selected terms. Where Army and joint definitions are different, (*Army*) follows the term. Terms for which FM 2-22.3 is the proponent manual (the authority) are marked with an asterisk (*) and followed by the number of the paragraph (¶) or page where they are defined. For other terms, refer to the manual listed. JP 1-02, Dictionary of Military and Associated Terms and FM 1-02 Operational Terms and Graphics are posted on the Joint Electronic Library, which is available online and on CD ROM.

- Use this URL to access JP 1-02 online:
  http://atiam.train.army.mil/soldierPortal/atia/adlsc/view/public/11444-1/FM/1-02/TOC.HTM
- Use this URL to access FM 1-02 online:
  http://atiam.train.army.mil/soldierPortal/atia/adlsc/view/public/11444-1/FM/1-02/TOC.HTM
- /jel/service_pubs/101_5_1.pdf
- Follow this path to access JP 1-02 on the Joint Electronic Library CD-ROM:
  Mainmenu>Joint Electronic Library>DOD Dictionary.
- Follow this path to access FM 1-02 on the Joint Electronic Library CD-ROM:
  Mainmenu>Joint Electronic Library>Service Publications>Multiservice Pubs>
  FM 101-5-1.

*2X	The *2X Staff conducts mission and RM for all HUMINT and CI entities located within the designated AOIR. It coordinates, deconflicts, and synchronizes all HUMINT and CI activities in the designated AOIR. ("*2X" indicates 2X functions at all levels.)
AC	Active Component
ACCO	Army Central Control Office
ACE	analysis and control element
ACT	Analysis Control Team
ADA	Air Defense Artillery
ADP	automated data processing
ADCON	administrative control
aka	also known as
AMHS	Automated Message Handling System
AMID	allied military intelligence battalion
AO	area of operations
AOI	area of interest
AOIR	area of intelligence responsibility
AOR	area of responsibility

approx	approximately
ARNG	Army Reserve National Guard
ASAS	All-Source Analysis System
ASCC	Army Service Component Command
ASD(C3I)	Assistant Secretary of Defense (Command, Control, Communications, and Intelligence)
ASI	additional skill identifier
assn	assassination
ATD	associated technical document
BAT	Biometric Automated Toolset
BCT	brigade combat team
BDA	battle damage assessment
BOS	Battlefield Operating System
BSC	Behavioral Science Consultant
BSD	basic source data
C2	command and control
CA	civil affairs
CAT	category
CCIR	commander's critical information requirement
CCNY	City College of New York
CDOC	captured document
CDR	commander
CGS	common ground station
CE	captured equipment (STANAG term)
CED	captured enemy document
CEE	captured enemy equipment
CENTCOM	US Central Command
CFSO	Counterintelligence Force Protection Source Operations
CHAMS	CI/HUMINT Automated Management System
CHATS	CI/HUMINT Automated Tool Set
CI	counterintelligence
CIA	Central Intelligence Agency
CICA	Counterintelligence Coordination Authority
CIAC	Counterintelligence Analysis Cell

CID	Criminal Investigation Division
CIFA	Counterintelligence Field Agency
C2X	Coalition Intelligence Staff Officer
C/J2X LNO	Coalition/Joint Intelligence Staff Liaison Officer
C/J/G2X	Coalition/Joint/Corps/Division Intelligence Staff Officer
C/J/G/S2	Coalition/Joint/Corps/Division/Brigade and Below Intelligence Staff Officer
C/J/G/S2X	Coalition/Joint/Corps/Division/Brigade and Below Intelligence Staff Officer
CMO	civil-military operations
COA	course of action
COCOM	Combatant Command
COE	common operating environment
COLISEUM	Community On-Line Intelligence System for End Users and Managers
COMMZ	communications zone
CONUS	continental United States
CONOP	contingency operations
COP	common operational picture
COR	contracting officer representative
COT	commercial off-the-shelf
COTR	Contracting Officer's Technical Representative
counterintelligence	Information gathered and activities conducted to protect against espionage, other intelligence activities, sabotage, or assassinations conducted by or on behalf of foreign governments or elements thereof, foreign organizations, or foreign persons, or international terrorist activities. (FM 2-0)
Counterintelligence Coordinating Authority	Subordinate to the J/G2X and coordinates all CI activities for a deployed force. It provides technical support to all CI assets and coordinates and deconflicts CI activities in the deployed AO. (FM 2-0)
CP	command post
CPERS	captured personnel (JP-2.5)
CPR	Common Point of Reference
CS	combat support
CSS	combat service support
CTF	coalition task force

DA	Department of the Army
DCGS-A	Distributed Common Ground System-Army
DCIS	Defense Criminal Investigative Service
DCISS	Defense Intelligence Agency CI Information System
DCP	detainee collection point
DEA	Drug Enforcement Agency
debriefing	The systematic questioning of individuals to procure information to answer specific collection requirements by direct and indirect questioning techniques. (FM 2-0)
DED	Data Element Dictionary
DH	Defense HUMINT
DHMO	Defense HUMINT Management Office
DHS	Department of Homeland Security
DIA	Defense Intelligence Agency
DII	DOD Information Infrastructure
DISCOM	Division Support Command
DOCEX	document exploitation
document exploitation	The systematic extraction of information from all media formats in response to collection requirements. (FM 2-0)
DOD	Department of Defense
DOE	Department of Energy
DP	displaced person
DRP	Detainee Reporting
DRS	Detainee Report System
DS	direct support
DSCA	defense support of civilian authorities
DSS	decision support software
DST	decision support template
DTG	date-time group
EAC	echelons above corps
EEFI	essential elements of friendly information
EPW	enemy prisoner of war
evaluating	In intelligence usage, appraisal of an item of information in terms of credibility, reliability, pertinence, and accuracy. (FM 2-0)

EW	electronic warfare
FBI	Federal Bureau of Investigation
FEO	forced entry operations
FFIR	friendly force information requirement
FHA	foreign humanitarian assistance
FISS	Foreign Intelligence Security Service
Five S's	search, seize, segregate, safeguard, and silence
FORSCOM	US Army Forces Command
FRAGO	fragmentary order
FRN	field reporting number
FSE	fire support element
Gb	gigabyte
GC	Geneva Convention Relative to the Protection of Civilian Persons in Time of War
GPS	Global Positioning System
GPW	Geneva Convention Relative to the Treatment of Prisoners of War
GRCS	Guardrail Common Sensor
GRIFN	Guardrail Information Node
GS	general support
GSR	general support-reinforcing
GUI	graphic user interface
GWS	Geneva Convention for the Amelioration of the Condition of the Wounded and Sick in Armed Forces in the Field
HAC	HUMINT analysis cell
HAT	HUMINT analysis team
HCR	HUMINT collection requirement
HCT	HUMINT collection team
HET	human exploitation team
HN	host nation
HOC	HUMINT operations cell
HSOC	Homeland Security Operations Center
HQ	headquarters
HQDA	Headquarters, Department of the Army

human intelligence	The collection by a trained HUMINT collector of foreign information from people and multimedia to identify elements, intentions, composition, strength, dispositions, tactics, equipment, and capabilities. (FM 2-0)
HUMINT	Human Intelligence
HUMINT Analysis Cell	The "fusion point" for all HUMINT reporting and operational analysis in the ACE and JISE. It determines gaps in reporting and coordinates with the requirements manager to cross-cue other intelligence sensor systems.
HUMINT Analysis Team	Sub-element of the G2 ACE that supports the G2 development of IPB products and developing and tailoring requirements to match HUMINT collection capabilities.
HUMINT Operations Cell	Assigned under the J/G2X to track all HUMINT activities in the area of intelligence responsibility (AOIR). It provides technical support to all HUMINT collection operations and deconflicts HUMINT collection operations in the AO. (FM 2-0)
I&WS	Deputy Under Secretary of Defense for Intelligence and Warfighting Support
IAFIS	Integrated Automated Fingerprint Identification System
IBS	Integrated Broadcast Services
ICF	intelligence contingency fund
ICRC	International Committee of the Red Cross
ID	identification
IED	improvised explosive device
IG	Inspector General
IIMG	Interagency Incident Management Group
IIR	intelligence information report
IMINT	imagery intelligence
INTREP	intelligence report
INTSUM	intelligence summary
IPB	intelligence preparation of the battlefield
IPF	Intelligence Processing Facility
IPSP	Intelligence Priorities for Strategic Planning
IR	information requirements
ISA	International Standardization Agreement
ISN	Internment Serial Number
ISR	intelligence, surveillance, and reconnaissance
IU	Interrogation Unit (AJP-2.5)

J2	Joint Intelligence Directorate/Staff
J2X	Joint Intelligence Staff Officer
J2X	Responsible for controlling, coordinating and deconflicting all HUMINT and CI collection activities and keeping the joint force J2 informed on all HUMINT and CI activities conducted in the joint force AOR. (*2X Staff Handbook) Umbrella organization consisting of human intelligence operation cell and the task force counterintelligence coordinating authority. The J2X is responsible for coordination and deconfliction of all human source-related activity. See also counterintelligence; human intelligence. (JP 2-01)
J5	Joint Staff Directorate, Civil Affairs
J/G2	Joint/Corps/Division Intelligence Staff Officer
J/G/S2	Joint/Corps/Division/Brigade and Below Intelligence Staff Officer
J/G/S2X	Joint/Corps/Division/Brigade and Below Intelligence Staff Officer
JAC	Joint Analysis Center
JAO	joint area of operations
JCMEC	Joint Captured Materiel Exploitation Center
JDEF	joint document exploitation facility
JDS	Joint Dissemination System
JFC	Joint Forces Commander
JIC	Joint Interrogation Center
JIDC	Joint Interrogation and Debriefing Center
JISE	Joint Intelligence Support Element
JTF	joint task force
JUMPS	job, unit, mission, PIR and IR, and supporting information
JWICS	Joint Worldwide Intelligence Communications System
KB	knowledgeability brief
LCC	Land Component Command
LDR	Lead Development Report
LEA	law enforcement agency
LNO	Liaison Officer
LRS	long-range surveillance
LTIOV	latest time information is of value
MAGTF	Marine Air-Ground Task Force
MASINT	measurement and signature intelligence
MDMP	Military Decisionmaking Process

MEDEVAC	medical evacuation
MEF	Marine expeditionary force
MEJA	Military Extraterritorial Jurisdiction Act
METT-TC	mission, equipment, terrain and weather, troops and support available, time available, and civil considerations
MI	Military Intelligence
MILO	mission, identification, location, and organization
MSO	military source operations
MOS	military occupation specialty
MP	Military Police
MTOE	modified table of organization and equipment
MTW	major theater war
NA	not applicable
NAC	national agency check
NAI	named area of interest
NATO	North Atlantic Treaty Organization
NBC	nuclear, biological, and chemical
NCO	noncommissioned officer
NCOIC	noncommissioned officer in charge
NGA	National Geospatial-Intelligence Agency
NGO	non-governmental organization
NIP	Notice of Intelligence Potential
NIST	national intelligence support team
NOFORN	no foreign dissemination
NRT	near-real time
NSA	National Security Agency
NVG	night vision goggles
OB	order of battle
OCONUS	outside continental United States
OCR	optical character recognition
OGA	other government agencies
O/I	operations and intelligence
OIC	officer in charge
OMT	operational management team

OPCON	operational control
OPLAN	operations plan
OPORD	operations order
OSC	operations support cell
OSD	Office of the Secretary of Defense
OSINT	open-source intelligence
OPTEMPO	operational tempo
OVOP	overt operational proposal
PDA	Personal Digital Assistant
PIR	priority intelligence requirement
PME	peacetime military engagement
PMO	Provost Marshal Office
POL	petroleum, oils, and lubricants
POW	prisoner of war
PRETECHREP	preliminary technical report
priority intelligence requirements	Those intelligence requirements for which a commander has an anticipated and stated priority in the task of planning and decisionmaking. (JP 1-02)
PSO	peacetime stability operation (STANAG term)
PSYOP	Psychological Operations
PVO	private volunteer organization
PW	Prisoner of War (as used in the GPW)
R	reinforcing
R&S	reconnaissance and surveillance
RC	Reserve Components
RFI	request for information
RM	requirements management
ROE	rules of engagement
RSR	Resource Status Report
SALUTE	Size, Activity, Location, Unit, Time, Equipment
SBI	special background investigation
SCC	Service Component Commands
SCI	sensitive compartmented information
SCO	Sub-Control Office
SDR	Source-Directed Requirement

security detainee	Those detainees who are not combatants, but who may be under investigation or pose a threat to US forces if released.
SE	southeast
SECDEF	Secretary of Defense
SERE	survival, evasion, resistance, and escape
S.F.P.	Students for Peace
SIGINT	signals intelligence
SII	statement of intelligence interest
SIO	senior intelligence officer
SIPRNET	Secret Internet Protocol Router Network
SIR	specific information requirement
SITMAP	situation map
SJA	Staff Judge Advocate
SME	subject matter expert
SOF	Special Operations Forces
SOFA	Status of Forces Agreement
SOI	signal operating instruction
SOP	standing operating procedure
SOW	statement of work
Source (DOD)	1. A person, thing, or activity from which information is obtained. 2. In clandestine activities, a person (agent), normally a foreign national, in the employ of an intelligence activity for intelligence purposes. 3. In interrogation activities, any person who furnishes information, either with or without the knowledge that the information is being used for intelligence purposes. In this context, a controlled source is in the employment or under the control of the intelligence activity and knows that the information is to be used for intelligence purposes. An uncontrolled source is a voluntary contributor of information and may or may not know that the information is to be used for intelligence purposes. See also agent; collection agency. (JP 1-02)
SQL	structured query language
SSC	small-scale contingency
Stability and Reconstruction Operations	Those operations that sustain and exploit security and control over areas, populations, and resources. They employ military capabilities to reconstruct or establish services and support civilian agencies. Stability and reconstruction operations involve both coercive and cooperative actions.
STANAG	Standardization Agreement

TACON	tactical control
tactical questioning	The expedient initial questioning for information of immediate tactical value. Soldiers conduct tactical questioning based on the unit's SOP, ROE, and the order for that mission. Unit leaders must include specific guidance for tactical questioning in the order for appropriate missions. The unit S3 and S2 must also provide specific guidance down to the unit level to help guide tactical questioning. (FM 2-0)
TCICA	Theater Counterintelligence Coordination Authority
TCP	traffic control point
TDA	table of distribution and allowances
TDRC	Theater Detainee Reporting Center
TECHDOC	technical document
TECHNINT	technical intelligence
TES	Tactical Exploitation System
TF	task force
THREATCON	threat condition
TOE	table of organization and equipment
TPFDDL	Time-Phased Forces Deployment Data List
TTP	tactics, techniques, and procedures
TS	Top Secret
tvl	travel
TWS	Trusted Workstation
UCMJ	Uniform Code of Military Justice
UNCLASS	unclassified
unk	unknown
USAR	US Army Reserve
USB	Universal Serial Bus
USCENTCOM	US Central Command
USD(I)	Under Secretary of Defense for Intelligence
UTM	universal transverse mercator (grid)
UW	unconventional warfare
WARNO	warning order
WMD	weapons of mass destruction
WO	warrant officer
WTC	World Trade Center

Bibliography

The bibliography lists field manuals by new number followed by old number, as indicated.

DOCUMENTS NEEDED

These documents must be available to the intended users of this publication.

JP 0-2. *Unified Action Armed Forces.* 10 July 2001.

JP 2-0. *Doctrine for Intelligence Support to Joint* Operations. 9 March 2000.

JP 2-01.2. *(U) Joint Doctrine and Tactics, Techniques, and Procedures for Counterintelligence Support to Operations (S//NF).* 13 June 2006.

JP 3-0. *Doctrine for Joint Operations.* 10 September 2001.

JP 3-63. *Joint Doctrine for Detainee Operations.* September 2005.

JP 4-0. *Doctrine for Logistic Support of Joint Operations.* 6 April 2000.

AR 40-25. *Nutrition Standards and Education.* 15 June 2001.

AR 715-9. *Contractors Accompanying the Force.* 29 October 1999.

DA Pam 715-16. *Contractor Deployment Guide.* 27 February 1998.

FM 1. *The Army.* 14 June 2005.

FM 1-02. *Operational Terms and Graphics.* 21 September 2004.

FM 2-0. *Intelligence.* 17 May 2004.

FM 3-0. *Operations.* 14 June 2001.

FM 3-100.21. *Contractors on the Battlefield.* 3 January 2003.

FM 5-0. *Army Planning and Orders Production.* 20 January 2005.

FM 6-0. *Mission Command: Command and Control of Army Forces.* 11 August 2003.

FM 27-10. *Law of Land Warfare.* July 1956.

FM 34-5. (U) *Human Intelligence and Related Counterintelligence Operations* (S//NF). 29 July 1994.

FM 34-130. *Intelligence Preparation of the Battlefield.* 8 July 1994.

FM 63-11. *Logistics Support Element Tactics, Techniques, and Procedures.* 8 October 1996.

FM 71-100. *Division Operations.* 28 August 1996.

FM 100-15. *Corps Operations.* 13 September 1989.

FM 101-5. *Staff Organization and Operations.* 31 May 1997.

AMC Pam 715-18. *AMC Contracts and Contractors Supporting Military Operations.* June 2000.

READINGS RECOMMENDED
These sources contain relevant supplemental information.

ARMY PUBLICATIONS
Most Army doctrinal publications are available online:
http://155.217.58.58/atdls.htm

AR 190-8. *Enemy Prisoners of War, Retained Personnel, Civilian Internees and Other Detainees.* 1 October 1997.

AR 190-40. *Serious Incident Report.* 15 July 2005.

AR 195-5. *Criminal Investigation Evidence Procedures.* 28 August 1982.

AR 380-5. *Department of the Army Information Security Program.* September 2000.

AR 380-10. *Technology Transfer, Disclosure of Information, and Contacts with Foreign Representatives.* 15 February 2001.

AR 380-15. *(C) Safeguarding Classified NATO Information.* 1 March 1984.

AR 380-28. *Department of the Army Special Security System.* 12 December 1991.

AR 380-40. *Policy for Controlling and Safeguarding Communications Security (COMSEC) Material.* 22 October 1990.

AR 380-49. *Industrial Security.* 15 April 1982.

AR 380-53. *Telecommunications Security Monitoring.* 29 April 1998.

AR 380-67. *Personnel Security Program.* 9 September 1988.

AR 381-1. *Security Controls on Dissemination of Intelligence Information.* 12 February 1990.

AR 381-10. *US Army Intelligence Activities.* 1 July 1984.

AR 381-12. *Subversion and Espionage Directed Against the US Army (SAEDA).* 15 January 1993.

AR 381-14. *Technical Counterintelligence (TCI).* 30 September 2002.

AR 381-20. *(U) US Army Counterintelligence Activities (S).* 26 September 1986.

AR 381-45. *Investigative Records Repository.* 25 August 1989.

AR 381-47. *(U) US Army Offensive Counterespionage Activities (S).* 30 July 1990.

AR 381-100. *(U) Army Human Intelligence Collection Program (S//NF).* 15 May 1988.

AR 381-102. *(U) US Army Cover Support Program (S).* 10 January 1991.

AR 381-143. *Military Intelligence Nonstandard Material Polices and Procedures.* 1 December 1986.

AR 381-172. (U) *Counterintelligence Force Protection Operations (CFSO) and Low Level Source Operations (LLSO) (S//NF).* 30 December 1994.

AR 385-10. *The Army Safety Program.* 23 May 1988.

AR 614-115. *Military Intelligence Excepted Career Program.* 12 April 2004.

AR 614-200. *Military Intelligence Civilian Excepted Career Program.* 30 September 2004.

AR 715-9. *Contractors Accompanying the Force.* 29 October 1999.

DA Pam 381-15. *Foreign Military Intelligence Collection Activities Program.* 1 June 1988.

DA Pam 385-1. *Small Unit Safety Officer/NCO Guide.* 29 November 2001.

TRADOC Reg 25-36. *The TRADOC Doctrinal Literature Program (DLP).* 1 October 2004.

FM 2-0. *Intelligence.* 17 May 2004.

FM 3-19.4. *Military Police Leader's Handbook.* March 2002.

FM 3-19-40. *Military Police Internment/Resettlement Operations.* August 2001.

FM 4-02.21. *Division and Brigade Surgeon's Handbook of Tactics, Techniques and Procedures.* November 2000.

FM 5-0. *Staff Organizations and Operations.* 31 May 1997.

FM 19-4. *Military Police Battlefield Circulation Control, Area Security, and Enemy Prisoner of War Operations.* 7 May 1993.

FM 27-10. *Law of Land Warfare.* July 1956.

FM 34-3. *Intelligence Analysis.* March 1990.

FM 34-7-1. *Tactical Human Intelligence and Counterintelligence Operations.* April 2002.

FM 34-8. *Combat Commander's Handbook for Intelligence.* September 1992.

FM 34-54. *Technical Intelligence.* January 1998.

FM 34-60. *Counterintelligence.* 3 October 1995.

FM 41-10. *Civil Affairs Operations.* 11 January 1993.

FMI 3-19.40. *Military Police Internment/Resettlement Operations.* 30 September 2005.

TC 25-20. *A Leader's Guide to After-Action Reviews.* 30 September 1993.

ST 2-22.7. *Tactical Human Intelligence and Counterintelligence Operations.* April 2002.

ST 2-33.5. *US Army Intelligence Reach Operations.* 1 June 2001.

ST 2-50. *Intelligence and Electronic Warfare Assets.* June 2002.

ST 2-91.6. *Small Unit Support to Intelligence.* March 2004.

OTHER PUBLICATIONS

JP 2-01. *Joint Intelligence Support to Military Operations.* November 1996.

AR 350-1. *Army Training and Education.* 9 April 2003.

DIAM 58-11. *(U) DOD HUMINT Policies and Procedures (S//NF).* August 1993.

DIAM 58-12. *(U) DOD HUMINT Management Systems (S//NF).* June 1997.

AJP-2.5. *Handling of Captured Personnel, Materiel, and Documents.* September 2001.

DOD Directive 2310.1. *DOD Program for Enemy Prisoners of War (POW) and Other Detainees (Short Title: DOD Enemy POW Detainee Program).* 18 August 1994.

DOD Directive 2310.1E. *The Department of Defense Detainee Program.*

DOD Directive 3115.09. *DOD Intelligence Interrogations, Detainee Debriefings, and Tactical Questioning.* 3 November 2005.

DOD Directive 5100.77. *DOD Law of War Program.* 9 December 1998.

DOD Directive 5240.1. *DOD Intelligence Activities.* 3 December 1982.

DOD Directive 5525.5. *DOD Cooperation with Civilian Law Enforcement Officials.* 15 January 1986.

DOD Instruction 3020.41. *Contractor Personnel Authorized to Accompany the U.S. Armed Forces.* 3 October 2005

DOD Regulation 5200.1-R. *Information Security Program,* 1977.

DOD Regulation 5240.0-R. *Procedures Governing the Activities of DOD Intelligence Components That Affect United States Persons.* December 1982.

DOD SOP for Collecting and Processing Detainee Biometric Data. 11 February 05

Protocol 1 Additional to the Geneva Conventions. *Part IV: Civilian Population, Section 1: General Protection Against Effects of Hostilities.* 1977.

Executive Order 12333, *United States Intelligence Activities.* 4 December 1981.

Under Secretary of Defense for Intelligence (USD(I)) Memorandum, *"Guidance for the Conduct and Oversight of Defense Human Intelligence (HUMINT) (U)."* 14 December 2004.

Implementing Instructions to the USD(I) Memorandum. *"Guidance for the Conduct and Oversight of Defense Human Intelligence (HUMINT) (U)."* 7 September 2004.

SECDEF Memorandum, *Policy on Assigning Detainee Serial Numbers (ISN) (U).* 20 September 2005.

18 U.S.C. *Posse Comitatus Act of 1878,* § 1385.

Homeland Security Act of 2002.

Detainee Treatment Act of 2005, Public Law No. 109-163, Title XIV.

DA Form 1132-R. *Prisoner's Personal Property List – Personal Deposit Fund (LRA).* April 1986.

DA Form 4137. *Evidence/Property Custody Document. July* 1976.

DA Form 4237-R. *Detainee Personal Record.* August 1995.

DD Form 2745. *Enemy Prisoner of War Capture Tag.* May 1996.

Standardization Agreements (STANAG):

STANAG 1059. *Distinguishing Letters for Geographic Entities for Use by NATO Armed Forces.* Edition 8. April 2003.

STANAG 2022. *Intelligence Reports.* 29 September 1988.

STANAG 2033. *Interrogation of Prisoners of War (PW)*. Edition 6. December 1994.

STANAG 2044. *Procedures for Dealing with Prisoners of War (PW)*. Edition 5. June 1994.

STANAG 2084. *Handling and Reporting of Captured Enemy Equipment and Documents*. June 1986.

Index

Entries are by paragraph number